Symbol	Description	Page
r	Coefficient of correlation	440
r_S	Rank-correlation coefficient	478
ρ (rho)	Population correlation coefficient	444
s	Sample standard deviation	62
s^2	Sample variance	62
σ (sigma)	Standard deviation of population, probability distribution, or continuous distribution	62, 223, 237
σ^2	Variance of population, probability distribution, or continuous distribution	62, 223, 237
Σ (sigma)	Summation	42
$\sigma_{\bar{x}}$	Standard deviation of sampling distribution of $\bar{x}$	271
s_e	Standard error of estimate	430
SK	Pearsonian coefficient of skewness	82
SSE	Error sum of squares	344
SST	Total sum of squares	344
$SS(Tr)$	Treatment sum of squares	344
t	t statistic	303
t_α	Critical value of t	304
$T_{i.}$	Total of values in ith sample	347
$T_{..}$	Grand total for all the samples	347
u	Total number of runs	472
$u_{\alpha/2}$ and $u'_{\alpha/2}$	Critical values of u	472
$U_1, U_2,$ or U	Statistics for U test	462
U'_α	Critical value of U	462
V	Coefficient of variation	69
$\frac{x}{n}$	Sample proportion	362
$\bar{x}$	Sample mean	43
$\bar{\bar{x}}$	Grand mean of combined data	46
$\tilde{x}$	Sample median	49
$\bar{x}_w$	Weighted mean	45
$\hat{y}$	Value of y on regression line	420
z	Standard unit	69
z	z statistic	322
z_α	Critical value of z	244

Statistics: A First Course

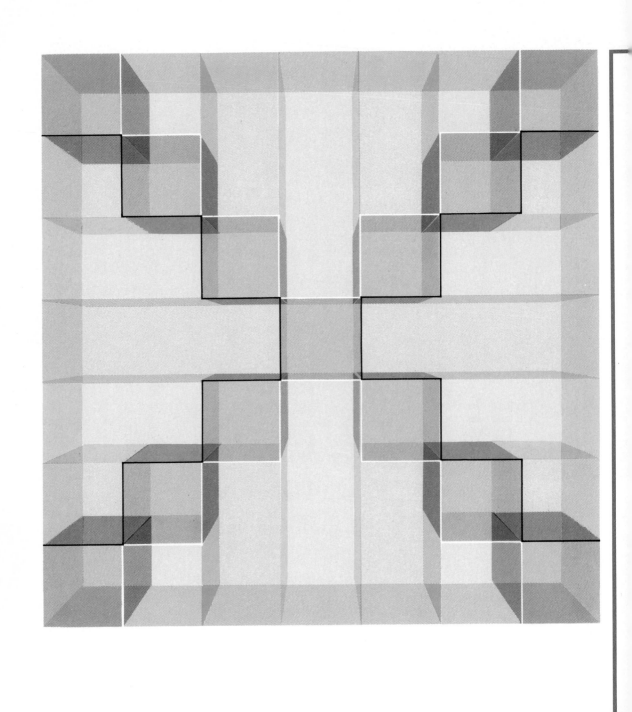

STATISTICS

4TH Edition

A FIRST COURSE

John E. Freund

Arizona State University

Richard Manning Smith

Bryant College

P R E N T I C E - H A L L ,   I N C . ,   *Englewood Cliffs, New Jersey 07632*

Library of Congress Cataloging in Publication Data

Freund, John E.
    Statistics: a first course.

    Bibliography: p.
Includes index.
1. Statistics. I. Smith, Richard Manning,
(date). II. Title.
QA276.12.F74 1986   519.5   85-9254
ISBN 0-13-845975-4

PHOTO CREDITS

Chapter 1 (page 2) Ken Karp
Chapter 2 (page 10) Laimute E. Druskis
Chapter 3 (page 38) Marc Anderson
Chapter 4 (page 102) Frederic Lewis, Inc.
Chapter 5 (page 138) Ken Karp
Chapter 6 (page 194) Maryland Academy of Sciences
Chapter 7 (page 232) USDA
Chapter 8 (page 262) H. Amstrong-Roberts
Chapter 9 (page 294) Irene Springer
Chapter 10 (page 360) H. Armstrong-Roberts
Chapter 11 (page 412) S. Shelton; Wide World Photos, Inc.
Chapter 12 (page 452) Irene Springer

Editorial/production supervision: Karen J. Clemments and
    Joan McCulley
Interior and cover design: Suzanne Behnke and Walter Behnke
Manufacturing buyer: John Hall

10 9 8 7 6 5 4

ISBN 0-13-845975-4 01

PRENTICE-HALL INTERNATIONAL (UK) LIMITED, *London*
PRENTICE-HALL OF AUSTRALIA PTY. LIMITED, *Sydney*
PRENTICE-HALL CANADA INC., *Toronto*
PRENTICE-HALL HISPANOAMERICANA, S.A., *Mexico*
PRENTICE-HALL OF INDIA PRIVATE LIMITED, *New Delhi*
PRENTICE-HALL OF JAPAN, INC., *Tokyo*
PRENTICE-HALL OF SOUTHEAST ASIA PTE. LTD., *Singapore*
EDITORA PRENTICE-HALL DO BRASIL, LTDA., *Rio de Janeiro*
WHITEHALL BOOKS LIMITED, *Wellington, New Zealand*

Dedicated to Doug and John

As is explained in the Preface, all sections marked ★ may be regarded as optional; that is, they are not prerequisites for any subsequent material.

Contents

Possibilities and Probabilities   103

Some Rules of Probability   139

Probability Distributions   **195**

The Normal Distribution   **233**

Sampling and Sampling Distributions   **263**

The Analysis of Measurements   295

The Analysis of Count Data   361

The Analysis of Paired Data   **413**

Nonparametric Tests   **453**

Preface

In the last few decades there have been many changes in the teaching of statistics. Not only has there been a pronounced shift in emphasis from descriptive methods to more and more inference, but there has also been a change in the level at which first courses in statistics are being taught. Whereas such courses used to be taught mostly to college juniors and seniors, now they are taught also to freshmen and sophomores, and even in advanced high school programs. The purpose of this book, like that of the first three editions, is to reach the student at this somewhat earlier level, as will be apparent from the organization of the material, the language of the book, its format, its notation, and above all, the examples and the exercises. The high level of mathematical precision of the previous three editions has been maintained without overcomplicating the material.

There have been some major changes in pedagogy. A greater attempt has been made to place exercises at the end of sections or after closely related sections, to make it easier for both teachers and students to find exercises relating to given topics. As before, the five review sections serve to reinforce concepts when they are out of context of a particular section or chapter. Also, there are more illustrations in the example–solution format and many of these are now given earlier in the respective sections than in previous editions.

Special care has been taken in revising the presentation of some of the more difficult concepts, so that they are now more readily accessible to the beginning student. Among others, these revisions involve the binomial distribution, sampling distributions, analysis of variance, and chi-square tests.

More formulas than before, and more definitions and rules, are boxed in or are highlighted with boldface color. Also, for the convenience of the student, the tables of the normal and t distributions are repeated on the end papers inside the back cover. The end papers inside the front cover now contain a list of symbols with descriptions and page references to their introduction in the text.

New topics in this edition include stem-and-leaf plots, box-and-whisker plots, and a separate section dealing with significance tests concerning the difference between the means of paired data.

This is the first edition of *Statistics: A First Course* in which there is mention of computers. The printouts shown in the text are intended to impress upon the reader that there are, indeed, packages of statistical programs which facilitate most of the techniques which are discussed, that these programs will sometimes provide information that is otherwise not easily accessible, and that they may be preferred, for instance, when given statistical tables are inadequate. These printouts were obtained with the widely used MINITAB package of statistical programs, but it is not necessary by any means that persons using the book have access to MINITAB or any other particular package of statistical programs.

The use of computers in conjunction with this text is entirely optional. However, to facilitate the use of computers, special exercises labeled ▱ are given at the end of relevant sets of exercises. For the most part, these exercises deal with problems that are not trivially solved without computers. An instructor may choose to assign them or not without loss of continuity.

As in the first three editions, controversial material (say, about the meaning of probability) has not been avoided. The reader is exposed to the strengths of statistical techniques as well as their weaknesses, and it is hoped that this honest approach will provide a stimulus as well as a challenge.

The authors are indebted to Professor E. S. Pearson and the *Biometrika* trustees for permission to reproduce parts of Tables 8 and 18 from their *Biometrika Tables for Statisticians;* to Prentice-Hall, Inc., to reproduce part of Table 2 of R. A. Johnson and D. W. Wichern's *Applied Multivariate Statistical Analysis;* to the Addison-Wesley Publishing Co. to base Table VII on Table 11.4 of D. B. Owen's *Handbook of Statistical Tables;* and to the editor of the *Annals of Mathematical Statistics* to reproduce the material in Table VIII.

The authors would like to express their appreciation to their many colleagues and students whose helpful criticisms and suggestions contributed greatly to the previous editions of this book and also to this fourth edition. In particular, they would like to thank Nancy Clemens Croll, Rita Ewer, Shirley Ann Lee, and Alan Olinsky. The authors would also like to express their appreciation to Bob Sickles, executive editor, and to Karen J. Clemments and Joan McCulley, production editors, for their courteous cooperation in the production of this book.

John E. Freund
Richard Manning Smith

Statistics: A First Course

Introduction

To see what statistics is and what statisticians do, let us refer to a recent edition of a popular dictionary.

> **STATISTICS** (stə-'tis-tiks) A branch of mathematics dealing with the collection, analysis, interpretation, and presentation of masses of numerical data.
>
> **STATISTICIAN** (‚stat-ə-'stish-ən) One versed in or engaged in compiling collections of data.

The first definition makes it clear that statistics covers a lot of ground—it concerns how data are obtained, how they are manipulated, and how they are put to use. A shortcoming of this definition is that statistics is not really limited to masses of data. Indeed, one of the major achievements of modern statistics is that it enables us to squeeze useful information out of relatively meager sets of data; for instance, limited data on the severity of a rare disease, difficult-to-obtain data on the environmental effects of supersonic transports, or very scarce data on the chemistry of distant stars.

The second definition leaves a great deal to be desired, for a statistician can be anyone from a clerk filing records on births and deaths or a person keeping track of batting averages and pass completions, to a consultant who applies sophisticated decision-making techniques on the managerial level or a scholar who develops the mathematical theory on which the ever-growing body of statistical methods is based. *Thus, statistics provides opportunities for persons with very little formal training and also for those with advanced college degrees.*

In Sections 1.1 and 1.3 we discuss the recent growth of statistics and its expected development in the future. Section 1.2 stresses the need for the study of statistics.

1.1

Statistics, Past and Present

The origin of the material we shall study in this book may be traced to two areas of interest which, on the surface, have very little in common: government (political science) and games of chance.

Governments have long used censuses to count persons and property. A famous example is the census reported in the *Domesday Book* of William of Normandy, completed in the year 1086, which covered most of England, listing its economic resources, including property owners and the land which they owned. In the first U.S. census, in 1790, government agents merely counted the population, but more recent U.S. censuses have become much wider in scope, providing a wealth of information about the population and the economy, and they are conducted every ten years. The most recent one was conducted in 1980.

The problem of describing, summarizing, and analyzing census data led to the development of methods which, until recently, constituted almost all there was to the subject of statistics. These methods, which originally consisted mainly of presenting data in the form of tables and charts, constitute what we now call **descriptive statistics.** This includes anything done to data which is designed to summarize, or describe, them without going any further, that is, without trying to infer anything that goes beyond the data themselves. For instance, if a newspaper reports net paid circulations of 172,316 in 1980 and 207,185 in 1985, and we perform the necessary calculations to show that there was an increase of 20.2%, our work belongs to the field of descriptive statistics. This would not be the case, however, if we used the given data to predict the newspaper's circulation in the year 1990.

Although descriptive statistics is an important branch of statistics and it continues to be widely used, statistical information usually arises from samples (from observations made on only part of a large set of items), and this means that its analysis will require generalizations which go beyond the data. As a result, an important feature of the recent growth of statistics has been the shift in emphasis from methods which merely describe to methods which serve to make generalizations, that is, a shift in emphasis from descriptive statistics to the methods of **statistical inference.**

Such methods are required, for instance, to predict the operating life span of a sewing machine (on the basis of the performance of several such machines);

to estimate the 1990 assessed value of all privately owned property in Orange County, California (on the basis of business trends, population projections, and so forth); to compare the effectiveness of two reducing diets (on the basis of the weight losses of persons who have been on the diets); to determine the optimum dose of a medication (on the basis of tests performed with volunteer patients from selected hospitals); or to predict the flow of traffic on a freeway which has not yet been built (on the basis of past traffic counts on alternate routes).

In each of the situations described in the preceding paragraph, there are uncertainties because there is only partial, incomplete, or indirect information, and it is with the use of the methods of statistical inference that we judge the merits of the results and, perhaps, suggest a "most profitable" choice, a "most promising" prediction, or a "most reasonable" course of action.

In view of the uncertainties, we handle problems like these with statistical methods which find their origin in games of chance. Although the mathematical study of games of chance dates back to the seventeenth century, it was not until the early part of the nineteenth century that the theory developed for "heads or tails," for example, or "red or black" or "even or odd," was applied also to real-life situations where the outcomes were "boy or girl," "life or death," "pass or fail," and so forth. Thus, **probability theory** was applied to many problems in the behavioral, natural, and social sciences, and nowadays it provides an important tool for the analysis of any situation (in science, in business, or in everyday life) which in some way involves an element of uncertainty or chance. In particular, it provides the basis for the methods which we use when we generalize from observed data, namely, when we use the methods of statistical inference.

1.2
The Study of Statistics

There are two reasons why the scope of statistics and the need to study statistics have grown enormously in the last few decades. One reason is the increasingly quantitative approach employed in all the sciences, as well as in business and in many other activities which directly affect our lives. This includes the use of mathematical techniques in the evaluation of antipollution controls, in inventory planning, in the analysis of cloud formations, in the study of diet and longevity, in the evaluation of teaching techniques, and so forth.

The other reason is that the amount of statistical information that is collected, processed, and disseminated to the public for one reason or another has increased almost beyond comprehension, and what part of it is "good" statistics and what part is "bad" statistics is anybody's guess. To act as watchdogs, more and more persons with some knowledge of statistics are needed to take an active part in the collection of the data, in the analysis of the data, and, what is equally important, in all the preliminary planning. Without the latter, it is frightening to think of all the things that can go wrong. The results of costly studies can be completely useless if questions are ambiguous or asked in the wrong way, for example, or if instruments are poorly adjusted or if all relevant factors are not taken into account.

In contrast to this text, which presents a general introduction to the subject of statistics, numerous books have been written on business statistics, educational statistics, medical statistics, psychological statistics, . . . , and even on statistics for historians. Although problems arising in these various disciplines will sometimes require special statistical techniques, none of the basic methods discussed in this text is restricted to any particular field of application. In the same way in which $3 + 3 = 6$ regardless of whether we are adding dollar amounts, horses, or trees, the methods we shall present provide appropriate **statistical models** regardless of whether the data are IQ's, tax payments, reaction times, test scores, or humidity readings. To emphasize this point, the examples and the exercises in this text were chosen to cover a wide spectrum of applications.

EXERCISES

(Exercises 1.1 and 1.6 are practice exercises; their complete solutions are given on page 9.)

1.1 In four history tests a student received grades of 45, 73, 77, and 86. Which of the following conclusions can be obtained from these figures by purely descriptive methods and which require generalizations? Explain your answers.
   (a) Only one of the grades exceeds 85.
   (b) The student's grades increased from each test to the next.
   (c) The student must have studied harder for each successive test.
   (d) The difference between the highest and lowest grades is 41.

1.2 Mary and Jean are real estate salespersons. In the first three months of 1984 Mary sold 3, 6, and 2 one-family homes and Jean sold 4, 0, and 5 one-family homes. Which of the following conclusions can be obtained from these figures by purely descriptive methods and which require generalizations? Explain your answers.
   (a) During the three months Mary sold more one-family homes than Jean.
   (b) Mary is a better real estate salesperson than Jean.
   (c) Mary sold at least two one-family homes during each of the three months.
   (d) Jean probably took her annual vacation during the second month.

1.3 The paid attendance of a minor league baseball team's first four home games was 5,308, 4,030, 6,386, and 5,770 in the year 1984 and 6,274, 5,883, 7,615, and 1,312 in the year 1985. Which of the following conclusions can be obtained from these figures by purely descriptive methods and which require generalizations? Explain your answers.

    (a) The fourth 1985 figure was probably recorded incorrectly and should have been 7,312 instead of 1,312.

    (b) Among the eight games, the paid attendance for any one game was highest in 1985.

    (c) Among the eight games, the paid attendance in 1985 exceeded 6,000 more often than in 1984.

    (d) Since the paid attendance at each of the first three home games was higher in 1985 than in 1984, the weather must have been better on those days.

1.4 Driving the same model car, five persons averaged 22.5, 21.7, 23.0, 22.5, and 21.8 miles per gallon. Which of the following conclusions can be obtained by purely descriptive methods and which require generalizations? Explain your answers.

    (a) More often than any of the other figures, the drivers averaged 22.5 miles per gallon.

    (b) The second and fifth persons must have done more city driving than the others.

    (c) None of the averages differs from 22.0 by more than 1.0.

    (d) If the whole experiment were repeated, none of the drivers would average less than 21.0 or more than 24.0 miles per gallon.

1.5 The three oranges which a person bought at a supermarket weighed 9, 8, and 13 ounces. Which of the following conclusions can be obtained from these data by purely descriptive methods and which require generalizations? Explain your answers.

    (a) The average weight of the three oranges is 10 ounces.

    (b) The average weight of oranges sold at that supermarket is 10 ounces.

1.6 "Bad" statistics may well result from asking questions in the wrong way or of the wrong persons. Explain why the following may lead to useless data:

    (a) To determine public sentiment about a certain foreign trade restriction, an interviewer asks voters: "Do you feel that this unfair practice should be stopped?"

    (b) In order to predict a municipal election, a public opinion poll telephones persons selected haphazardly from the city's telephone directory.

    (c) In a study of art appreciation, persons are asked whether they like Indian art.

1.7 "Bad" statistics may also result from asking questions in the wrong place or at the wrong time. Explain why the following may lead to useless data:

    (a) A house-to-house survey is made during weekday mornings to study consumer reaction to certain convenience foods.

    (b) To predict an election, a poll taker interviews persons coming out of a building which houses the national headquarters of a political party.

(c) To determine what the average person spends on a vacation, a researcher interviews the passengers on a luxury cruise.

1.8 Explain why the following may lead to useless data:

(a) To determine the proportion of improperly sealed cans of coffee, a quality-control inspector examines every 50th can coming off a production line.

(b) To determine the average annual income of its graduates 10 years after graduation, a college's alumni office sent questionnaires in 1985 to all members of the class of 1975, and the estimate was based on the questionnaires returned.

(c) To study executives' reaction to its copying machines, the Xerox corporation hires a research organization to ask executives the question: How do you like using Xerox copies?

1.9 A statistically minded lawyer has his office on the third floor of a very tall office building, and whenever he leaves his office he records whether the first elevator which stops at his floor is going up or coming down. Having done this for some time, he discovers that the vast majority of the time the first elevator which stops is going down. Comment on his conclusion that fewer elevators are going up than are coming down.

1.3

Statistics, What Lies Ahead

In Section 1.1 we indicated how the emphasis in statistics has shifted from summarizing data by means of charts and tables to making inferences (that is, generalizations) on the basis of partial, incomplete, or indirect information. This is not meant to imply, however, that the subject of statistics has now become stable and inflexible, and that it has ceased to grow. Aside from the fact that new statistical techniques are constantly being developed to meet particular needs, the whole philosophy of statistics continues to be in a state of change. Most recently, attempts have been made to treat all problems of statistical inference within the framework of a unified theory called **decision theory,** which, so to speak, covers everything "from cradle to grave." One of the main features of this theory is that we must account for all the consequences which can arise when we base decisions on statistical data. This poses serious problems, as it is generally difficult, if not impossible, to put cash values on the consequences of one's actions. For instance, how can we put a cash value on the consequences of the decision whether or not to market a new medication, especially if the wrong decision may well involve the loss of human lives?

There are also statisticians who suggest that the emphasis has swung too far from descriptive statistics to statistical inference; rightly so, they feel that the solution of many problems requires only descriptive methods. To accommodate their needs, some new descriptive techniques have recently been developed under the general heading of **exploratory data analysis.** Two of these will be introduced in Sections 2.3 and 3.4.

We have mentioned all this mainly to impress upon the reader that statistics, like most other fields of learning, is not static. Indeed, it is difficult to picture what a beginning course in statistics will be like twenty years hence. Certain aspects will probably still be the same, and that includes the role of probability theory in the foundations of statistics as well as certain "bread and butter" techniques which have been very useful in the past and will undoubtedly continue to be widely used in the future.

SOLUTIONS OF PRACTICE EXERCISES

1.1 (a) The conclusion merely describes the data, as it can be seen that 86 exceeds 85, but 45, 73, and 77 do not.

(b) The conclusion merely describes the data, since 73 exceeds 45, 77 exceeds 73, and 86 exceeds 77.

(c) This is a generalization, as there can be many other reasons for the increases in the grades. For instance, the student may have felt better physically when he got the higher grades, or he may have been just lucky in studying the exact material asked for in the tests.

(d) This conclusion merely describes the data; the highest grade is 86, the lowest grade is 45, and their difference is $86 - 45 = 41$.

1.6 (a) This is called "begging the question," because the interviewer suggests to the voters that the practice is, in fact, unfair.

(b) Persons selected from a telephone directory will generally not provide a satisfactory cross section of all persons eligible to vote, because ownership of telephones is related to economic status and, hence, often to political judgments.

(c) The term "Indian art" is ambiguous; some persons may respond with reference to the work of American Indians, while others may be thinking about art produced in India.

Frequency Distributions

In recent years the collection of statistical data has grown at such a rate that it would be impossible to keep up with even a small part of the things which directly affect our lives unless this information is disseminated in "predigested" or summarized form. The whole matter of putting large masses of data into a usable form has always been important, but it has multiplied greatly in the last few decades. This has been due partly to the development of electronic computers which have made it possible to accomplish in minutes what previously had to be left undone because it would have taken months or years, and partly to the deluge of data generated by the increasingly quantitative approach of the sciences, especially the behavioral and social sciences, where nearly every aspect of human life is nowadays measured in one way or another.

The most common method of summarizing data is to present them in condensed form in tables or charts, and the key word here is

> **DISTRIBUTION** (ˌdis-trə-ˈbyü-shən) An arrangement of statistical data that shows how many items, or what parts of the data, go into the different intervals or categories into which the data are grouped.

This definition does not come from a dictionary, not word for word, as we could not find one which actually reflects the usage of this term in statistics.

Sections 2.1 and 2.2 deal with problems related to the grouping of data and the presentation of such groupings in graphical form; in Section 2.3 we discuss a relatively new way of presenting grouped data, and in Section 2.4 we present the grouping of paired data.

11

2.1
Frequency Distributions

To illustrate the importance of grouping, or classifying, data, let us consider the problem of a social scientist who wants to study the ages of persons in the United States with incomes below the poverty level. Not even considering the possibility of conducting a survey of his own—the cost would be staggering—he turns to the most logical source for this kind of information, the *Statistical Abstract of the United States* (published annually by the U.S. Department of Commerce). It is conceivable that the Bureau of the Census, which compiles such data, might make its **raw** (untreated) **data** available to the social scientist, but this would put him in the unenviable position of having to look at printouts containing thousands of figures. Thus, he turns to the *Statistical Abstract of the United States* with the hope of finding the desired information in a "more usable" form.

In connection with large sets of data, a good overall picture and sufficient information can often be conveyed by grouping the data into a number of classes (intervals or categories), and our social scientist finds that the information he seeks is presented as follows for the year 1981:

Persons with Income below Poverty Level	
Age (years)	Number of persons (thousands)
Under 16	11,223
16 to 21	3,867
22 to 44	8,754
45 to 64	4,125
65 and over	3,853
Total	31,822

A table like this is called a **frequency distribution** (or simply a **distribution**)—it shows how the ages of the more than 31 million persons are distributed among the chosen classes. Since the data are grouped according to their numerical size, we also refer to such tables as **numerical** or **quantitative distributions.**

In the example above, each class covered more than one possible value; for instance, the second class covered the values 16, 17, 18, 19, 20, and 21. Each class may also cover a single value, as is illustrated by the following example, based on a study in which 400 persons were asked how many full-length movies they had seen on television during the preceding week:

Number of movies	Number of persons
0	72
1	106
2	153
3	40
4	18
5	7
6	3
7	0
8	1
Total	400

If data are grouped into nonnumerical categories, the resulting table is called a **categorical** or **qualitative distribution.** This kind of distribution is illustrated by the following table pertaining to the college plans of a high school's 548 seniors:

	Number of seniors
Plan to attend college	240
May attend college	146
Plan to or may attend a vocational school	57
Will not attend any school	105
Total	548

Since we said on page 3 that statistics deals with numerical data, the inclusion of categorical data, such as the four alternatives in the example above, requires some explanation. The answer is simple, because we can always code categorical alternatives by means of numbers. The four choices of the high school seniors might be labeled 1, 2, 3, and 4, "yes or no" responses to a question might be recorded as 1 and 2 (or, perhaps, as 29 and 30 if we are referring to the 15th "yes or no" question on a questionnaire), and a person's marital status might be recorded as 0, 1, 2, or 3, depending on whether he or she is single, married, widowed, or divorced.

Frequency distributions present data in a relatively compact form, give a good overall picture, and contain information that is adequate for many purposes, but there are usually some things which can be determined only from the original data. For instance, given only the distribution on page 12, we cannot tell how many of the persons are 19 years old, or how many of them are over 62. Similarly, the study of the high school seniors' plans might have contained information about which colleges they plan to attend, or what they intend to do if they will not attend any school. Nevertheless, frequency distributions present raw data in a more readily usable form, and the price we pay for this—the loss of certain information—is usually a fair exchange.

The construction of numerical distributions consists essentially of four steps: (1) choosing the classes, (2) sorting (or tallying) the data into these classes, (3) counting the number of items in each class, and (4) displaying the results in the form of a chart or table. Since the second and third steps are purely mechanical and the fourth step is a matter of craftsmanship and taste, we shall concentrate here on the first step, namely, that of choosing suitable classifications.

This involves choosing the number of classes and the range of values each class should cover, namely, from where to where each class should go. Both of these choices are arbitrary to some extent, but they depend on the nature of the data and on the purpose the distribution is to serve. The following are some rules that are generally observed:

> **We seldom use fewer than 6 or more than 15 classes; the exact number we use in a given situation depends mainly on the number of measurements or observations we have to group.**

Clearly, we would lose more than we gain if we group 6 observations into 12 classes with many of them empty, and we would probably give away too much information if we group 10,000 measurements into 3 classes.

> **We always make sure that each item (measurement or observation) goes into one and only one class.**

To this end we must make sure that the smallest and largest values fall within the classification, that none of the values can fall into possible gaps between successive classes, and that the classes do not overlap, namely, that successive classes have no values in common.

Whenever possible, we make the classes cover equal ranges (or intervals) of values.

If we can, we also make these ranges numbers that are easy to work with, such as 5, 10, or 100, for this will facilitate constructing, reading, and using the distribution.

In connection with these rules, the age distribution on page 12 is not a good example, but, presumably, the government statisticians had good reasons for choosing the classes as they did. There are only five classes, none of the classes cover equal ranges of values, and the "65 and over" class is **open;** for all we know, the oldest person may be 87 or 105.

In general, we refer to classes of the "less than," "or less," "more than," and "or more" type as **open classes,** and they are used to reduce the number of classes that are required when a few of the values are much smaller (or much greater) than the rest. However, they should be avoided when possible, for as we shall see, they make it difficult, or even impossible, to calculate certain further descriptions that may be of interest.

So far as the second rule is concerned, it is important to watch whether the data are given to the nearest dollar or to the nearest cent, whether they are given to the nearest inch or to the nearest tenth of an inch, whether they are given to the nearest ounce or to the nearest hundredth of an ounce, and so forth. For instance, to group the weights of certain animals, we could use the first of the following three classifications if the weights are given to the nearest kilogram, the second if the weights are given to the nearest tenth of a kilogram, and the third if the weights are given to the nearest hundredth of a kilogram:

Weight (kilograms)	Weight (kilograms)	Weight (kilograms)
10–14	10.0–14.9	10.00–14.99
15–19	15.0–19.9	15.00–19.99
20–24	20.0–24.9	20.00–24.99
25–29	25.0–29.9	25.00–29.99
30–34	30.0–34.9	30.00–34.99

To illustrate what we have been discussing in this section, let us now go through the actual steps of grouping a given set of data into a frequency distribution.

EXAMPLE Construct a distribution of the following data on the amount of time (in hours) that 80 college students devoted to leisure activities during a typical school week:

```
23  24  18  14  20  24  24  26  23  21
16  15  19  20  22  14  13  20  19  27
29  22  38  28  34  32  23  19  21  31
16  28  19  18  12  27  15  21  25  16
30  17  22  29  29  18  25  20  16  11
17  12  15  24  25  21  22  17  18  15
21  20  23  18  17  15  16  26  23  22
11  16  18  20  23  19  17  15  20  10
```

Solution Since the smallest value is 10 and the largest is 38, we might choose the six classes 10–14, 15–19, 20–24, 25–29, 30–34, and 35–39; or we might choose the eight classes 10–13, 14–17, 18–21, 22–25, 26–29, 30–33, 34–37, and 38–41, to mention two possibilities. Note that in each case the classes accommodate all of the data, they do not overlap, and they are all of the same size.

Deciding upon the first of these classifications, we now tally the 80 observations and get the results shown in the following table:

Hours	Tally	Frequency
10–14	⊬⊬ ///	8
15–19	⊬⊬ ⊬⊬ ⊬⊬ ⊬⊬ ⊬⊬ ///	28
20–24	⊬⊬ ⊬⊬ ⊬⊬ ⊬⊬ ⊬⊬ //	27
25–29	⊬⊬ ⊬⊬ //	12
30–34	////	4
35–39	/	1
	Total	80

The numbers given in the right-hand column of this table, which show how many items fall into each class, are called the **class frequencies.** Also, the smallest and largest values that can go into any given class are referred to as its **class limits,** and in our example they are 10 and 14, 15 and 19, 20 and 24, . . . , and 35 and 39. More specifically, 10, 15, 20, . . . , and 35 are called the **lower class limits,** and 14, 19, 24, . . . , and 39 are called the **upper class limits.**

The amounts of time which we grouped in our example were all given to the nearest hour, so that the first class actually covers the interval from 9.5

hours to 14.5 hours, the second class covers the interval from 14.5 hours to 19.5 hours, and so forth. It is customary to refer to these numbers as the **class boundaries** or the **"real" class limits.** In actual practice, class limits are used much more widely than class boundaries, and we have mentioned them here mainly because they will be needed in the next chapter for calculating certain descriptive measures of a distribution.

Numerical distributions also have what we call **class marks** and **class intervals.** Class marks are simply the midpoints of the classes, and they are obtained by adding the upper and lower limits of a class (or its upper and lower boundaries) and dividing by 2. A class interval is the length of a class, or the range of values it can contain, and it is given by the difference between its class boundaries. If the classes of a distribution are all of equal length, their common class interval, which we refer to as the **class interval of the distribution,** is also given by the difference between any two successive class marks.

EXAMPLE   Find the class marks and the class interval of the distribution obtained in the preceding example.

Solution   The class marks are $\dfrac{10 + 14}{2} = 12$, $\dfrac{15 + 19}{2} = 17, \ldots$, and $\dfrac{35 + 39}{2} =$ 37, and the class interval of the distribution is $17 - 12 = 5$. Note that class intervals are *not* given by the differences between the respective class limits, which in our example all equal 4 and not 5.

There are essentially two ways in which frequency distributions can be modified to suit particular needs. One way is to convert a distribution into a **percentage distribution** by dividing each class frequency by the total number of items grouped, and then multiplying by 100.

EXAMPLE   Convert the distribution of the amounts of time the 80 college students devoted to leisure activities into a percentage distribution.

Solution   The first class contains $\dfrac{8}{80} \cdot 100 = 10\%$ of the data, the second class contains $\dfrac{28}{80} \cdot 100 = 35\%$ of the data, the third class contains $\dfrac{27}{80} \cdot 100 = 33.75\%$ of the data, $\ldots$, and the sixth class contains $\dfrac{1}{80} \cdot 100 = 1.25\%$

of the data. These results are shown in the following table:

Hours	Percentage
10–14	10
15–19	35
20–24	33.75
25–29	15
30–34	5
35–39	1.25

The other way of modifying a frequency distribution is to convert it into a "less than," "or less," "more than," or "or more" **cumulative distribution.** To construct a cumulative distribution we simply add the class frequencies, starting either at the top or at the bottom of the distribution.

EXAMPLE Convert the distribution on page 16 into a cumulative "less than" distribution.

Solution Since none of the values is less than 10, 8 are less than 15, $8 + 28 = 36$ are less than 20, $8 + 28 + 27 = 63$ are less than 25, . . . , the result is shown in the following table:

Hours	Cumulative frequency
Less than 10	0
Less than 15	8
Less than 20	36
Less than 25	63
Less than 30	75
Less than 35	79
Less than 40	80

In the same way, we can also convert a percentage distribution into a **cumulative percentage distribution** by adding the percentages, starting either at the top or at the bottom of the distribution.

So far we have discussed only the construction of numerical distributions, but the general problem of constructing categorical (or qualitative) distributions is very much the same. Again we must decide how many categories (classes) to use and what kind of items each category is to contain, making sure that all the data can be accommodated and that there are no ambiguities. Since the categories are often chosen before any data are collected, it is prudent to include a category labeled "others" or "miscellaneous."

For categorical distributions, we do not have to worry about such mathematical details as class limits, class boundaries, and class marks. On the other hand, there is often a serious problem with ambiguities and we must be very careful and explicit in defining what each category is to contain. For instance, in the example on page 13, the one dealing with the college plans of a group of high school seniors, it is not clear where we should put one of the seniors who may attend college or a vocational school.

EXERCISES   (Exercises 2.1, 2.4, 2.7, 2.11, and 2.14 are practice exercises; their complete solutions are given on pages 36 and 37.)

2.1 The weights of a city's firemen (to the nearest pound) vary from 148 to 236 pounds. Indicate the limits of ten classes into which these weights might be grouped.

2.2 Measurements of the boiling point of a substance, measured to the nearest degree Celsius, vary from 136° to 168°. Indicate the limits of seven classes into which these measurements might be grouped.

2.3 The wages paid to the employees of a company in a given week varied from \$215.26 to \$313.53. Indicate the limits of six classes into which these wages might be grouped.

2.4 The number of empty seats on flights from Dallas to New Orleans are grouped into a table with the classes 0–9, 10–19, 20–29, 30–39, and 40 or more. Will it be possible to determine from this table the number of flights on which there were
   (a) at least 20 empty seats;
   (b) more than 20 empty seats;
   (c) more than 19 empty seats;
   (d) at least 19 empty seats;
   (e) exactly 19 empty seats?

2.5 The declared values of packages mailed from a foreign country are grouped into a distribution with the classes \$0.00–\$49.99, \$50.00–\$99.99, \$100.00–\$149.99, \$150.00–\$199.99, and \$200.00 or more. Will it be possible to determine from the distribution the number of packages valued at
   (a) more than \$100.00;
   (b) at least \$100.00;
   (c) exactly \$100.00;
   (d) more than \$99.99?

2.6 The following is the distribution of the weights of 125 mineral specimens collected on a field trip:

Weight (grams)	Number of specimens
0– 19.9	19
20.0– 39.9	38
40.0– 59.9	35
60.0– 79.9	17
80.0– 99.9	11
100.0–119.9	3
120.0–139.9	2
Total	125

If possible, find how many of the specimens weigh
  (a) at most 59.9 grams;
  (b) less than 40.0 grams;
  (c) more than 100.0 grams;
  (d) 80.0 grams or less;
  (e) exactly 20.0 grams;
  (f) anywhere from 40.0 to 80.0 grams.

2.7 The number of students who are absent each day from a certain school are grouped into a distribution having the classes 0–14, 15–29, 30–44, and 45–59. Determine
  (a) the lower class limits;
  (b) the upper class limits;
  (c) the class marks;
  (d) the class interval of the distribution.

2.8 The number of suitcases which are lost each week by an airline on flights between Honolulu and San Francisco are grouped into a distribution having the classes 0–2, 3–5, 6–8, 9–11, 12–14, and 15–17. Find
  (a) the lower class limits;
  (b) the upper class limits;
  (c) the class marks;
  (d) the class interval of the distribution.

2.9 The number of congressmen absent each day during a session of Congress are grouped into a distribution having the classes 0–19, 20–39, 40–59, 60–79, 80–99, 100–119, 120–139, and 140–159. Determine
  (a) the lower class limits;
  (b) the upper class limits;

(c) the class marks;

(d) the class interval of the distribution.

2.10 The number of nurses on duty each day at a hospital are grouped into a distribution having the classes 20–34, 35–49, 50–64, 65–79, and 80–94. Find

    (a) the class limits;

    (b) the class boundaries;

    (c) the class marks;

    (d) the class interval of the distribution.

2.11 The class marks of a distribution of the daily number of traffic accidents reported in a county are 4, 11, 18, and 25. Find

    (a) the class boundaries;

    (b) the class limits.

2.12 The class marks of a distribution of the daily number of calls received by a small cab company are 22, 27, 32, 37, 42, 47, and 52. Find

    (a) the class boundaries;

    (b) the class limits.

2.13 The class marks of a distribution of temperature readings, given to the nearest degree Fahrenheit, are 36, 45, 54, 63, 72, 81, and 90. Find

    (a) the class boundaries;

    (b) the class limits.

2.14 To group sales invoices ranging from $5.00 to $30.00, a clerk uses the following classification: $5.00–$9.99, $10.00–$15.99, $15.00–$19.99, $20.00–$24.90, and $25.00–$29.99. Explain where difficulties might arise.

2.15 To group data on the number of rainy days reported by a weather station for the month of May during the last fifty years, a meteorologist uses the classes 0–5, 6–10, 12–17, 18–23, and 23–30. Explain where difficulties might arise.

2.16 The following are the body weights (in grams) of 50 rats used in a study of vitamin deficiencies:

136	92	115	118	121	137	132	120	104	125
119	115	101	129	87	108	110	133	135	126
127	103	110	126	118	82	104	137	120	95
146	126	119	119	105	132	126	118	100	113
106	125	117	102	146	129	124	113	95	148

    (a) Group these weights into a table having the classes 80–89, 90–99, 100–109, 110–119, 120–129, 130–139, and 140–149.

    (b) Convert the distribution obtained in part (a) into a cumulative "or more" distribution.

2.17 Convert the distribution obtained in part (a) of the preceding exercise into a

    (a) percentage distribution;

    (b) cumulative "or less" percentage distribution.

2.18 The following are the grades which 50 students obtained in an accounting test:

73	65	82	70	45	50	70	54	32	75
75	67	65	60	75	87	83	40	72	64
58	75	89	70	73	55	61	78	89	93
43	51	59	38	65	71	75	85	65	85
49	97	55	60	76	75	69	35	45	63

(a) Group these grades into a distribution having the classes 30–39, 40–49, 50–59, . . . , and 90–99.

(b) Convert the distribution obtained in part (a) into a cumulative "less than" distribution, beginning with "less than 30."

2.19 The following is a distribution of the ages of the members of a video dating service for single persons:

Age (years)	Frequency
20–24	129
25–29	221
30–34	310
35–39	163
40–44	105
45–49	62
50–54	10

(a) Convert the distribution into a cumulative "less than" distribution.

(b) Convert the distribution into a cumulative "more than" distribution.

2.20 The following are the miles per gallon obtained with 40 tankfuls of gas:

24.1	25.0	24.8	24.3	24.2	25.3	24.2	23.6	24.5	24.4
24.5	23.2	24.0	23.8	23.8	25.3	24.5	24.6	24.0	25.2
25.2	24.4	24.7	24.1	24.6	24.9	24.1	25.8	24.2	24.2
24.8	24.1	25.6	24.5	25.1	24.6	24.3	25.2	24.7	23.3

(a) Group these figures into a distribution having the classes 23.0–23.4, 23.5–23.9, 24.0–24.4, 24.5–24.9, 25.0–25.4, and 25.5–25.9.

(b) Convert the distribution obtained in part (a) into a cumulative "more than" distribution, beginning with "more than 22.9" and ending with "more than 25.9."

(c) Convert the distribution obtained in part (b) into a cumulative "more than" percentage distribution.

2.21 In the forty lectures of a psychology class, 2, 1, 5, 0, 0, 3, 2, 1, 1, 4, 3, 1, 0, 1, 2, 1, 3, 6, 2, 2, 1, 0, 2, 1, 1, 3, 4, 1, 0, 2, 1, 0, 1, 2, 4, 1, 3, 1, 2, and 3 students were absent. Construct a distribution showing how many times 0, 1, 2, 3, 4, 5, or 6 students were absent.

2.22 A survey made at a resort city showed that 50 tourists arrived by the following means of transportation: car, train, plane, plane, plane, bus, train, car, car, car, plane, car, plane, train, car, car, bus, car, plane, plane, train, train, plane, plane, car, car, train, car, car, plane, car, car, plane, bus, plane, bus, car, plane, car, car, train, train, car, plane, bus, plane, car, car, train, and bus. Construct a categorical distribution showing the frequencies corresponding to the different means of transportation.

2.23 Asked to rate the maneuverability of a car as excellent, very good, good, fair, poor, or very poor, 40 drivers responded as follows: very good, good, good, fair, excellent, good, good, good, very good, poor, good, good, good, good, very good, good, fair, good, good, very poor, very good, fair, good, good, excellent, very good, good, good, good, fair, fair, very good, good, very good, excellent, very good, fair, good, good, and very good. Construct a distribution showing the frequencies corresponding to the different ratings of the maneuverability of the car.

2.2

Graphical Presentations

When frequency distributions are constructed primarily to condense large sets of data and display them in an "easy to digest" form, they are often presented graphically. The most common form of graphical presentation of a frequency distribution is the **histogram,** an example of which (pertaining to the amounts of time that 80 college students devoted to leisure activities) is shown in Figure 2.1. A histogram is constructed by representing the measurements or observations that are grouped on a horizontal scale, the class frequencies on a vertical scale, and drawing rectangles whose bases equal the class intervals and whose heights are determined by the corresponding class frequencies. The markings on the horizontal scale can be the class limits as in Figure 2.1, the class boundaries, the class marks, or arbitrary key values. For easy readability it is usually preferable to indicate the class limits, although the rectangles actually go from one class boundary to the next. Histograms cannot be used in connection with frequency distributions having open classes, and they must be used with extreme care when the class intervals are not all equal. In that case, it is best to represent the class frequencies by the areas of the rectangles instead of their heights.

For large sets of data, it may be convenient to construct histograms directly from raw data by using a computer package designed specially for data analysis.

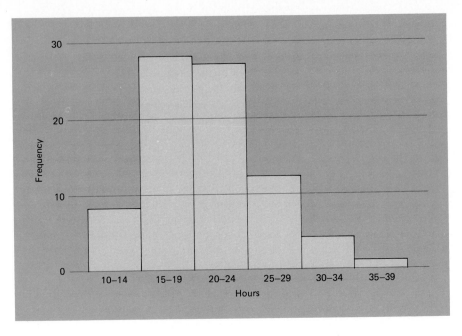

2.1

Histogram of the distribution of the amounts of time that 80 students engaged in leisure activities.

2.2

Computer printout for histogram of the amounts of time that 80 students engaged in leisure activities.

```
MTB > SET C1
DATA> 23   24   18   14   20   24   24   26   23   21
DATA> 16   15   19   20   22   14   13   20   19   27
DATA> 29   22   38   28   34   32   23   19   21   31
DATA> 16   28   19   18   12   27   15   21   25   16
DATA> 30   17   22   29   29   18   25   20   16   11
DATA> 17   12   15   24   25   21   22   17   18   15
DATA> 21   20   23   18   17   15   16   26   23   22
DATA> 11   16   18   20   23   19   17   15   20   10
MTB > HIST C1 12 5

  C1

  MIDDLE OF      NUMBER OF
  INTERVAL       OBSERVATIONS
    12.00            8      ********
    17.00           28      ****************************
    22.00           27      ***************************
    27.00           12      ************
    32.00            4      ****
    37.00            1      *
```

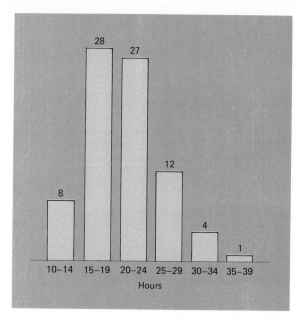

2.3

Bar chart of the distribution of the amounts of time that 80 students engaged in leisure activities.

For instance, the computer printout of Figure 2.2 shows a histogram of a distribution with the class marks 12, 17, 22, 27, 32, and 37 (as on page 16 for the data dealing with the amounts of time that 80 students devoted to leisure activities.[†] Here, "SET C1" instructs the computer to place the 80 numbers in column 1, and "HIST C1 12 5" instructs it to print a histogram with 12 as the first class mark and 5 as the class interval for the data in column 1. Note that the frequency scale is horizontal, so that the histogram is, so to speak, on its side. Strictly speaking, the result shown in Figure 2.2 is not a histogram; at least, not as we have defined this term. However, combining some of the features of Figures 2.3 and 2.7, it conveys the same idea.

Similar to histograms are **bar charts,** such as the one shown in Figure 2.3. The heights of the rectangles, or bars, again represent the class frequencies, but there is no pretense of having a continuous horizontal scale.

Another, less widely used, form of graphical presentation is the **frequency polygon** (see Figure 2.4). In these, the class frequencies are plotted at the class marks and the successive points are connected by means of straight lines. Note that we added classes with zero frequencies at both ends of the distribution to

---

[†] In the printout of Figure 22 and in others appearing in this text, words and numbers appear which relate to the technical aspects of operating the particular computer program employed. If a computer is available, the reader should refer to the appropriate manuals for operating instructions and for a list of problems which can be solved with existing programs. Trained users can create additional programs as needed.

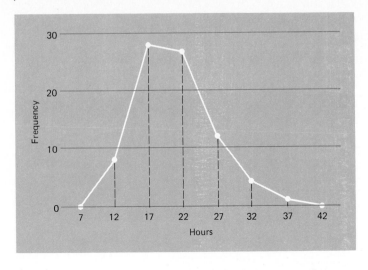

2.4

Frequency polygon of the distribution of the amounts of time that 80 students engaged in leisure activities.

"tie down" the graph to the horizontal scale. If we apply the same technique to a cumulative distribution, we obtain what is called an **ogive.** However, the cumulative frequencies are plotted at the corresponding class limits or class boundaries instead of the class marks—it stands to reason that "less than 20," for example, should be plotted at 20, or at 19.5, the class boundary, since "less than 20" actually includes everything up to 19.5. Figure 2.5 shows an ogive of the cumulative "less than" distribution which we constructed on page 18.

Although the visual appeal of histograms, bar charts, frequency polygons, and ogives exceeds that of tables, there are various ways in which distributions can be presented more dramatically and more effectively. Two kinds of such

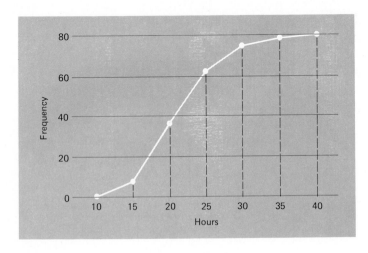

2.5

Ogive of the distribution of the amounts of time that 80 students engaged in leisure activities.

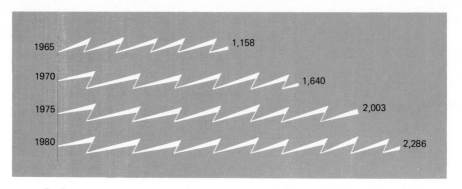

2.6

Electric energy production in the United States (billions of kilowatt-hours).

pictorial presentations (often seen in newspapers, magazines, and reports of various sorts) are illustrated by the **pictograms** of Figures 2.6 and 2.7.

Categorical (or qualitative) distributions are often presented graphically as **pie charts** such as the one shown in Figure 2.8, where a circle is divided into sectors (pie-shaped pieces) which are proportional in size to the corresponding frequencies or percentages. To construct a pie chart we first convert the distribution into a percentage distribution. Then, since a complete circle corresponds to 360 degrees, we obtain the central angles of the various sectors by multiplying the percentages by 3.6.

2.7

Population of the United States.

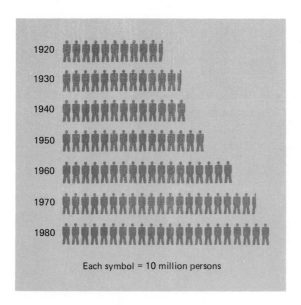

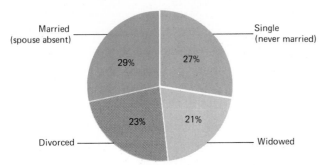

Married
(spouse absent)

Single
(never married)

29%

27%

23%

21%

Divorced

Widowed

2.8

Marital status of female family householders with no spouse present, 1981.

EXAMPLE   To study their attitudes toward social issues, 1,200 persons were asked (among other things) whether we are spending "Too little," "About right amount," or "Too much" on social welfare programs. Draw a pie chart to display the results shown in the following table:

	Number of persons
Too little	296
About right amount	360
Too much	544

Solution   The percentages corresponding to the three categories are $\dfrac{296}{1,200} \cdot 100 = 24.7\%$, $\dfrac{360}{1,200} \cdot 100 = 30.0\%$, and $\dfrac{544}{1,200} \cdot 100 = 45.3\%$, and multiplying by 3.6, we find that the central angles of the three sectors are $(24.7)(3.6) = 88.92$ degrees, $(30.0)(3.6) = 108.00$ degrees, and $(45.3)(3.6) = 163.08$ degrees. Rounding the angles to the nearest degree and using a protractor, we get the pie chart shown in Figure 2.9.

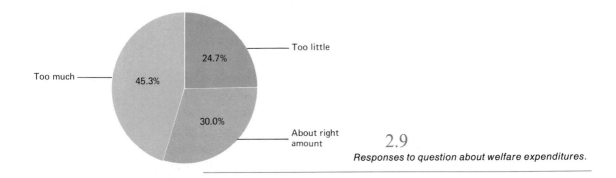

Too little

24.7%

Too much

45.3%

30.0%

About right
amount

2.9

Responses to question about welfare expenditures.

Many computers are preprogrammed so that, upon entering the data, pie charts can be produced directly on their display screens.

2.24 The following is the distribution of the total finance charges which 200 customers paid on their budget accounts at a department store:

Amount (dollars)	Frequency
0–19	18
20–39	62
40–59	63
60–79	43
80–99	14

(a) Draw a histogram of this distribution.
(b) Draw a bar chart of this distribution.

2.25 Convert the distribution of the preceding exercise into a cumulative "less than" distribution and draw an ogive.

2.26 The following is the distribution of the weights of the 140 freshmen women entering a certain college:

Weight (pounds)	Frequency
90–99	4
100–109	23
110–119	49
120–129	38
130–139	17
140–149	6
150–159	3

(a) Draw a histogram of this distribution.
(b) Draw a frequency polygon of this distribution.

2.27 Convert the distribution of the preceding exercise into a cumulative "less than" percentage distribution and draw an ogive.

2.28 The following table shows how workers in Phoenix, Arizona, get to work:

Means of transportation	Percentage
Ride alone	81
Car pool	14
Ride bus	3
Varies or work at home	2

Construct a pie chart of this percentage distribution.

2.29 Construct a pie chart of the following distribution, which shows how the dogs entered in a dog show are distributed according to AKC classifications:

Group	Number
Sporting dogs	39
Hounds	50
Working dogs	112
Terriers	24
Toys	34
Nonsporting dogs	41

2.30 Construct a pie chart of the categorical distribution on page 13 (representing the 548 high school seniors' college plans).

2.31 Here, again, are the grades (from Exercise 2.18) which 50 students obtained in an accounting test:

73	65	82	70	45	50	70	54	32	75
75	67	65	60	75	87	83	40	72	64
58	75	89	70	73	55	61	78	89	93
43	51	59	38	65	71	75	85	65	85
49	97	55	60	76	75	69	35	45	63

Use a computer package to construct a histogram with the classes 30–39, 40–49, 50–59, . . . , and 90–99.

2.32 Use a computer package to construct a histogram with the classes 80–89, 90–99, 100–109, . . . , and 140–149 of the weights of Exercise 2.16 on page 21.

2.33 Use a computer package to construct a histogram with the classes 23.0–23.4, 23.5–23.9, 24.0–24.4, . . . , and 25.5–25.9 of the mileages of Exercise 2.20 on page 22.

2.3

Stem-and-Leaf Plots

So far in this chapter we have directed our attention to the grouping of mass data, with the objective of putting such data into a manageable form. As we saw, this entailed some loss of information. In recent years, similar techniques have been proposed for the preliminary exploration of relatively small sets of data which yield a good overall picture of the data without any loss of information.

To illustrate, consider the following scores on a test of physical coordination given to 20 students who had consumed the legal limit amount of alcohol permitted for driving:

$$69 \quad 84 \quad 52 \quad 93 \quad 61 \quad 74 \quad 79 \quad 65 \quad 88 \quad 63$$
$$57 \quad 64 \quad 67 \quad 72 \quad 74 \quad 55 \quad 82 \quad 61 \quad 68 \quad 77$$

Proceeding as in Section 2.1, we might group these data into the following distribution:

Scores	Tally	Frequency
50–59	///	3
60–69	//// ///	8
70–79	////	5
80–89	///	3
90–99	/	1

where the tally pictures the overall pattern of the data like a histogram (or bar chart) lying on its side.

If we wanted to avoid the loss of information inherent in the above table, we could replace the tally marks with the last digits of the corresponding scores, getting

50–59	2	7	5				
60–69	9	1	5	3	4	7	1	8
70–79	4	9	2	4	7		
80–89	4	8	2				
90–99	3						

This can also be written as

5∗	2	7	5				
6∗	9	1	5	3	4	7	1	8
7∗	4	9	2	4	7		
8∗	4	8	2				
9∗	3						

where ∗ is a placeholder for 0, 1, 2, 3, 4, 5, 6, 7, 8, or 9, or simply as

```
5 │ 2 7 5
6 │ 9 1 5 3 4 7 1 8
7 │ 4 9 2 4 7
8 │ 4 8 2
9 │ 3
```

In either of these final forms, the table is referred to as a **stem-and-leaf plot**—each line is a **stem** and each digit on a stem to the right of the vertical line is a **leaf.** To the left of the vertical line are the **stem labels,** which, in our example, are 5∗, 6∗, . . . , and 9∗, or 5, 6, . . . , and 9.

Essentially, a stem-and-leaf plot presents the same picture as the corresponding tally, yet it retains all the original information. For instance, if a stem-and-leaf plot has the stem

$$12* \quad | \quad 3 \quad 5 \quad 2 \quad 0 \quad 8$$

the corresponding data are 123, 125, 122, 120, and 128, and if a stem-and-leaf plot has the stem

$$3** \quad | \quad 17, 03, 55, 89$$

with two-digit leaves, the corresponding data are 317, 303, 355, and 389.

There are various ways in which stem-and-leaf plots can be modified to meet particular needs (see Exercise 2.38), but we shall not go into this here in any detail as it has been our objective only to present one of the relatively new techniques which come under the general heading of **exploratory data analysis.** Let us point out, though, that in a stem-and-leaf plot the data are grouped with respect to the stem labels—and as we shall see in Chapter 3, this simplifies the determination of certain further descriptions.

EXERCISES

(Exercise 2.38 is a practice exercise; its complete solution is given on page 37.)

2.34 The following are the IQ's of sixteen high school students: 120, 105, 112, 108, 102, 117, 100, 108, 103, 107, 115, 143, 98, 126, 103, and 114. Construct a stem-and-leaf plot with the stem labels 9, 10, . . . , and 14.

2.35 The following are the weights of twenty-four applicants for jobs with a city's fire department: 216, 170, 194, 212, 194, 205, 186, 190, 181, 198, 204, 223, 169, 226, 196, 175, 207, 183, 199, 187, 203, 218, 187, and 192. Construct a stem-and-leaf plot with one-digit leaves.

2.36 The following are the weekly earnings (in dollars) of fifteen salespersons: 305, 255, 319, 167, 270, 291, 512, 283, 334, 362, 188, 217, 440, 195, and 408. Construct a stem-and-leaf plot with the stem labels 1, 2, 3, 4, and 5 (and, hence, with two-digit leaves).

2.37 List the data which correspond to the following stems of stem-and-leaf plots:

(a) 1 *	$\vert$	0	2	7	5	1	1	8
(b) 12	$\vert$	5	3	3	0	2		
(c) 3 **	$\vert$	45,	18,	66,	01			
(d) 1.5	$\vert$	0	7	2	2	9		

2.38 If we want to construct a stem-and-leaf plot with more stems than there would be otherwise, we might use * as a placeholder for 0, 1, 2, 3, and 4, and · as a placeholder for 5, 6, 7, 8, and 9. For the data on page 31 we would thus get the **double-stem plot**

5*	$\vert$	2			
5·	$\vert$	7	5		
6*	$\vert$	1	3	4	1
6·	$\vert$	9	5	7	8
7*	$\vert$	4	2	4	
7·	$\vert$	9	7		
8*	$\vert$	4	2		
8·	$\vert$	8			
9*	$\vert$	3			

where we doubled the number of stems by cutting the interval covered by each class in half. Construct a double-stem plot with one-digit leaves for the data of Exercise 2.35.

2.39 The following are the ages of thirty heads of household in a retirement community: 68, 81, 61, 62, 76, 65, 69, 73, 78, 60, 64, 74, 57, 70, 68, 66, 83, 71, 59, 66, 61, 60, 85, 72, 76, 65, 67, 73, 72, and 67. Construct a double-stem plot (see Exercise 2.38) with one-digit leaves.

2.4

Two-Way Tables ★ [†]

There are many problems of a statistical nature where we deal with **paired data**—the ages of husbands and wives, the supply and the demand for crude oil, rainfall and the per acre yield of wheat, grade-point averages and IQ's, and so on. The decisions we must make when grouping paired data are almost identical with the ones we met in the construction of an ordinary frequency distribution. We must decide how many classes to use for each variable and from where to where each class is to go.

[†] As is explained in the Preface, all sections marked ★ may be regarded as optional. Exercises relating to optional sections are marked in the same way.

Group the following paired data on the number of chicken dinners, x, and the number of steak dinners, y, which were served by a restaurant on 24 Sundays:

x	y	x	y
65	41	73	53
71	37	75	46
74	48	67	36
75	46	63	38
72	45	70	48
78	43	74	47
64	35	76	41
65	40	60	43
78	51	71	45
76	49	66	48
67	43	76	46
72	42	72	41

Solution   Since the smallest number of chicken dinners is 60 and the largest is 78, the four classes 60–64, 65–69, 70–74, and 75–79 provide a suitable, though arbitrary, classification for x; similarly, since the smallest number of steak dinners is 35 and the largest is 53, the four classes 35–39, 40–44, 45–49, and 50–54 provide a suitable, though arbitrary, classification for y. With these classes, we get the following **two-way classification:**

	x			
y	60–64	65–69	70–74	75–79
35–39				
40–44				
45–49				
50–54				

Altogether there are $4 \cdot 4 = 16$ classes, or **cells,** and this is why we chose a small number of classes, only four, for each of the two variables.

Next, we tally the data by observing that the first pair, 65 and 41, goes into the second row of the second column; the second pair, 71 and

37, goes into the first row of the third column; . . . ; and the last pair, 72 and 41, goes into the second row of the third column. Thus, we arrive at the following **two-way table:**

	60–64	65–69	70–74	75–79
35–39	2	1	1	
40–44	1	3	2	2
45–49		1	5	4
50–54			1	1

x above columns, *y* beside rows.

EXERCISES

⋆ 2.40 The following are the final examination grades which 36 students obtained in psychology, *x*, and economics, *y*:

x	y	x	y	x	y
53	70	31	72	35	57
24	38	80	86	56	72
62	55	45	46	65	63
90	78	78	57	78	76
18	35	71	71	49	53
94	91	84	72	82	100
73	69	58	59	22	38
85	83	9	14	90	82
25	51	16	42	77	82
71	53	97	93	35	19
81	60	65	61	52	43
52	58	42	58	93	79

Tally these data into a two-way table having the classes 1–20, 21–40, 41–60, 61–80, and 81–100 for both *x* and *y*, and show the frequencies corresponding to the 25 cells.

⋆ 2.41 In twenty-four consecutive weeks, police officers Jones and Brown issued, respectively, 31 and 35, 38 and 43, 42 and 38, 32 and 37, 39 and 33, 26 and 28, 33 and 27, 40 and 38, 33 and 39, 29 and 34, 45 and 33, 34 and 31, 37 and 39, 27 and 26, 43 and 39, 31 and 33, 34 and 34, 40 and 38, 33 and 35, 36 and 37,

29 and 27, 41 and 37, 32 and 30, and 41 and 24 speeding tickets. Tally these figures into the following table:

	Officer Jones			
	26–30	31–35	36–40	41–45
Officer Brown 21–25				
26–30				
31–35				
36–40				
41–45				

Also show the frequencies associated with the twenty cells of the table.

SOLUTIONS OF PRACTICE EXERCISES

2.1 Since the weights cover a range of $236 - 148 = 88$ (89 values including 148 and 236), each of the ten classes must cover at least nine values, and it suggests, itself, that we use the ten classes 140–149, 150–159, 160–169, 170–179, 180–189, 190–199, 200–209, 210–219, 220–229, and 230—239.

2.4 (a) Yes; (b) no; (c) yes; (d) no; (e) no.

2.7 (a) The lower class limits are 0, 15, 30, and 45.
 (b) The upper class limits are 14, 29, 44, and 59.
 (c) The class marks are $\dfrac{0 + 14}{2} = 7, \dfrac{15 + 29}{2} = 22, \dfrac{30 + 44}{2} = 37$, and $\dfrac{45 + 59}{2} = 52$.
 (d) The class interval of the distribution is $22 - 7 = 15$.

2.11 (a) The boundary between the first two classes is $\dfrac{4 + 11}{2} = 7.5$, the boundary between the second and third classes is $\dfrac{11 + 18}{2} = 14.5$, and the boundary between the third and fourth classes is $\dfrac{18 + 25}{2} = 21.5$; since the differences between successive class boundaries is $14.5 - 7.5 = 7$, the lower boundary of the first class is $7.5 - 7 = 0.5$, and the upper boundary of the fourth class is $21.5 + 7 = 28.5$.
 (b) The limits of the first class are the smallest and largest integers falling between 0.5 and 7.5, namely, 1 and 7; the limits of the second class are the smallest and largest integers falling between 7.5 and 14.5, namely, 8 and 14;

CHAP. 2: Frequency Distributions

similarly, the limits of the third class are 15 and 21, and the limits of the fourth class are 22 and 28.

2.14 There is an ambiguity because values from \$15.00 to \$15.99 can be put into the second or third class; also, there is no place to put values from \$24.91 to \$24.99 and no place to put \$30.00.

2.38

16·	9			
17∗	0			
17·	5			
18∗	1	3		
18·	6	7	7	
19∗	4	4	0	2
19·	8	6	9	
20∗	4	3		
20·	5	7		
21∗	2			
21·	6	8		
22∗	3			
22·	6			

Jan.	38.0
Feb.	40.1
Mar.	48.6
Apr.	61.1
May	71.5
June	80.1
July	85.3
Aug.	83.7
Sep.	76.4
Oct.	65.6
Nov.	53.6
Dec.	42.1

The average daily maximum temperature of New York City is a "comfortable" 62.2 degrees.

Statistical Descriptions

In the introduction to the first chapter we said that statistics is a branch of mathematics dealing with the collection, analysis, interpretation, and presentation of masses of numerical data. Actually, "statistics" has several other meanings. It is also used to refer to the data themselves—for instance, when we refer to birth and death records as vital statistics—and as the plural of

STATISTIC (stə-'tis-tik) A quantity (such as an average) that is computed from observed data.

In this sense, a statistic is also referred to as a statistical measure or simply as a statistical description.

It is customary to classify statistical measures according to the particular features of a set of data which they are supposed to describe. In this chapter we shall be concerned with measures of location, with special emphasis on measures of central location, and measures of variation; others will be taken up later, as needed.

Measures of location are discussed in Sections 3.2 through 3.5, and measures of variation in Sections 3.6 through 3.8. The description of grouped data is treated in Section 3.9, and some further kinds of descriptions are presented in Section 3.10.

3.1

Samples and Populations

Actually, the definition of "statistic" on the preceding page pertains only to sample data, so let us make the following distinction:

> **If a set of data consists of all conceivably possible (or hypothetically possible) observations of a certain phenomenon, we call it a population; if a set of data consists of only a part of these observations, we call it a sample.**

We added the phrase "hypothetically possible" to take care of such clearly hypothetical situations where we look at the outcomes (heads or tails) of 12 flips of a coin as a sample from the population of all possible flips of the coin, where we look at the weights of ten 30-day-old calves as a sample of the weights of all (past, present, and future) 30-day-old calves, or we look at four determinations of the copper content of an ore as a sample of all possible determinations of the copper content of the ore. In fact, we often look at the results of an experiment as a sample of what we might obtain if the experiment were repeated over and over again.

The term "population" comes from the origin of statistics in describing human populations. Even though it may sound strange to refer to the heights of all the trees in a forest as a "population of heights" or to the speeds of all the cars passing a check point as a "population of speeds," in statistics "population" is a technical term with a meaning of its own.

Although we are free to call any set of data a population, what we do in practice depends on the context in which the data are to be viewed. For instance, the complete figures for a recent year, giving the weekly amounts of sales tax collected by the 32 restaurants of a fast-food chain, can be looked upon as either a population or a sample. If a tax collector, auditing the records of this fast-food chain, is interested only in the figures for the given year, they constitute a population; on the other hand, if the management of the chain wants to estimate future collections of sales tax or guess at the amounts of sales tax that was collected by its competitors, the original data must be looked upon as a sample.

EXAMPLE   The social committee of a fraternal order knows how much money it has raised at each of its last six benefit luncheons. Give one example each of a problem in which these data would be looked upon as (a) a population; (b) a sample.

Solution   (a) The data would constitute a population if the committee has to decide how to invest the money raised at the six luncheons.

(b) The data would constitute a sample if the committee wants to estimate how much money might be raised at a future luncheon.

As we have used it here, the word "sample" has very much the same meaning as it has in everyday life. A newspaper considers the attitudes of 200 readers toward a political issue to be a sample of the attitudes of all its readers toward the issue; and a consumer considers a trial tube of toothpaste as a sample of all the toothpaste made by the given firm. Later, we shall use the word "sample" only in connection with data that can reasonably be used to make generalizations about the population from which they came. In this more technical sense, many sets of data which are popularly called samples are not acceptable.

Since the term "statistic" was introduced in connection with sample data, let us add that there is also a name for statistical descriptions of populations—we call them **parameters.** As we shall see, the distinction between statistics and parameters will serve to simplify and clarify our language. Indeed, we shall even use different symbols for statistical measures, depending on whether they are used to describe samples or populations.

3.2

Measures of Location: The Mean

Measures of central location may be described crudely as "averages" in the sense that they are indicative of the "center," "middle," or the "most typical" of a set of data. Indeed, the most popular measure of central location is what the layman calls an "average" and what the statistician calls a **mean.**[†] We put the word "average" in quotes because in everyday language it has all sorts of connotations—we speak of a baseball player's batting average, we talk about

---

[†] The mean is also referred to as the **arithmetic mean,** to distinguish it from the **geometric mean** and the **harmonic mean,** two other kinds of "averages" which are used only in very special situations and will not be discussed in this text.

the average suburban family, we describe a holdup man's appearance as average, and so on. Formally, then, the mean is defined as follows:

The mean of *n* numbers is their sum divided by *n*.

EXAMPLE   On a certain day, nine students received 1, 3, 2, 0, 1, 5, 2, 1, and 3 pieces of mail. Find the mean.

Solution   The total number of pieces of mail which the nine students received is $1 + 3 + 2 + 0 + 1 + 5 + 2 + 1 + 3 = 18$. Since $\dfrac{18}{9} = 2$, the mean number of letters per student is 2.

EXAMPLE   A supermarket manager, who wants to study the "traffic" in her store, finds that 295, 1,002, 941, 768, and 1,283 persons entered the store during the past five days. Find the mean number of persons who entered the supermarket during these five days.

Solution   Altogether, the number of persons who entered the supermarket during the past five days is $295 + 1,002 + 941 + 768 + 1,283 = 4,289$. Since $\dfrac{4,289}{5} = 857.8$, this is the mean (or average) number of persons who entered the store per day.

Since we shall have the occasion to calculate the means of many different sets of data, it will be convenient to have a formula which is always applicable. This requires that we represent the numbers to be averaged by means of symbols such as x, y, or z; the number of values in a sample, the **sample size,** is usually denoted by the letter n. Choosing the letter x, we can refer to the n values in a sample as x_1 (which is read "x sub-one"), x_2 (which is read "x sub-two"), $x_3, \ldots$, and x_n, and write

$$\text{sample mean} = \frac{x_1 + x_2 + x_3 + \cdots + x_n}{n}$$

This formula will take care of any set of sample data, but it can be made more compact by assigning the sample mean the symbol $\bar{x}$ (which is read "x bar") and by using the $\sum$ **notation.** The symbol $\sum$ is capital *sigma*, the Greek letter for

S. In this notation, we let $\sum x$ stand for "the sum of the x's" (that is, $\sum x = x_1 + x_2 + x_3 + \cdots + x_n$), so that we can write

Sample mean

$$\bar{x} = \frac{\sum x}{n}$$

If we refer to the measurements as y's or z's, we write their mean as $\bar{y}$ or $\bar{z}$ and substitute $\sum y$ or $\sum z$ for $\sum x$.

In the formula for $\bar{x}$, the expression $\sum x$ does not state explicitly which x's are to be added; let it be understood, therefore, that $\sum x$ always stands for the sum of all the x's under consideration in a given situation. In Section 3.10, the technical note near the end of this chapter, the use of the sigma notation is discussed in some detail.

The mean of a population of N items is defined in the same way. It is the sum of the N items, $x_1 + x_2 + x_3 + \cdots + x_N$, or $\sum x$, divided by the **population size** N. Assigning the population mean the symbol μ (lowercase *mu*, the Greek letter for m), we write

Population mean

$$\mu = \frac{\sum x}{N}$$

with the reminder that $\sum x$ is now the sum of all N values which constitute the population. To distinguish between parameters and statistics, namely, descriptions of populations and samples, it is common practice to denote the former with Greek letters.

The popularity of the mean as a measure of the "middle" or "center" of a set of data is not accidental. Any time we use a single number to describe some aspect of a set of data, there are certain requirements, or desirable features, that should be kept in mind. Aside from the fact that the mean is a simple and familiar measure, the following are some of its noteworthy properties:

It can be calculated for any set of numerical data, so it always exists.

A set of numerical data has one and only one mean, so it is always unique.

It lends itself to further statistical treatment; as we shall see, for example, the means of several sets of data can always be combined into the overall mean of all the data.

It is relatively reliable in the sense that means of many samples drawn from the same population generally do not fluctuate, or vary, as widely as other statistics used to estimate the mean of a population.

This last property is of fundamental importance in statistical inference, and we we shall study it in some detail in Chapter 8.

There is another property of the mean which, on the surface, seems desirable:

It takes into account every item of a set of data.

However, sometimes a set of data may contain very small or very large values which are so far removed from the main body of the data that the appropriateness of including them when describing the data is questionable. Such values may be due to chance, or they may be due to gross errors in recording the data, gross errors in calculations, malfunctioning of equipment, or other identifiable sources of contamination. When such values are "averaged in" with the other values, they can affect the mean to such an extent that it is debatable whether it really provides a useful description of the "middle" of the data.

EXAMPLE   Five light bulbs burned out after lasting, respectively, for 867, 849, 840, 852, and 822 hours of continuous use. Find the mean and also determine what the mean would have been if the second value had been recorded incorrectly as 489 instead of 849.

Solution   For the original data we get

$$\bar{x} = \frac{867 + 849 + 840 + 852 + 822}{5} = \frac{4,230}{5} = 846$$

and with 489 instead of 849 we get

$$\bar{x} = \frac{867 + 489 + 840 + 852 + 822}{5} = \frac{3,870}{5} = 774$$

This shows that a very small or very large value, here due to a careless error in recording the data, can have a pronounced effect on the mean.

To avoid the possibility of being misled by very small or very large values, we sometimes describe the "middle" or "center" of a set of data with other kinds of statistical measures, for instance, those given in Sections 3.4 and 3.5.

3.3

Measures of Location: The Weighted Mean

When averaging quantities, it is often necessary to account for the fact that not all of them are equally important in the phenomenon being described. For instance, in 1980 the percentage of housing units that were owner occupied was 38.9 in Chicago, 49.1 in San Diego, and 56.3 in Memphis. The mean of these three percentages is $\dfrac{38.9 + 49.1 + 56.3}{3} = 48.1$, but we cannot very well say that this is the average owner occupancy rate for the three cities. The three figures do not carry equal weight because there are not equally many housing units in the three cities (see Exercise 3.19).

In order to give quantities being averaged their proper degree of importance, it is necessary to assign them (relative importance) **weights,** and then calculate a **weighted mean.** In general, the weighted mean $\bar{x}_w$ of a set of numbers $x_1, x_2, x_3, \ldots$, and x_n, whose relative importance is expressed numerically by a corresponding set of numbers $w_1, w_2, w_3, \ldots$, and w_n, is given by

Weighted mean

$$\bar{x}_w = \frac{w_1 x_1 + w_2 x_2 + \cdots + w_n x_n}{w_1 + w_2 + \cdots + w_n} = \frac{\sum w \cdot x}{\sum w}$$

Here $\sum w \cdot x$ stands for the sum of the products obtained by multiplying each x by the corresponding weight, and $\sum w$ is simply the sum of the weights. Note that if the weights are all equal, the formula for the weighted mean reduces to that of the (ordinary) mean.

EXAMPLE   In 1981, cod, flounder, haddock, and ocean perch brought commercial fishermen 33.0, 57.9, 39.4, and 28.3 cents per pound. Given that they caught 100 million pounds of cod, 201 million pounds of flounder, 55 million pounds of haddock, and 19 million pounds of ocean perch, what is the overall average price that they received per pound?

Solution   Substituting $x_1 = 33.0, x_2 = 57.9, x_3 = 39.4, x_4 = 28.3, w_1 = 100, w_2 = 201, w_3 = 55$, and $w_4 = 19$ into the formula for $\bar{x}_w$, we get

$$x_w = \frac{(100)(33.0) + (201)(57.9) + (55)(39.4) + (19)(28.3)}{100 + 201 + 55 + 19}$$

$$= \frac{17,642.6}{375}$$

$$= 47.0$$

Note that the figure in the denominator, 375, is the total catch in millions of pounds, and that the figure in the numerator, 17,642.6, is the total value of the catch in millions of cents, that is, in units of $10,000. Also, if we had averaged 33.0, 57.9, 39.4, and 28.3 without using weights, we would have obtained

$$\bar{x} = \frac{33.0 + 57.9 + 39.4 + 28.3}{4} = 39.6$$

Can you explain why this is much less than the actual average price of 47.0 cents?

A special application of the formula for the weighted mean arises when we must find the overall mean, or **grand mean,** of k sets of data having the means $\bar{x}_1, \bar{x}_2, \bar{x}_3, \ldots,$ and $\bar{x}_k,$ and consisting, respectively, of $n_1, n_2, n_3, \ldots,$ and n_k measurements or observations. The result is given by

Grand mean
of combined data

$$\bar{\bar{x}} = \frac{n_1\bar{x}_1 + n_2\bar{x}_2 + \cdots + n_k\bar{x}_k}{n_1 + n_2 + \cdots + n_k} = \frac{\sum n \cdot \bar{x}}{\sum n}$$

where the weights are the sizes of the respective samples, the numerator is the total of all the measurements or observations, and the denominator is the sample size of the combined data.

EXAMPLE   In a biology class there are 20 freshmen, 18 sophomores, and 12 juniors. If the freshmen averaged 68 in an examination, the sophomores averaged 75, and the juniors averaged 86, find the mean grade for the entire class.

Solution   Substituting $n_1 = 20$, $n_2 = 18$, $n_3 = 12$, $\bar{x}_1 = 68$, $\bar{x}_2 = 75$, and $\bar{x}_3 = 86$ into the formula for the grand mean of combined data, we get

$$\bar{\bar{x}} = \frac{20 \cdot 68 + 18 \cdot 75 + 12 \cdot 86}{20 + 18 + 12}$$

$$= \frac{3,742}{50}$$

$$= 74.84$$

or 75 rounded to the nearest integer.

(Exercises 3.1, 3.4, 3.10, 3.17, and 3.18 are practice exercises; their complete solutions are given on page 88.)

3.1 A producer of television commercials knows exactly how much money was spent on the production of each of four commercials. Give one example each of a problem in which these data would be looked upon as

(a) a population;

(b) a sample.

3.2 The final election returns from a county show that the three candidates for a certain office received 16,255, 11,278, and 3,455 votes. What office might these candidates be running for so that these figures would constitute

(a) a population;

(b) a sample?

3.3 Suppose that we are given complete information about the number of lunches which the executives of an insurance company charged to their expense accounts during the first six months of 1985. Give one illustration each of a situation where these data would be looked upon as

(a) a population;

(b) a sample.

3.4 The following are the ages of twenty persons empaneled for jury duty by a court: 48, 58, 33, 42, 57, 31, 52, 25, 46, 60, 61, 49, 38, 53, 30, 47, 52, 63, 41, and 34. Find their mean age.

3.5 The following are the weight losses (in pounds) of eight persons following a prescribed diet for four weeks: 5.3, 2.4, 1.9, 3.5, 4.0, 2.6, 2.5, and 1.8. Find their mean weight loss.

3.6 At their first inauguration, the first ten presidents of the United States were 57, 61, 57, 57, 58, 57, 61, 54, 68, and 51 years old. Calculate the mean.

3.7 Trying to determine the calories per serving of lasagna in a laboratory assignment in nutrition, fifteen students obtained 329, 335, 347, 318, 316, 322, 330, 351, 362, 315, 342, 346, 353, 327, and 333. Find the mean.

3.8 A week's records of a cab company show the following amounts (in dollars) spent on gasoline by each of its 32 cabs:

37.87	24.75	38.67	32.45	31.55	50.55	64.50	39.01
69.49	52.83	53.41	60.75	21.45	47.82	40.58	30.56
37.51	34.69	41.88	30.24	15.25	27.63	24.65	45.14
20.11	31.22	41.35	26.27	36.00	38.76	25.68	23.65

Find the mean expenditure of these cabs.

3.9 The records of 15 persons convicted of various crimes showed that, respectively, 4, 3, 0, 0, 2, 4, 4, 3, 1, 0, 2, 0, 2, 1, and 4 of their grandparents were foreign-born. Find the mean and discuss whether it can be used to support the contention that the "average criminal" has two foreign-born grandparents.

3.10 An elevator in a hotel is designed to carry a maximum load of 2,000 pounds. Is it overloaded if at one time it carries 8 women whose mean weight is 123 pounds and 5 men whose mean weight is 174 pounds?

3.11 A bridge is designed to carry a maximum load of 180,000 pounds. If at a given moment it is loaded with 36 vehicles having a mean weight of 4,630 pounds, is there any real danger that it might collapse?

3.12 By mistake, an instructor has erased the grade which one of ten students in her class received in a final examination. However, she knows that the students averaged (had a mean grade of) 71 in the examination, and that the other nine students received grades of 96, 44, 82, 70, 47, 74, 94, 78, and 56. What must have been the grade which the teacher erased?

3.13 The hours of sleep that a student has had during each of the last ten nights are 6, 8, 2, 13, 9, 7, 6, 5, 10, and 7. Find the mean and discuss its usefulness in describing the "middle" of the data.

3.14 A bill was introduced in a state legislature to repeal the sales tax on prescription drugs. Comment on the argument of the state finance director that the average (mean) per capita prescription bill for the past four years was a trifling $5.20, which is not really a burden to anyone.

3.15 Records show that in Phoenix, Arizona, the normal daily maximum temperature for each month is 65, 69, 74, 84, 93, 102, 105, 102, 98, 88, 74, and 66 degrees Fahrenheit. Verify that the mean of these figures is 85 and comment on the claim that, in Phoenix, the average daily maximum temperature is a very comfortable 85 degrees.

3.16 Careful measurements show that the actual coffee content of six jars of instant coffee is 6.03, 5.98, 6.04, 6.00, 5.99, and 6.02 ounces of coffee. What would have been the error in the mean coffee content of the six jars if the third value had been recorded incorrectly as 6.40?

3.17 If somebody invests $1,000 at 7%, $3,000 at $7\frac{1}{2}$%, and $16,000 at 8%, what is the overall percentage yield of these investments?

3.18 In a recent year, the average salaries of elementary school teachers in Oregon, Washington, and Alaska were $13,300, $14,500, and $21,000. Given that there were 13,000, 17,200, and 2,400 elementary school teachers in these states, find the average salary of all the elementary school teachers in the three states.

3.19 On page 45 we stated that in 1980 the percentage of housing units that were owner occupied was 38.9 in Chicago, 49.1 in San Diego, and 56.3 in Memphis. Given that there were 1,175 thousand housing units in Chicago, 342 thousand in San Diego, and 244 thousand in Memphis, what is the corresponding percentage for the three cities combined?

3.20 An instructor counts the final examination in a course four times as much as each of the four one-hour examinations. What is the average grade of a student who received grades of 71, 77, 58, and 74 in the four one-hour examinations and 80 in the final examination?

3.21 In a recent season, a baseball team's five best hitters had batting averages of .307, .299, .297, .291, and .283. What is their combined batting average if they had, respectively, 488, 137, 646, 533, and 502 at bats?

3.22 During a special promotion, a discount chain sold 575, 410, and 520 microwave ovens in three of its stores at average prices of $495, $525, and $500. What is the mean price of the ovens sold?

3.23 In a study of home replacement costs, it was found that 12 houses in one city were underinsured on the average by $5,000, 24 homes in another city were underinsured on the average by $6,200, and 14 homes in a third city were underinsured on the average by $6,900. On the average, by how much were these 50 homes underinsured?

3.24 The average IQ's of the students in three high schools are 108, 112, and 102. If there are, respectively, 1,024, 686, and 790 students in these schools, what is the average IQ of all the students in the three schools?

3.4

Measures of Location: The Median and Other Fractiles

To avoid the possibility of being misled by a few very small or very large values, as in the example on page 44, it is sometimes preferable to describe the "middle" or "center" of a set of data with statistical measures other than the mean. The definition of one of these, the **median** on n values, requires that we arrange the data according to size. Then,

When n is odd, the median is the value of the item that is in the middle.

When n is even, the median is the mean of the two items that are nearest to the middle.

The symbol which we use for the median of a set of x's is $\tilde{x}$ (and, hence, $\tilde{y}$ or $\tilde{z}$ if we refer to the measurements as y's or z's).

EXAMPLE   In a recent month, a state's Game and Fish Department reported 53, 31, 67, 53, and 36 hunting or fishing violations for five different regions. Find the median number of violations for these months.

Solution   The median is not 67, the third (or middle) item, because the figures must first be arranged according to size. Thus, we get

$$31 \quad 36 \quad 53 \quad 53 \quad 67$$

and it can be seen that the median is 53.

Note that in this example there are two 53's among the data and that we do not refer to either of them as *the* median—the median is a number and not necessarily a particular measurement or observation.

EXAMPLE   Ten persons, sent out to interview 50 students at each of ten different campuses, found that 18, 13, 15, 12, 8, 21, 7, 11, 16, and 3 of the students sampled jog regularly. Find the median.

Solution   Arranging these figures according to size, we get

$$3 \quad 7 \quad 8 \quad 11 \quad 12 \quad 13 \quad 15 \quad 16 \quad 18 \quad 21$$

and it can be seen that the median is $\dfrac{12 + 13}{2} = 12.5$, namely, the mean of the two items nearest the middle.

We cannot give a general formula for the value of the median, but we can give one for the **median position.** With data ranked from low to high or high to low, and counting from either end,

The median is the value of the $\dfrac{n + 1}{2}$th item.

When n is odd, $\dfrac{n + 1}{2}$ is an integer and it gives the position of the median; when n is even, $\dfrac{n + 1}{2}$ is midway between two integers and the median is the mean of the values of the corresponding items.

EXAMPLE   Find the median position for (a) $n = 11$, (b) $n = 25$, and (c) $n = 75$.

Solution   With the data arranged according to size and counting from either end,

(a) $\dfrac{n + 1}{2} = \dfrac{11 + 1}{2} = 6$, so that for $n = 11$ the median is the value of the 6th item;

(b) $\dfrac{n + 1}{2} = \dfrac{25 + 1}{2} = 13$, so that for $n = 25$ the median is the value of the 13th item;

(c) $\dfrac{n + 1}{2} = \dfrac{75 + 1}{2} = 38$, so that for $n = 75$ the median is the value of the 38th item.

EXAMPLE   Find the median position for (a) $n = 8$, (b) $n = 20$, and (c) $n = 100$.

Solution   With the data arranged according to size and counting from either end,

(a) $\dfrac{n+1}{2} = \dfrac{8+1}{2} = 4.5$, so that for $n = 8$ the median is the mean of the values of the 4th and 5th items;

(b) $\dfrac{n+1}{2} = \dfrac{20+1}{2} = 10.5$, so that for $n = 20$ the median is the mean of the values of the 10th and 11th items;

(c) $\dfrac{n+1}{2} = \dfrac{100+1}{2} = 50.5$, so that for $n = 100$ the median is the mean of the values of the 50th and 51st items.

It is important to remember that $\dfrac{n+1}{2}$ is a formula for the median position, and not a formula for the median, itself.

To simplify the determination of a median, it sometimes helps to utilize the grouping provided by a stem-and-leaf plot.

EXAMPLE   The following are the numbers of passengers on 25 runs of a ferryboat: 52, 84, 40, 57, 61, 65, 77, 64, 62, 35, 82, 58, 50, 78, 103, 71, 75, 41, 53, 66, 60, 95, 58, 49, and 89. Construct a stem-and-leaf plot with one-digit leaves and use it to find the median.

Solution   First constructing the stem-and-leaf plot, we get

3	5
4	0 1 9
5	2 7 8 0 3 8
6	1 5 4 2 6 0
7	7 8 1 5
8	4 2 9
9	5
10	3

Since the median position is $\dfrac{25+1}{2} = 13$ and ten of the values fall on the first three stems, we must find the third smallest value on the fourth stem. As can be seen by inspection, it is 62.

51

If we had calculated the mean in the example on page 49 which dealt with the hunting and fishing violations, we would have obtained

$$\frac{53 + 31 + 67 + 53 + 36}{5} = \frac{240}{5} = 48$$

and it should not come as a surprise that it differs from the median, which was 53. Each of these averages describes the "middle" or "center" of the data in its own way. The median is average in the sense that it splits the data into two parts so that as many of the values are to the left of the median position as there are to its right. The mean, on the other hand, is typical in the sense that if each value in a set of data is replaced by the same number while the total remains unchanged, this number will have to be the mean. This follows directly from the formula for $\bar{x}$, which, multiplied by n, yields $n \cdot \bar{x} = n \cdot \dfrac{\sum x}{n} = \sum x$.

The median shares the first two properties of the mean listed on page 43, that is, it can be determined for any set of numerical data and it is always unique. Also like the mean, the median is simple enough to find once the data have been arranged according to size, but it should be kept in mind that ordering large sets of data manually can be a very tedious job. So far as the third and fourth properties are concerned, the medians of several sets of data can generally not be combined into the overall median of all the data, and in problems of inference the median is generally not as reliable as the mean; that is, it is subject to greater chance fluctuations, as is illustrated by Exercises 3.37 and 3.38 on pages 56 and 57. When it comes to the fifth property, the median is actually preferable to the mean, as it is not so easily affected by extreme (very large or very small) values.

EXAMPLE In the example on page 44 we showed that the mean lifetime of five light bulbs is $\bar{x} = 846$ hours, but if one of the values, 849, is recorded incorrectly as 489, the mean lifetime is $\bar{x} = 774$, which is off by $846 - 774 = 72$ hours. What would have been the corresponding error if we had used the medians?

Solution Since the original values are 867, 849, 840, 852, and 822, the correct value of the median is $\tilde{x} = 849$; with 489 substituted for 849, the values are 867, 489, 840, 852, and 822, and $\tilde{x} = 840$, which is off by $849 - 840 = 9$ hours. This is only an eighth of the error we made when we used the mean.

Also, unlike the mean, the median can be used to define the "middle" of a number of objects, properties, or attributes. It is possible, for example, to rank a number of tasks according to their difficulty and then describe the middle (or median) one as being of "average" difficulty; also, we might rank samples of chocolate fudge according to their consistency and then describe the middle (or median) one as having "average" consistency.

The median is but one of many different **fractiles** which divide a set of data into two or more equal parts. Also of importance in statistics are **quartiles** and **percentiles,** but since percentiles are used mainly in connection with large sets of data, we shall discuss them for grouped data in Section 3.9. Thus, let us treat here the three quartiles, Q_1, Q_2, and Q_3, which are defined as follows: With the data arranged from left to right in an increasing order of magnitude,

> The first quartile, Q_1, is the median of all the values to the left of the median position for the whole set of data.
>
> The second quartile, Q_2, is the median.
>
> The third quartile, Q_3, is the median of all the values to the right of the median position for the whole set of data.

Generally, we find Q_3 by counting as many places as for Q_1, starting at the other end.

Note that with this definition there are always as many values to the left of the Q_1 position as there are between the Q_1 position and the median position, between the median position and the Q_3 position, and to the right of the Q_3 position.[†]

EXAMPLE   Verify this assertion for $n = 23$.

Solution   The median position is $\dfrac{23 + 1}{2} = 12$, so that there are eleven values to

its left and eleven values to its right. Thus, the Q_1 position is $\dfrac{11 + 1}{2} = 6$

and, correspondingly, Q_3 is the 6th value counting from the other end;

[†] Some recent texts define **hinges** instead of the first and third quartiles, and there may be a difference for odd values of n. For instance, the **lower hinge** is defined as the median of all the values to the left of or at the median position, so that for $n = 23$, where the median position is $\dfrac{23 + 1}{2} = 12$, the position of the lower hinge is $\dfrac{12 + 1}{2} = 6.5$ and the position of the first quartile is $\dfrac{11 + 1}{2} = 6$. Note that there are six values to the left of the position of the lower hinge and only five between the positions of the lower hinge and the median.

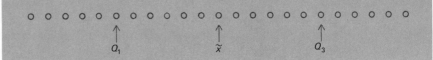

3.1

The median and quartile positions for n = 23.

that is, the Q_3 position is 18 (see Figure 3.1). It follows that there are five values to the left of the Q_1 position, five values between the Q_1 position and the median position (between 6 and 12), five values between the median position and the Q_3 position (between 12 and 18), and five values to the right of the Q_3 position.

EXAMPLE   Use the stem-and-leaf plot constructed on page 51 to find Q_1 and Q_3 for the data on the passengers of the ferryboat.

Solution   Since we have already shown that the median position is 13, we find that the Q_1 position is $\frac{12 + 1}{2} = 6.5$. Thus, Q_1 is the mean of the 6th and 7th values, and it can be seen from the stem-and-leaf plot that it equals $\frac{52 + 53}{2} = 52.5$. To find Q_3 we count 6.5 values starting at the other end. Since the 6th value is 78 and the 7th value is 77, we get $\frac{78 + 77}{2} = 77.5$.

EXAMPLE   The following are the numbers of minutes which a person, on her way to work, had to wait for the bus on fourteen working days: 10, 2, 17, 6, 8, 3, 10, 2, 9, 5, 9, 13, 1, and 10. Find the median, Q_1, and Q_3.

Solution   For $n = 14$, the median position is $\frac{14 + 1}{2} = 7.5$, so that the Q_1 position is $\frac{7 + 1}{2} = 4$, and Q_3 is the fourth value from the other end. Since the data, arranged according to size, are

$$1 \quad 2 \quad 2 \quad 3 \quad 5 \quad 6 \quad 8 \quad 9 \quad 9 \quad 10 \quad 10 \quad 10 \quad 13 \quad 17$$

it can be seen that the median is $\frac{8 + 9}{2} = 8.5$, Q_1 is 3, and $Q_3 = 10$.

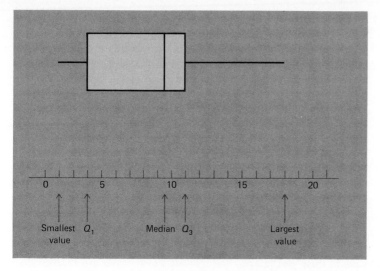

3.2

Box-and-whisker plot for data on passengers of ferryboat.

The information provided by the median, the two quartiles, and also the smallest and largest values, is sometimes presented in the form of a **box-and-whisker plot.** Such a plot is shown in Figure 3.2 for the data of the preceding example. It provides useful information for the exploratory analysis of the overall shape of the distribution of a set of data (see also Section 3.10).

EXERCISES

(Exercises 3.25, 3.28, 3.37, 3.42, and 3.45 are practice exercises; their complete solutions are given on pages 89 and 90.)

3.25 Find the median position for (a) $n = 15$, and (b) $n = 40$.

3.26 Find the median position for (a) $n = 17$, and (b) $n = 30$.

3.27 Find the median position for (a) $n = 39$, and (b) $n = 150$.

3.28 In 1985, twelve used-car salespersons sold, respectively, 58, 70, 85, 42, 64, 46, 66, 89, 44, 93, 58, and 79 used cars. Find the median.

3.29 The following are the numbers of restaurant meals which thirteen persons ate during a given week: 3, 10, 5, 1, 8, 5, 6, 12, 15, 1, 0, 6, and 5. Find the median.

3.30 Copying a report, the sixteen students in a typing class made 0, 1, 3, 0, 0, 0, 0, 4, 0, 0, 1, 2, 0, 1, 0, and 2 mistakes. Find
  (a) the mean;
  (b) the median.
Discuss the relative merits of the two results.

3.31 Twenty power failures lasted 18, 125, 44, 96, 31, 26, 80, 49, 125, 63, 45, 33, 89, 12, 103, 75, 40, 80, 61, and 28 minutes. Find the median.

3.32 In Exercise 3.4 on page 47 we gave the ages of twenty persons empaneled for jury duty. Find the median.

3.33 Find the median of the distribution on page 13 concerning the number of full-length movies that 400 persons had seen on television during a given week.

3.34 Twenty-five NBA games lasted, respectively, 138, 142, 113, 126, 135, 142, 159, 157, 140, 157, 121, 128, 142, 164, 155, 139, 143, 158, 140, 118, 142, 146, 123, 130, and 137 minutes. Find the median length of these games
  (a) directly;
  (b) by first constructing a stem-and-leaf plot.

3.35 Use the stem-and-leaf plot on page 32 to determine the median of the scores which the twenty students obtained on the physical coordination test.

3.36 Each of 15 persons soliciting funds for a charitable organization was assigned a quota (amount of money) he or she should raise, and the following are the percentages of their respective quotas which they actually attained: 92, 107, 353, 90, 78, 80, 74, 92, 102, 86, 106, 109, 95, 102, and 91. Calculate the mean and the median of these percentages, and indicate which of the two measures is a better indication of these persons' "average" performance.

3.37 To verify the claim that the mean is generally more reliable than the median (namely, that it is subject to smaller chance fluctuations), a student conducted an experiment consisting of fifteen times drawing three cards from an ordinary deck of 52 playing cards after having removed the jacks, queens, and kings. The following are the results: 8, 3, and 2; 6, 6, and 7; 3, 5, and 9; 4, 4, and 10; 8, 3, and 8; 10, 1, and 4; 7, 10, and 4; 1, 3, and 1; 5, 1, and 9; 8, 10, and 3; 4, 5, and 9; 5, 3, and 8; 3, 6, and 7; 1, 8, and 8; and 10, 5, and 6.
  (a) Calculate the fifteen medians and the fifteen means.
  (b) Group the medians and the means obtained in part (a) into separate distributions having the classes 0.5–2.5, 2.5–4.5, 4.5–6.5, and 6.5–8.5. (There can be no ambiguities since the medians and the means of three whole numbers cannot equal 2.5, 4.5, or 6.5.)
  (c) Draw histograms of the two distributions obtained in part (b) and explain how they illustrate the claim that the mean is generally more reliable than the median.

3.38 To verify the claim that the mean is generally more reliable than the median (namely, that it is subject to smaller chance fluctuations), a student conducted an experiment consisting of 12 tosses of three dice. The following are his results: 2, 4, and 6; 5, 3, and 5; 4, 5, and 3; 5, 2, and 3; 6, 1, and 5; 3, 2, and 1; 3, 1, and 4; 5, 5, and 2; 3, 3, and 4; 1, 6, and 2; 3, 3, and 3; and 4, 5, and 3.
  (a) Calculate the twelve medians and the twelve means.
  (b) Group the medians and the means obtained in part (a) into separate distributions having the classes 1.5–2.5, 2.5–3.5, 3.5–4.5, and 4.5–5.5. (Note that there will be no ambiguities since the medians of three whole numbers and the means of three whole numbers cannot equal 2.5, 3.5, or 4.5.)

(c) Draw histograms of the two distributions obtained in part (b) and explain how they illustrate the claim that the mean is generally more reliable than the median.

3.39 Repeat the preceding exercise with your own data by repeatedly rolling three dice (or one die three times) and construct corresponding distributions for the twelve medians and the twelve means. (If no dice are available, simulate the experiment mentally or by drawing numbered slips of paper out of a hat.)

3.40 A consumer testing service obtained the following miles per gallon in five test runs performed with each of three compact cars:

Car A: 27.9, 30.4, 30.6, 31.4, 31.7
Car B: 31.2, 28.7, 31.3, 28.7, 31.3
Car C: 28.6, 29.1, 28.5, 32.1, 29.7

(a) If the manufacturers of car A want to advertise that their car performed best in this test, which of the "averages" discussed in this text could they use to substantiate their claim?

(b) If the manufacturers of car B want to advertise that their car performed best in this test, which of the "averages" discussed in this text could they use to substantiate their claim?

★ 3.41 Suppose that the manufacturers of car C of the preceding exercise hire an unscrupulous statistician and instruct him to find some kind of "average" which will show that their car performed best in the test. Show that the **midrange,** which is defined as the mean of the smallest and largest values, will serve their purpose.

3.42 Find the positions of the median, Q_1, and Q_3 when $n = 21$, and verify that there are as many values to the left of the Q_1 position as there are between the Q_1 position and the median position, between the median position and the Q_3 position, and to the right of the Q_3 position.

3.43 Rework the preceding exercise with $n = 18$.

3.44 Rework Exercise 3.42 with $n = 20$.

3.45 With reference to Exercise 3.29, find Q_1 and Q_3 for the numbers of restaurant meals which the thirteen persons ate during the given week.

3.46 With reference to Exercise 3.31, find Q_1 and Q_3 for the lengths of the power failures.

3.47 With reference to Exercise 3.34, find Q_1 and Q_3 for the lengths of the NBA games
(a) directly from the data;
(b) by first constructing a stem-and-leaf plot.

3.48 Use the results of Exercises 3.31 and 3.46 to construct a box-and-whisker plot for the lengths of the power failures.

3.49 Use the results of Exercises 3.34 and 3.47 to construct a box-and-whisker plot for the lengths of the NBA games.

57

3.5

Measures of Location: The Mode

The **mode** is another measure of location which is sometimes used to describe the "middle" of a set of data. It is defined as follows:

The mode is the value which occurs with the highest frequency.

In this sense it is "most typical" of a set of data; its two main advantages are that it requires no calculations, only counting, and that it can be determined for qualitative as well as quantitative data.

EXAMPLE   The 20 meetings of a square dance club were attended by 26, 25, 28, 23, 25, 24, 24, 21, 23, 26, 28, 26, 24, 23, 24, 32, 25, 27, 24, and 22 of its members. Find the mode.

Solution   Among these numbers, 21, 22, 27, and 32 each occurs once, 28 occurs twice; 23, 25, and 26 each occurs three times; and 24 occurs five times. Thus, 24 is the **modal attendance.**

EXAMPLE   Asked to name the best collegiate basketball team in the country, twenty-five sportswriters named: Virginia, Houston, UCLA, Memphis State, North Carolina, Indiana, Houston, Louisville, Houston, UCLA, Houston, Virginia, Michigan, North Carolina, Louisville, De Paul, UCLA, North Carolina, Memphis State, Houston, Houston, Virginia, UCLA, Louisville, and North Carolina. Find the mode.

Solution   Since Indiana, Michigan, and De Paul are named once, Memphis State is named twice, Virginia and Louisville are named three times, UCLA and North Carolina are named four times, and Houston is named six times, Houston is the **modal choice.**

Two definite disadvantages of the mode are that it may not exist, which is the case when no two items are alike, and it may not be unique. When there is more than one mode, this is sometimes indicative of the fact that the data are not homogeneous, namely, that they can be looked upon as a combination of several sets of data.

A sample from the records of a motor vehicle bureau shows that sixteen drivers in a certain age group received 2, 3, 3, 1, 0, 0, 2, 1, 0, 3, 4, 0, 3, 2, 3, and 0 tickets in a recent year. Find the mode.

Solution Since 0 occurs five times, 1 occurs two times, 2 occurs three times, 3 occurs five times, and 4 occurs once, 0 and 3 each occurs with the maximum frequency of five. Thus, there are two modes. We might infer from this that there are many very good drivers, many very poor drivers, and fewer drivers in the categories between these two extremes.

There are many other measures of location besides the mean, the median, and the mode, and the question of what particular "average" should be used in a given situation is not always easily answered. The fact that the selection of statistical descriptions is to some extent arbitrary, has led some persons to believe that the magic of statistics can be used to prove almost anything. A famous nineteenth-century British statesman is often quoted as having said that there are three kinds of lies: lies, damned lies, and statistics. Exercises 3.40 and 3.41 on page 57 described a situation where this kind of criticism would well be justified.

EXERCISES (Exercise 3.50 is a practice exercise; its complete solution is given on page 90.)

3.50 The following are the amounts of time (in minutes) which sixteen persons spent standing in line waiting to buy tickets for a concert: 8, 2, 9, 1, 16, 5, 7, 11, 9, 1, 14, 12, 9, 0, 8, and 4. Find the mode.

3.51 An inspection of eighteen newly completed home plumbing jobs in one city turned up 0, 1, 3, 0, 0, 0, 1, 0, 4, 1, 0, 1, 1, 2, 0, 1, 0, and 2 violations of the city's building code. Find the mode.

3.52 With reference to Exercise 3.29, find the mode of the number of restaurant meals which the persons ate.

3.53 With reference to Exercise 3.34, find the mode of the lengths of the basketball games.

3.54 With reference to Exercise 2.18 on page 22, find the mode of the accounting grades of the fifty students.

3.55 Asked for their favorite color, fifty persons said: red, blue, blue, green, yellow, blue, brown, red, blue, red, red, green, white, blue, red, green, blue, red, green, green, purple, white, yellow, blue, blue, blue, red, red, brown, orange, white, green, blue, blue, black, red, blue, red, yellow, green, yellow, blue, blue, orange, red, green, white, purple, blue, and red. What was their modal choice?

3.56 Forty registered voters were asked whether they considered themselves Democrats, Republicans, or Independents. Use the following results to determine their modal choice: Democrat, Republican, Independent, Independent, Democrat, Independent, Republican, Republican, Independent, Democrat, Democrat, Independent, Democrat, Independent, Republican, Independent, Independent, Independent, Democrat, Democrat, Republican, Independent, Independent, Republican, Republican, Democrat, Republican, Democrat, Independent, Independent, Democrat, Democrat, Independent, Republican, Independent, Independent, Democrat, Independent, Republican, Democrat.

3.6

Measures of Variation: The Range

An important feature of most any kind of data is that the values are not all alike, and the extent to which they are unalike or vary among themselves is of basic importance in statistics. The following are some examples which illustrate the importance of measuring the variability of statistical data.

The chocolate chip ice cream produced by one company averages 360 chocolate chips per quart, with all quarts containing anywhere from 340 to 380 chocolate chips. Another company's chocolate chip ice cream also averages 360 chocolate chips per quart, but some quarts contain as few as 20 chips and others as many as 740. If the ice cream produced by the two companies is the same in all other respects, it stands to reason that most persons would prefer the chocolate chip ice cream made by the first company—this should give them a better chance of getting ice cream with a "desirable" number of chocolate chips.

Suppose that in a hospital each patient's pulse rate is taken three times a day, and that on a certain day the records of two patients show

> *Patient A:* 72, 76, 74
> *Patient B:* 59, 92, 71

The mean pulse rates are the same, as can easily be checked, but observe the difference in variability. Whereas patient A's pulse rate is quite stable, that of patient B fluctuates widely, and this should be taken into account when treatments are prescribed.

Finally, suppose that we have a coin which is slightly bent and we wonder whether it is still balanced or "fair." So, we toss the coin 200 times, and if we get 100 heads and 100 tails there would be no doubt. What if we get 130 heads, however, or only 92? Are the discrepancies due to chance or due to the coin's being bent? Clearly, we need some indication of how much variability can be attributed to chance.

These examples show the need for statistical descriptions which measure the extent to which data are dispersed, or spread out, namely, the need for **measures of variation.**

To introduce one of the simplest ways of measuring variability, let us refer to the second of the preceding examples, and let us observe that the pulse rate of patient A varied from 72 to 76, while that of patient B varied from 59 to 92. These extreme (smallest and largest) values tell us something about the variability of the respective sets of data, and so do their differences. Thus, we make the following definition:

The range of a set of data is the largest value minus the smallest.

For patient A of the example above we get a range of $76 - 72 = 4$, and for patient B we get a range of $92 - 59 = 33$.

The range is easy to calculate and easy to understand, but despite these advantages it is generally not a useful measure of variation. Its main shortcoming is that it does not tell us anything about the dispersion of the values which fall between the two extremes. Each of the following sets of data

$$
\begin{array}{llllllllll}
\textit{Set 1:} & 5, & 20, & 20, & 20, & 20, & 20, & 20, & 20 \\
\textit{Set 2:} & 5, & 5, & 5, & 5, & 20, & 20, & 20, & 20 \\
\textit{Set 3:} & 5, & 7, & 9, & 12, & 15, & 17, & 19, & 20
\end{array}
$$

has a range of $20 - 5 = 15$, but the dispersion is entirely different in each case. Thus, the range is used mainly as a "quick and easy" indication of variability, for instance, in industrial quality control to keep a close check on raw materials or products by observing, and charting, the range of small samples taken at regular intervals of time.

3.7

Measures of Variation: The Standard Deviation

To introduce the **standard deviation,** by far the most generally useful measure of variation, let us observe that the dispersion of a set of data is small if the values are closely bunched about their mean, and that it is large if the values are scattered widely about their mean. It would seem reasonable, therefore, to measure the variation of a set of data in terms of the amounts by which the individual values differ from their mean. If a set of numbers $x_1, x_2, x_3, \ldots$, and x_N, constituting a population, has the mean μ, the differences $x_1 - \mu, x_2 - \mu, x_3 - \mu, \ldots$, and $x_N - \mu$ are called the **deviations from the mean,** and it suggests itself that we might use their average (namely, their mean) as a measure of the variation

61

in the population. Unfortunately, this will not do. Unless the x's are all equal, some of the deviations will be positive, some will be negative, and as the reader will be asked to show in Exercise 3.118 on page 88, the sum of the deviations from the mean, $\sum (x - \mu)$, and consequently also their mean, is always zero.

Since we are really interested in the magnitude of the deviations, and not in whether they are positive or negative, we might simply ignore the signs and define a measure of variation in terms of the absolute values of the deviations from the mean. Indeed, if we add the deviations from the mean as if they were all positive or zero and divide by n, we obtain the statistical measure which is called the **mean deviation.** This measure has intuitive appeal, but because of the absolute values it leads to theoretical difficulties in problems of inference, and it is rarely used.

An alternative approach is to work with the squares of the deviations from the mean, as this will also eliminate the effect of the signs. Squares of real numbers cannot be negative; in fact, squares of the deviations from a mean are always positive unless a value happens to coincide with the mean, in which case both $x - \mu$ and $(x - \mu)^2$ are equal to zero. Then, if we average the squared deviations from the mean and take the square root of the result (to compensate for the fact that the deviations were squared), we get the **population standard deviation**

Population standard deviation

$$\sigma = \sqrt{\frac{\sum (x - \mu)^2}{N}}$$

This measure of variation is denoted by σ (lowercase *sigma,* the Greek letter for s), and, expressing literally what we have done here mathematically, it is also called the **root-mean-square deviation.** The square of σ is called the **population variance.**

It may seem logical to use the same formula, with n and $\bar{x}$ substituted for N and μ, for the standard deviation of a sample, but this is not quite what we do. Instead of dividing the sum of the squared deviations from the mean by n, we divide it by $n - 1$ and define the **sample standard deviation,** denoted by s, as

Sample standard deviation

$$s = \sqrt{\frac{\sum (x - \bar{x})^2}{n - 1}}$$

Its square, s^2, is called the **sample variance.**

In dividing by $n - 1$ instead of n we are not just being arbitrary. There is a good theoretical reason for doing this; if we divided by n and used s^2 as an estimate of σ^2 (namely, used the variance of a sample to estimate the variance of

CHAP. 3: Statistical Descriptions

the population from which it came), our result would tend to be too small, and we correct for this by dividing by $n - 1$ instead of n. If n is large, it generally will not matter much whether we divide by $n - 1$ or n, but it is common practice to define σ and s as we did.

In calculating the sample standard deviation using the formula by which it is defined, we must (1) find $\bar{x}$, (2) determine the n deviations from the mean $x - \bar{x}$, (3) square these deviations, (4) add all the squared deviations, (5) divide by $n - 1$, and (6) take the square root of the result obtained in step 5. In actual practice, this formula is rarely used, but we shall illustrate it here to emphasize what is really measured by σ and s.

EXAMPLE   On six consecutive Sundays, a tow-truck operator received 9, 7, 11, 10, 13, and 7 service calls. Calculate s.

Solution   First calculating the mean, we get

$$\bar{x} = \frac{9 + 7 + 11 + 10 + 13 + 7}{6} = \frac{57}{6} = 9.5$$

and the work required to find $\sum (x - \bar{x})^2$ may be arranged as in the following table:

x	$x - \bar{x}$	$(x - \bar{x})^2$
9	-0.5	0.25
7	-2.5	6.25
11	1.5	2.25
10	0.5	0.25
13	3.5	12.25
7	-2.5	6.25
	0.0	27.50

Then, dividing by $6 - 1 = 5$ and taking the square root using Table XII or a calculator, we get

$$s = \sqrt{\frac{27.50}{5}} = \sqrt{5.5} = 2.3$$

rounded to one decimal. Note in the table above that the total for the middle column is zero; since this must always be the case, it provides a check on the calculations.

It was easy to calculate s in this example because the data were whole numbers and the mean was exact to one decimal. Otherwise, the calculations required by the formula defining s can be quite tedious, and, unless we can get s directly with a statistical calculator or a computer, it helps to use the computing formula

Computing formula for the sample standard deviation

$$s = \sqrt{\frac{n(\sum x^2) - (\sum x)^2}{n(n-1)}}$$

EXAMPLE   Use this computing formula for s to rework the preceding example.

Solution   First we calculate $\sum x$ and $\sum x^2$, getting

x	x^2
9	81
7	49
11	121
10	100
13	169
7	49
57	569

Then, substituting $\sum x = 57$ and $\sum x^2 = 569$, together with $n = 6$, into the formula, we find that

$$s = \sqrt{\frac{6(569) - (57)^2}{6 \cdot 5}} = \sqrt{\frac{165}{30}} = \sqrt{5.5} = 2.3$$

This agrees, as it should, with the result we obtained before.

Since beginners often seem to confuse $\sum x^2$ with $(\sum x)^2$, let us emphasize the point that for $\sum x^2$ we first square each x and then add all the squares; for $(\sum x)^2$, on the other hand, we first add all the x's and then square their sum.

EXERCISES   (Exercises 3.57 and 3.62 are practice exercises; their complete solutions are given on page 90.)

3.57 In 1982, the daily average hospital room charges in the eight Mountain states were $151, $151, $130, $167, $152, $145, $146, and $169. What is their range?

3.58 For a recent ten-year period, the production of natural (sun-dried) raisins in California yielded 715, 825, 640, 900, 790, 965, 895, 700, 915, and 945 trays per acre. Find the range.

3.59 The following are the closing prices of two stocks on five consecutive Fridays:

$$Stock\ A: \quad 18, 17\tfrac{5}{8}, 18\tfrac{1}{4}, 17\tfrac{3}{8}, 18\tfrac{1}{8}$$
$$Stock\ B: \quad 20\tfrac{1}{4}, 20, 19\tfrac{7}{8}, 20\tfrac{1}{2}, 20\tfrac{3}{8}$$

Calculate the range for each stock and decide which one is more stable, that is, less variable.

3.60 The 42 employees of a company are given an intensive course in CPR. The following are the numbers of correct answers they gave in a test administered after the completion of the course: 13, 9, 18, 15, 14, 21, 7, 10, 11, 20, 5, 18, 23, 16, 17, 8, 11, 18, 20, 9, 18, 17, 14, 21, 21, 18, 9, 15, 10, 21, 12, 14, 20, 18, 19, 7, 9, 13, 11, 20, 22, and 16. Find the range of these figures.

3.61 With reference to Exercise 2.16 on page 21, find the range of the weights of the 50 rats.

3.62 In five attempts, it took a person 11, 15, 12, 8, and 14 minutes to change a tire on a car. Calculate s using
(a) the formula which defines s,
(b) the computing formula.

3.63 The following are the wind velocities reported at an airport at 6 P.M. on eight consecutive days: 13, 8, 15, 11, 3, 10, 12, and 8 miles per hour. Calculate s using
(a) the formula which defines s;
(b) the computing formula.

3.64 A filling machine in a high-production bakery is set to fill open-face pies with 16 fluid ounces of fill. A sample of four pies from a large production lot shows fills of 16.2, 15.9, 15.8, and 16.1 fluid ounces. Calculate s using
(a) the formula which defines s;
(b) the computing formula.

3.65 The following are the times it took eight cars to accelerate from 0 to 60 mph: 15, 12, 15, 18, 19, 14, 17, and 15 seconds. Use the computing formula to determine the value of s.

3.66 The following are the numbers of hours that twelve students studied for a final examination: 7, 14, 22, 19, 20, 13, 25, 28, 32, 11, 20, and 24. Use the computing formula to determine s^2 and s.

3.67 With reference to Exercise 3.29 on page 55, find s for the numbers of restaurant meals which the thirteen persons ate.

3.68 With reference to Exercise 3.31, find s^2 for the lengths of the power failures.

3.69 In 1981 there were 29, 26, 21, 118, 13, and 47 institutions of higher learning in the six New England states. Modify the computing formula for the standard deviation so that it applies to populations (that is, replace the $n-1$ by n, and then replace each n by N), and then use it to calculate σ for the given data.

3.70 With reference to Exercise 3.60, determine s for the numbers of correct answers which the company's 42 employees got on the test.

3.71 It has been claimed that for samples of size $n = 4$, the range should be roughly twice as large as the standard deviation. Check this claim with reference to the following data, representing the number of times that four students were absent from a course in anthropology: 4, 7, 3, and 7.

3.72 It has been claimed that for samples of size $n = 10$, the range should be roughly three times as large as the standard deviation. Check this claim with reference to the following test scores which ten students obtained in an examination: 77, 90, 72, 71, 91, 84, 64, 80, 75, and 86.

 3.73 The following are the interest rates which a bank paid on its *18-Month Variable-Rate Individual Retirement Accounts* in 1983: 8.343, 8.327, 8.242, 8.180, 8.201, 8.260, 8.413, 8.580, 8.536, 8.489, 8.477, 8.620, 8.813, 8.934, 8.910, 8.804, 8.665, 8.501, 8.494, 8.513, 8.594, 8.754, 8.886, 9.051, 9.188, 9.303, 9.424, 9.513, 9.568, 9.579, 9.647, 9.778, 9.841, 9.854, 9.833, 9.739, 9.636, 9.587, 9.442, 9.314, 9.258, 9.164, 9.146, 9.064, 9.103, 9.143, 9.272, 9.316, 9.356, 9.416, and 9.426%. Use a computer package or a preprogrammed statistical calculator to find $\bar{x}$ and s for these percentages.

 3.74 The following are the ignition times of certain upholstery materials exposed to a flame (in seconds):

2.50	4.50	5.11	9.70	5.62	6.77	3.49	4.90
10.21	8.76	9.33	4.12	3.85	4.97	5.04	2.97
3.81	10.60	7.95	7.41	8.64	5.33	3.90	11.25
1.92	1.42	12.80	9.45	6.25	4.71	7.86	2.65
4.79	6.20	1.52	1.38	3.87	4.54	5.12	5.15
11.75	7.35	2.80	6.85	1.20	9.20	1.76	5.21
3.40	7.29	8.66	5.04	10.25	6.43	2.97	4.45
5.50	5.92	4.56	2.46	6.90	1.47	2.11	2.32
4.19	2.20	4.32	1.58	6.43	4.04	2.51	2.58
3.78	3.75	3.10	6.43	1.70	6.40	3.24	1.79
8.75	2.46	3.62	4.72	7.40	8.81	5.83	6.75
7.65	8.79	10.92	9.65	5.09	4.11	6.37	5.40
2.51	10.28	5.49	3.76				

Use a computer package or a preprogrammed statistical calculator to find the mean and the variance of these sample data.

3.8

Some Applications of the Standard Deviation

After this chapter, we shall use sample standard deviations primarily as estimates of population standard deviations in problems of inference. To get more of a feeling for what a standard deviation really measures, let us devote this section to some applications.

In the argument which led to the definition of the standard deviation, we observed that the dispersion of a set of data is small if the values are bunched closely about their mean, and that it is large if the values are scattered widely about their mean. Correspondingly, we can now say that if the standard deviation of a set of data is small, the values are concentrated near the mean, and if the standard deviation is large, the values are scattered widely about the mean. This idea is expressed more formally by the following theorem, named **Chebyshev's theorem** after the Russian mathematician P. L. Chebyshev (1821–1894):

Chebyshev's theorem

> *For any set of data (population or sample) and any constant k greater than 1, at least $1 - 1/k^2$ of the data must lie within k standard deviations on either side of the mean.*

Thus, we can be sure that at least $1 - \dfrac{1}{2^2} = 1 - \dfrac{1}{4} = \dfrac{3}{4}$, or 75%, of the values in any set of data must lie within two standard deviations on either side of the mean; at least $1 - \dfrac{1}{3^2} = 1 - \dfrac{1}{9} = \dfrac{8}{9}$, or 88.9%, must lie within three standard deviations on either side of the mean; and at least $1 - \dfrac{1}{10^2} = 1 - \dfrac{1}{100} = \dfrac{99}{100}$, or 99% must lie within ten standard deviations on either side of the mean.

EXAMPLE   If all the 1-pound cans of coffee filled by a food processor have a mean weight of 16.00 ounces with a standard deviation of 0.02 ounce, at least what percentage of the cans must contain anywhere from 15.95 to 16.05 ounces of coffee?

Solution   Since k standard deviations, or $k(0.02)$, equals $16.05 - 16.00 = 16.00 - 15.95 = 0.05$, we find that $k = \dfrac{0.05}{0.02} = 2.5$. Thus, at least $1 - \dfrac{1}{2.5^2} = 1 - \dfrac{1}{6.25} = 0.84$, or 84%, of the cans must contain anywhere from 15.95 to 16.05 ounces of coffee.

Chebyshev's theorem applies to any kind of data, but it tells us only "at least what percentage" must lie between certain limits. For some distributions

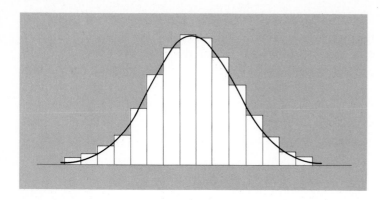

3.3

Bell-shaped distribution.

we can make the following much stronger statement:

> **For distributions having the general shape of the cross section of a bell (see Figure 3.3),**
>
> **(1) about 68% of the values will lie within one standard deviation of the mean, that is, between $\bar{x} - s$ and $\bar{x} + s$;**
> **(2) about 95% of the values will lie within two standard deviations of the mean, that is, between $\bar{x} - 2s$ and $\bar{x} + 2s$;**
> **(3) about 99.7% of the values will lie between three standard deviations of the mean, that is, between $\bar{x} - 3s$ and $\bar{x} + 3s$.**

In some books, this result is referred to as the **empirical rule,** although actually it is a theoretical result based on the normal distribution, which we shall discuss in Chapter 7.

In Section 3.6 we showed that there are many ways in which knowledge of the variability of a set of data can be of importance. Another application arises in the comparison of numbers belonging to different sets of data.

EXAMPLE   In an English class the final examination grades average 60 with a standard deviation of 16, and in a mathematics class the final examination grades average 58 with a standard deviation of 10. If a student gets a 72 in the English examination and a 68 in the mathematics examination, how many standard deviations is each of her grades above the average of the respective class? What does this tell us about her performance in the two subjects?

Solution   In the English examination her grade is $72 - 60 = 12$ points and, hence, $\frac{12}{16} = 0.75$ standard deviation above average; in the mathematics examination her grade is $68 - 58 = 10$ points and, hence, $\frac{10}{10} = 1.00$ standard deviation above average. Even though her English grade is higher than her mathematics grade and her English grade is 12 points above average while her mathematics grade is only 10 points above average, her performance relative to the two classes is better in mathematics than it is in English. Being 1.00 standard deviation above average in mathematics puts her in a relatively better position than being 0.75 standard deviation above average in English.

What we have done here consisted of converting the grades into **standard units** or **z-scores.** In general, if x is a measurement belonging to a set of data having the mean $\bar{x}$ (or μ) and the standard deviation s (or σ), then its value in standard units, denoted by z, is

Formula for converting to standard units

$$z = \frac{x - \bar{x}}{s} \quad or \quad z = \frac{x - \mu}{\sigma}$$

depending on whether the data constitute a sample or a population. In these units, z tells us how many standard deviations a value lies above or below the mean of the set of data to which it belongs. Standard units will be used frequently in later chapters.

One disadvantage of the standard deviation as a measure of variation is that it depends on the units of measurement. For instance, the weights of certain objects may have a standard deviation of 0.1 ounce or 2,835 milligrams, which is the same, but neither value really tells us whether it reflects a great deal of variation or very little variation. If the objects we are weighing are the eggs of small birds, either figure would reflect a great deal of variation, but this would not be the case if the objects we are weighing are 100-pound bags of potatoes. What we need in a situation like this is a **measure of relative variation,** such as the **coefficient of variation**

Coefficient of variation

$$V = \frac{s}{\bar{x}} \cdot 100 \quad or \quad V = \frac{\sigma}{\mu} \cdot 100$$

69

which expresses the standard deviation as a percentage of what is being measured, at least on the average.

EXAMPLE   During the past few months, one runner averaged 12 miles per week with a standard deviation of two miles, while another runner averaged 25 miles per week with a standard deviation of three miles. Which of the two runners is relatively more consistent in his weekly running habits?

Solution   The two coefficients of variation are, respectively,

$$\frac{2}{12} \cdot 100 = 16.7\% \quad \text{and} \quad \frac{3}{25} \cdot 100 = 12.0\%$$

Thus the second runner is relatively more consistent in his weekly running habits.

EXERCISES   (Exercises 3.75, 3.76, 3.82, and 3.86 are practice exercises; their complete solutions are given on page 90.)

3.75 According to Chebyshev's theorem, what can we assert about the percentage of any set of data that must lie within k standard deviations on either side of the mean when
(a) $k = 4$;
(b) $k = 12$?

3.76 An airline's records show that its flights between two cities arrive on the average 5.4 minutes late with a standard deviation of 1.4 minutes. At least what percentage of its flights between the two cities arrive anywhere between
(a) 2.6 minutes late and 8.2 minutes late;
(b) 1.6 minutes early and 12.4 minutes late?

3.77 According to Chebyshev's theorem, what can we assert about the percentage of any set of data that must lie within k standard deviations on either side of the mean when
(a) $k = 5$;
(b) $k = 8$.

3.78 According to Chebyshev's theorem, what can we assert about the percentage of any set of data that must lie within k standard deviations on either side of the mean when
(a) $k = 7$;
(b) $k = 20$.

3.79 With reference to Exercise 3.76, at least what percentage of the flights between the two cities must arrive anywhere between

(a) 1.2 minutes late and 9.6 minutes late;

(b) 0.5 minute late and 10.3 minutes late?

3.80 A study of the nutritional value of a certain kind of bread shows that on the average one slice contains 0.260 milligram of thiamine (vitamin B_1) with a standard deviation of 0.005 milligram. According to Chebyshev's theorem, between what values must be the thiamine content of

(a) at least $\dfrac{35}{36}$ of all slices of this bread;

(b) at least $\dfrac{80}{81}$ of all slices of this bread?

3.81 With reference to the preceding exercise, at least what percentage of the slices of the given kind of bread must have a thiamine content between 0.245 and 0.275 milligram? What can we say about this percentage if it can be assumed that the distribution of the thiamine content of the slices of bread is bell-shaped?

3.82 In a large city the average retail price of a head of lettuce is $0.71 (with a standard deviation of $0.05), the average retail price of a pound of tomatoes is $0.40 (with a standard deviation of $0.03), and the average retail price of a cucumber is $0.19 (with a standard deviation of $0.02). If a certain food market charges $0.78 for a head of lettuce, $0.45 for a pound of tomatoes, and $0.21 for a cucumber, which of these food items is relatively the most overpriced?

3.83 An investment service reports for each stock that it lists the price at which it is currently selling, its average price over a period of time, and a measure of its variability. Stock C, it reports, has a normal (mean) price of $58 with a standard deviation of $11, and it is currently selling at $76.50; stock D sells normally for $38, has a standard deviation of $4, and is currently selling at $50. If a person owns both stocks and wants to dispose of one, which one should he sell and why?

3.84 Between two persons on a reducing diet, the first belongs to an age group for which the mean weight is 146 pounds with a standard deviation of 14 pounds, and the second belongs to an age group for which the mean weight is 160 pounds with a standard deviation of 17 pounds. If their respective weights are 178 pounds and 193 pounds, which of the two is more seriously overweight for his or her age group?

3.85 The applicants to one state university have an average ACT mathematics score of 21.4 with a standard deviation of 3.1, while the applicants to another state university have an average ACT mathematics score of 22.1 with a standard deviation of 2.8. With respect to which of these two universities is a student in a relatively better position, if he or she scores

(a) 26 on this test;

(b) 31 on this test?

3.86 To compare the precision of two measuring instruments, a laboratory technician studies recent measurements made with both instruments. The first was recently used to measure the diameter of a ball bearing and the measurements had a mean of 4.92 mm with a standard deviation of 0.018 mm; the second was recently used to measure the unstretched length of a spring and the measuremeans had a mean of 2.54 in. with a standard deviation of 0.012 in. Which of the two measuring instruments is relatively more precise?

3.87 In five tests, one student averaged 63.2 with a standard deviation of 3.3, while another student averaged 78.8 with a standard deviation of 5.3. Which student is relatively more consistent?

3.88 One patient's blood pressure, measured daily over several weeks, averaged 202 with a standard deviation of 12.5, while that of another patient averaged 124 with a standard deviation of 8.1. Which patient's blood pressure is relatively more variable?

3.9

The Description of Grouped Data ★

In the past, considerable attention was paid to the description of grouped data, because it was generally advantageous to group data before calculating the appropriate descriptions. This was true, particularly, in connection with large sets of data, where the manual determination of quantities such as $\sum x$ and $\sum x^2$ entailed a considerable amount of work. Today, this is no longer the case, since such quantities can be determined in a matter of seconds, even with a hand-held calculator. Nevertheless, we shall devote this section to the description of grouped data, since some data (published government figures, for example) are available only in the form of frequency distributions.

As we saw in Chapter 2, the grouping of data entails some loss of information. Each item loses its identity, so to speak; we only know how many items there are in each class, so we must be satisfied with approximations. In the case of the mean and the standard deviation, we can usually get good approximations by proceeding as follows:

> **We assign to each item falling into a class the value of the class mark.**

For instance, to calculate the mean or the standard deviation of the data on page 16, those pertaining to the amount of time that 80 students devoted to leisure activities, we treat the eight values falling into the class 10–14 as if they were all equal to 12, the twenty-eight values falling into the class 15–19 as if they were all equal to 17, . . . , and the value falling into the class 35–39 as if it were equal to 37. This procedure is usually quite satisfactory, since the errors which are thus introduced into the calculations will tend to "average out."

To give general formulas for the mean and the standard deviation of a distribution with k classes, let us denote the successive class marks by x_1, x_2, $x_3, \ldots$, and x_k, and the corresponding class frequencies by $f_1, f_2, f_3, \ldots$, and f_k. Then, the sum of all the measurements or observations is represented by $x_1 f_1 + x_2 f_2 + x_3 f_3 + \cdots + x_k f_k = \sum x \cdot f$, the sum of their squares is represented by $x_1^2 f_1 + x_2^2 f_2 + x_3^2 f_3 + \cdots + x_k^2 f_k = \sum x^2 \cdot f$, and the formula for $\bar{x}$ and the computing formula for s can be written as

$$\bar{x} = \frac{\sum x \cdot f}{n} \quad \text{and} \quad s = \sqrt{\frac{n(\sum x^2 \cdot f) - (\sum x \cdot f)^2}{n(n-1)}}$$

To get corresponding formulas for the mean and the standard deviation of a population, we substitute N for n in the formula for the mean, and also in the formula for the standard deviation after first replacing the $n - 1$ by n.

EXAMPLE Find the mean and the standard deviation of the following distribution, which we obtained on page 16 for the amounts of time that 80 college students devoted to leisure activities during a typical school week:

Hours	Frequency
10–14	8
15–19	28
20–24	27
25–29	12
30–34	4
35–39	1

Solution To obtain $\sum x \cdot f$ and $\sum x^2 \cdot f$, we perform the calculations shown in the following table:

Class mark x	x^2	Frequency f	$x \cdot f$	$f x^2 \cdot f$
12	144	8	96	1,152
17	289	28	476	8,092
22	484	27	594	13,068
27	729	12	324	8,748
32	1,024	4	128	4,096
37	1,369	1	37	1,369
		80	1,655	36,525

```
MTB > SET C1
DATA> 23   24   18   14   20   24   24   26   23   21
DATA> 16   15   19   20   22   14   13   20   19   27
DATA> 29   22   38   28   34   32   23   19   21   31
DATA> 16   28   19   18   12   27   15   21   25   16
DATA> 30   17   22   29   29   18   25   20   16   11
DATA> 17   12   15   24   25   21   22   17   18   15
DATA> 21   20   23   18   17   15   16   26   23   22
DATA> 11   16   18   20   23   19   17   15   20   10

MTB > MEAN C1
    MEAN    =        20.613
MTB > STDEV C1
    ST.DEV. =        5.5632
```

3.4

Computer printout for the mean and the standard deviation of the amounts of time that 80 students engaged in leisure activities.

Then, substitution into the formulas yields

$$\bar{x} = \frac{1,655}{80} = 20.6875$$

or $\bar{x} = 20.69$ rounded to two decimals, and

$$s = \sqrt{\frac{80(36,525) - (1,655)^2}{80 \cdot 79}}$$

$$= \sqrt{\frac{182,975}{6,320}}$$

$$= 5.38$$

rounded to two decimals. To check on the **grouping error,** namely, the error introduced by first grouping the data, let us refer to the computer printout of Figure 3.4. It shows that the mean of the ungrouped data is 20.612 rounded to three decimals and that the standard deviation is 5.56 rounded to two decimals. Thus, both errors are fairly small.

In the preceding example, the calculations were rather tedious, and we went through them mainly to dramatize the simplification that can be attained by **coding** the class marks so that we can work with smaller numbers. When the class intervals are all equal, and only then, this coding consists of assigning the value 0 to one of the class marks (preferably at or near the center of the distribution), and representing all the class marks by means of successive integers. For instance, if a distribution has nine classes and the class mark of the middle class is assigned the value 0, the successive class marks of the whole distribution are assigned the values $-4, -3, -2, -1, 0, 1, 2, 3,$ and 4.

Of course, when we code the class marks like this, we must account for it in the formulas for the mean and the standard deviation. Referring to the new (coded) class marks as u's, we write

Mean of grouped data (with coding)

$$\bar{x} = x_0 + \frac{\sum u \cdot f}{n} \cdot c$$

where x_0 is the class mark in the original scale to which we assign 0 in the new scale, c is the class interval, n is the number of items grouped, and $\sum u \cdot f$ is the sum of the products obtained by multiplying each of the new class marks by the corresponding class frequency. Similarly, we write

Standard deviation of grouped data (with coding)

$$s = c \sqrt{\frac{n(\sum u^2 \cdot f) - (\sum u \cdot f)^2}{n(n-1)}}$$

where $\sum u^2 \cdot f$ is the sum of the products obtained by multiplying the squares of the new class marks by the corresponding class frequencies. If a set of data constitutes a population rather than a sample, we make the same modifications in the formulas as before; that is, we substitute N for n in the formula for the mean, and also in the formula for the standard deviation after first replacing the $n-1$ by n.

EXAMPLE   To demonstrate the simplification brought about by coding, recalculate $\bar{x}$ and s for the distribution of the amounts of time that the 80 college students devoted to leisure activities during a typical school week.

Solution   Arranging the work, as before, in a table, we get

Original class mark x	New class mark u	f	$u \cdot f$	$u^2 \cdot f$
12	-2	8	-16	32
17	-1	28	-28	28
22	0	27	0	0
27	1	12	12	12
32	2	4	8	16
37	3	1	3	9
		80	-21	97

where the class mark 22 is taken to be 0 in the new scale. Of course, we could have used any class mark, for instance, 27, as the zero of the u scale, *but the objective of coding is to make the arithmetic as simple as possible.*

Substituting $c = 5$, $x_0 = 22$, $n = 80$, $\sum u \cdot f = -21$, and $\sum u^2 \cdot f = 97$ into the foregoing formulas for $\bar{x}$ and s, we get

$$\bar{x} = 22 + \frac{-21}{80} \cdot 5 = 20.69$$

and

$$s = 5 \sqrt{\frac{80(97) - (-21)^2}{80 \cdot 79}} = 5.38$$

These results are, as they should be, identical with the ones which we obtained earlier without coding.

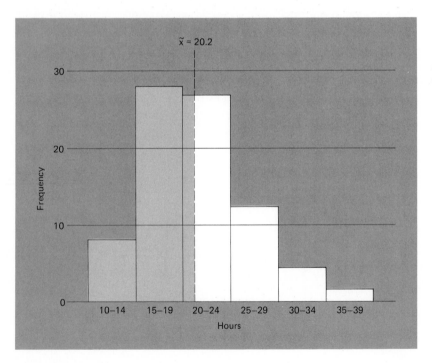

3.5

The median of the distribution of the amounts of time that the students engaged in leisure activities.

Once a set of data has been grouped, we can still determine most other statistical measures besides the mean and the standard deviation, but we may have to modify their definitions. For instance, we can no longer determine the exact value of the median and we define it as follows:

> **For grouped data, the median is such that half the total area of the rectangles of a histogram of their distribution lies to its left and the other half lies to its right (see Figure 3.5).**

This definition is equivalent to the assumption that the values in the class containing the median are distributed evenly (that is, spread out evenly) throughout that class.

To find the dividing line between the two halves of a histogram (each of which represents $\frac{n}{2}$ of the items grouped), we must somehow count $\frac{n}{2}$ of the items starting at either end of the distribution. How this is done is illustrated by the following example:

EXAMPLE   Find the median of the distribution of the amounts of time that the students devoted to leisure activities.

Solution   Since $\frac{n}{2} = \frac{80}{2} = 40$, we must count 40 of the items, starting at either end. Starting at the bottom of the distribution (beginning with the smallest values), we find that $8 + 28 = 36$ of the values fall into the first two classes, and that $8 + 28 + 27 = 63$ of the values fall into the first three classes. Therefore, we must count $40 - 36 = 4$ of the values beyond the 36 which fall into the first two classes, and on the assumption that the 27 values in the third class are spread evenly throughout that class, we add $\frac{4}{27}$ of the class interval of 5 to the lower boundary of the third class. This gives us

$$\tilde{x} = 19.5 + \frac{4}{27} \cdot 5 = 20.2$$

for the median of the distribution.

In general, if L is the lower boundary of the class into which the median must fall, f is its frequency, and j is the number of items we still lack when we reach L, then the median of the distribution is given by

Median of grouped data

$$\tilde{x} = L + \frac{j}{f} \cdot c$$

If we prefer, we can find the median of a distribution by starting to count at the other end (beginning with the largest values) and subtracting an appropriate fraction of the class interval from the upper boundary U of the class into which the median must fall. The corresponding formula is

Alternative formula for the median of grouped data

$$\tilde{x} = U - \frac{j'}{f} \cdot c$$

where j' is the number of items we still lack when we reach U.

EXAMPLE   Use the alternative formula to find the median of the distribution of the amounts of time that the 80 students devoted to leisure activities.

Solution   Since $1 + 4 + 12 = 17$ of the values fall into the three classes at the top of the distribution, we need $40 - 17 = 23$ of the 27 values in the next class to reach the median, and we write

$$\tilde{x} = 24.5 - \frac{23}{27} \cdot 5 = 20.2$$

The result is, of course, the same.

Note that the median of a distribution can be found by this method regardless of whether the class intervals are all equal; in fact, it can usually be found even when either or both of the classes at the top and at the bottom of a distribution are open (see Exercise 3.98).

The method by which we found the median of a distribution can also be used to determine other fractiles. For instance, the quartiles Q_1 and Q_3 of a distribution are defined so that 25% of the total area of the rectangles of the histogram lies to the left of Q_1 and 25% to the right of Q_3. To find them, we use either of the two formulas above.

EXAMPLE   Find Q_1 and Q_3 for the distribution of the amounts of time that the 80 students devoted to leisure activities.

Solution   Since $\dfrac{n}{4} = \dfrac{80}{4} = 20$, we must count 20 of the items, starting at the bottom of the distribution, to find Q_1. Since there are eight values in the first class, we must count $20 - 8 = 12$ of the 28 values in the second class to reach Q_1, and we write

$$Q_1 = 14.5 + \frac{12}{28} \cdot 5 = 16.6$$

rounded to one decimal. To find Q_3 we must count 20 of the items starting at the other end of the distribution. Since $1 + 4 + 12 = 17$ of the values fall into the three classes at the top of the distribution, we must count $20 - 17 = 3$ of the 27 values in the next class to reach Q_3, and we write

$$Q_3 = 24.5 - \frac{3}{27} \cdot 5 = 23.9$$

The results we have obtained here for the median, Q_1, and Q_3, together with the fact that the smallest value is 10 and the largest value is 38 (see page 16), are summarized in the box-and-whisker plot of Figure 3.6.

Another set of useful fractiles are the **percentiles,** which are defined so that 1% of the total area of the rectangles of the histogram lies to the left of the first percentile, P_1, 2% to the left of the second percentile, P_2, ..., and 99% to the left of the ninety-ninth percentile, P_{99}. Again, they are determined by using either of the two formulas on page 78.

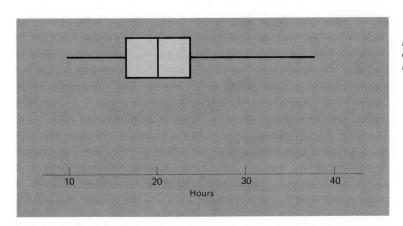

3.6

Box-and-whisker plot for the amounts of time that the 80 students engaged in leisure activities.

EXAMPLE   Find P_{33} and P_{85} for the distribution of the amounts of time that the 80 students devoted to leisure activites.

Solution   To find P_{33} we must count $0.33(80) = 26.4$ of the items starting at the bottom of the distribution, and we get

$$P_{33} = 14.5 + \frac{18.4}{28} \cdot 5 = 17.8$$

To find P_{85} we must count $0.15(80) = 12$ of the items starting at the top of the distribution, and we get

$$P_{85} = 29.5 - \frac{7}{12} \cdot 5 = 26.6$$

3.10
Some Further Descriptions ★

So far we have discussed statistical descriptions which come under the general heading of "measures of location" and "measures of variation." Actually, there is no limit to the number of ways in which statistical data can be described, and statisticians continually develop new methods of describing characteristics of numerical data that are of interest in particular problems. In this section we shall consider briefly the problem of describing the overall shape of a distribution.

Although frequency distributions can take on almost any shape or form, most of the distributions we meet in practice can be described fairly well by one or another of a few standard types. Among these, foremost in importance is the aptly described **symmetrical bell-shaped distribution** shown at the top of Figure 3.7. Indeed, there are theoretical reasons why, in many instances, distributions of actual data can be expected to follow this general pattern. The other two distributions of Figure 3.7 can still, by a stretch of the imagination, be called bell-shaped, but they are not symmetrical. Distributions like these, having a "tail" on one side or the other, are said to be **skewed;** if the tail is on the left we say that they are **negatively skewed,** and if the tail is on the right we say that they are **positively skewed.** Distributions of incomes or wages are often positively skewed because of the presence of some relatively high values that are not offset by correspondingly low values.

The symmetry or skewness of a distribution can also be judged visually by inspection of a box-and-whisker plot. For a symmetrical distribution, the me-

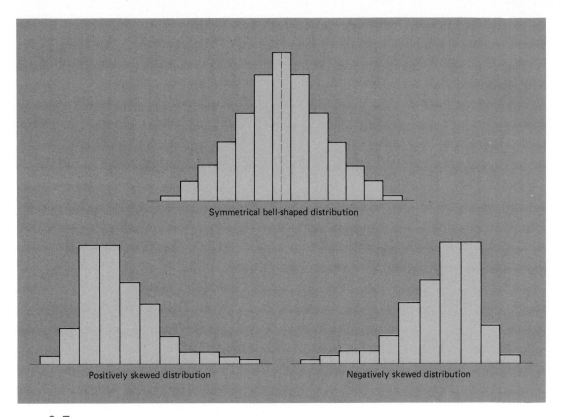

Symmetrical bell-shaped distribution

Positively skewed distribution

Negatively skewed distribution

3.7

Bell-shaped distributions.

dian line divides the box into equal halves. It is moved to the left of center when a distribution is positively skewed, and to the right of center when a distribution is negatively skewed. So far as Figure 3.6 is concerned, the median line is slightly to the left of center, and this reflects the mild positive skewness which is apparent from the histogram of Figure 3.5. Note also the long "whisker" extending from Q_3 to the largest value, which was 38.

There are several ways of actually measuring the extent to which a distribution is skewed. A relatively easy one is based on the fact that for a perfectly symmetrical bell-shaped distribution such as the one at the top of Figure 3.7, the values of the median and the mean coincide. Since the presence of some relatively high values that are not offset by correspondingly low values will tend to make the mean greater than the median (and the presence of some relatively low values that are not offset by correspondingly high values will tend to make the mean less than the median), we can use this relationship between the mean and the median to define a relatively simple measure of the extent to which a distribution is skewed. It is called the **Pearsonian coefficient of skewness** and it is given by

Pearsonian coefficient of skewness

$$SK = \frac{3(mean - median)}{standard\ deviation}$$

For a perfectly symmetrical distribution the value of *SK* is 0, and in general its values must fall between -3 and 3.

EXAMPLE   Find the Pearsonian coefficient of skewness for the distribution of the amounts of time the 80 students devoted to leisure activities.

Solution   Substituting into the formula the values obtained earlier for the mean, the median, and the standard deviation, namely, $\bar{x} = 20.7$, $\tilde{x} = 20.2$, and $s = 5.4$, we get

$$SK = \frac{3(20.7 - 20.2)}{5.4} = 0.3$$

rounded to one decimal. This reflects the fact that the distribution has a slight positive skewness.

Two other kinds of distributions which sometimes arise in practice are the **reverse J-shaped** and **U-shaped distributions** shown in Figure 3.8. As can be seen from the diagram, the names of such distributions quite literally describe

3.8
Reverse J-shaped and U-shaped distributions.

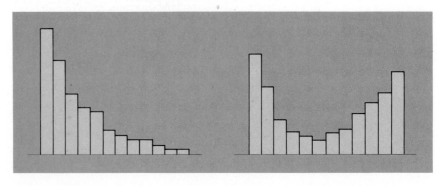

CHAP. 3: Statistical Descriptions

their shape. Examples of such distributions may be found in Exercises 3.108 and 3.109 on page 85.

(Exercises 3.89, 3.93, 3.97, 3.102, and 3.105 are practice exercises; their complete solutions are given on pages 91 and 92.)

EXERCISES

★ **3.89** Find $\bar{x}$ and s for the following distribution of the weekly earnings of 125 wage earners:

Weekly earnings (dollars)	Frequency
120.00–129.99	9
130.00–139.99	20
140.00–149.99	36
150.00–159.99	30
160.00–169.99	15
170.00–179.99	11
180.00–189.99	4

★ **3.90** Find the mean and the standard deviation of the age distribution of Exercise 2.19 on page 22.

★ **3.91** Find the mean and the standard deviation of the following distribution of the total travel costs of 200 students on a two-week vacation in Europe:

Travel cost (dollars)	Frequency
1,600–1,799	12
1,800–1,999	31
2,000–2,199	73
2,200–2,399	57
2,400–2,599	16
2,600–2,799	7
2,800–2,999	4

★ **3.92** The following are the numbers of minutes that a doctor kept 60 patients waiting beyond their appointment times:

12.1	9.8	10.5	5.6	8.2	0.5	6.8	10.1	17.2	4.2
8.3	1.3	7.9	11.3	6.3	7.2	9.3	9.9	7.2	12.7
1.2	4.6	10.3	8.5	10.0	12.8	9.6	13.5	10.8	5.1
12.7	11.5	3.8	12.9	13.0	3.9	7.5	16.1	11.1	8.3
9.6	6.4	15.7	5.8	9.7	11.9	2.4	5.2	8.4	16.7
2.5	13.0	4.8	10.7	11.4	9.3	4.7	6.0	9.5	14.6

SEC. 3.10: Some Further Descriptions

Group these data into a distribution with the classes 0.0–2.9, 3.0–5.9, 6.0–8.9, 9.0–11.9, 12–14.9, and 15.0–17.9, and then find the mean and the standard deviation.

★ 3.93 Find the median of the distribution of Exercise 3.89.

★ 3.94 Find the median of the following distribution of the numbers of raffle tickets sold by the 70 members of a social-service organization:

Number of raffle tickets	Frequency
less than 30	28
30–44	19
45–59	10
60–74	8
75–89	5

★ 3.95 Find the median of the distribution of Exercise 2.19 on page 22.

★ 3.96 Find the median of the distribution of Exercise 3.91.

★ 3.97 If possible, find Q_1 and Q_3 for the distribution of Exercise 3.94.

★ 3.98 Is it possible to find the mean and the median of each of the following distributions? Explain your answers.

(a) Grade	Frequency
40–49	5
50–59	18
60–69	27
70–79	15
80–89	6

(b) IQ	Frequency
less than 90	3
90–99	14
100–109	22
110–119	19
more than 119	7

(c) Weight	Frequency
100 or less	41
101–110	13
111–120	8
121–130	3
131–140	1

★ 3.99 Find Q_1 and Q_3 for the distribution of Exercise 2.19 on page 22.

★ 3.100 With reference to Exercise 3.92, find the shortest and the longest waiting times, and use the distribution obtained in that exercise to determine Q_1, the median, and Q_3. Also, use these results to draw a box-and-whisker plot.

★ **3.101** With reference to Exercise 2.19 on page 22, suppose that the youngest member of the video dating service is 20 and the oldest is 53. Use this information together with the results of Exercises 3.95 and 3.99 to draw a box-and-whisker plot.

★ **3.102** If possible, find P_{10}, P_{40}, and P_{90} for the distribution of Exercise 3.94.

★ **3.103** With reference to Exercise 3.91, find P_{20} and P_{80}.

★ **3.104** With reference to Exercise 3.92, find P_{10} and P_{90}.

★ **3.105** Use the results of Exercises 3.89 and 3.93 to calculate the Pearsonian coefficient of skewness for the distribution of the weekly earnings.

★ **3.106** Use the results of Exercises 3.90 and 3.95 to calculate the Pearsonian coefficient of skewness for the distribution of the ages of the members of the video dating service.

★ **3.107** Use the values of $\bar{x}$ and s obtained in Exercise 3.92 and the value of the median obtained in Exercise 3.100 to calculate SK for the distribution of the lengths of the waiting times.

★ **3.108** Roll a pair of dice 120 times and construct a distribution showing how many times there were 0 sixes, how many times there was 1 six, and how many times there were 2 sixes. Draw a histogram of this distribution and describe its shape.

★ **3.109** If a coin is flipped five times, the result may be represented by means of a sequence of H's and T's (for example, HHTTH), where H stands for *heads* and T for *tails*. Having obtained such a sequence of H's and T's, we can then check after each successive flip whether the number of heads exceeds the number of tails. For example, for the sequence HHTTH, heads is ahead after the first flip, after the second flip, after the third flip, not after the fourth flip, but again after the fifth flip; altogether, it is ahead four times. Repeat this experiment 50 times, and construct a histogram showing in how many cases heads was ahead altogether 0 times, 1 time, 2 times, ..., and 5 times. Explain why the resulting distribution should be U-shaped.

 ★ **3.110** Use a computer package or a preprogrammed statistical calculator to find the mean and the standard deviation of the raw data of Exercise 3.92. Determine the respective grouping errors by comparing the results with those of Exercise 3.92.

3.11

Technical Note (Summations)

In the abbreviated notation introduced on page 43, the expression $\sum x$ does not make it clear which, or how many, values of x we have to add. This is taken care of by the more explicit notation

$$\sum_{i=1}^{n} x_i = x_1 + x_2 + \cdots + x_n$$

where it is made clear that we are adding the x's whose subscripts i are 1, 2, ..., and n. Generally, we shall not use the more explicit notation in this text to simplify the overall appearance of the formulas, assuming that it is clear in each case what x's we are referring to and how many there are.

Using the $\sum$ notation, we shall also have occasion to write such expressions as $\sum x^2, \sum xy, \sum x^2 f, \ldots$, which (more explicitly) represent the sums

$$\sum_{i=1}^{n} x_i^2 = x_1^2 + x_2^2 + x_3^2 + \cdots + x_n^2$$

$$\sum_{j=1}^{m} x_j y_j = x_1 y_1 + x_2 y_2 + \cdots + x_m y_m$$

$$\sum_{i=1}^{n} x_i^2 f_i = x_1^2 f_1 + x_2^2 f_2 + \cdots + x_n^2 f_n$$

Working with two subscripts, we shall also have occasion to evaluate **double summations** such as

$$\sum_{j=1}^{3} \sum_{i=1}^{4} x_{ij} = \sum_{j=1}^{3} (x_{1j} + x_{2j} + x_{3j} + x_{4j})$$

$$= x_{11} + x_{21} + x_{31} + x_{41} + x_{12} + x_{22} + x_{32} + x_{42}$$

$$+ x_{13} + x_{23} + x_{33} + x_{43}$$

To verify some of the formulas involving summations that are stated but not proved in the text, the reader will find it helpful to use the following rules:

$$\text{Rule A:} \quad \sum_{i=1}^{n} (x_i \pm y_i) = \sum_{i=1}^{n} x_i \pm \sum_{i=1}^{n} y_i$$

$$\text{Rule B:} \quad \sum_{i=1}^{n} k \cdot x_i = k \cdot \sum_{i=1}^{n} x_i$$

$$\text{Rule C:} \quad \sum_{i=1}^{n} k = k \cdot n$$

Rules for summations

The first of these rules states that the summation of the sum (or difference) of two terms equals the sum (or difference) of the individual summations, and it can be extended to the sum or difference of more than two terms. The second rule states that we can, so to speak, factor a constant out of a summation, and the third rule states that the summation of a constant is simply n times that constant. All these rules can be proved by actually writing out in full what each of the summations represents.

(Exercise 3.113 is a practice exercise; its complete solution is given on page 92.)

3.111 Write each of the following in full, that is, without summation signs:

(a) $\sum_{i=1}^{6} x_i$;

(d) $\sum_{j=1}^{8} x_j f_j$;

(b) $\sum_{i=1}^{5} y_i$;

(e) $\sum_{i=3}^{7} x_i^2$;

(c) $\sum_{i=1}^{3} x_i y_i$;

(f) $\sum_{j=1}^{4} (x_j + y_j)$.

3.112 Write each of the following as summations:

(a) $z_1 + z_2 + z_3 + z_4 + z_5$;
(b) $x_5 + x_6 + x_7 + x_8 + x_9 + x_{10} + x_{11} + x_{12}$;
(c) $x_1 f_1 + x_2 f_2 + x_3 f_3 + x_4 f_4 + x_5 f_5 + x_6 f_6$;
(d) $y_1^2 + y_2^2 + y_3^2$.

3.113 Given $x_1 = 1$, $x_2 = 3$, $x_3 = 5$, $x_4 = 7$, $x_5 = 9$, $f_1 = 1$, $f_2 = 5$, $f_3 = 10$, $f_4 = 3$, and $f_5 = 2$, find

(a) $\sum_{i=1}^{5} x_i$;

(c) $\sum_{i=1}^{5} x_i \cdot f_i$;

(b) $\sum_{i=1}^{5} f_i$;

(d) $\sum_{i=1}^{5} x_i^2 \cdot f_i$.

3.114 Given $x_1 = -2$, $x_2 = 3$, $x_3 = 1$, and $x_4 = 4$, find

(a) $\sum_{i=1}^{4} x_i$;

(b) $\sum_{i=1}^{4} x_i^2$.

3.115 Given $x_1 = 2$, $x_2 = 3$, $x_3 = 4$, $x_4 = 5$, $x_5 = 6$, $x_6 = 7$, $f_1 = 3$, $f_2 = 12$, $f_3 = 10$, $f_4 = 6$, $f_5 = 3$, and $f_6 = 1$, find

(a) $\sum_{i=1}^{6} x_i$;

(c) $\sum_{i=1}^{6} x_i \cdot f_i$;

(b) $\sum_{i=1}^{6} f_i$;

(d) $\sum_{i=1}^{6} x_i^2 \cdot f_i$.

3.116 Given $x_1 = 2$, $x_2 = 1$, $x_3 = 5$, $x_4 = 3$, $y_1 = 1$, $y_2 = 3$, $y_3 = 2$, and $y_4 = 1$, find

(a) $\sum_{i=1}^{4} x_i$;

(d) $\sum_{i=1}^{4} y_i^2$;

(b) $\sum_{i=1}^{4} y_i$;

(e) $\sum_{i=1}^{4} x_i \cdot y_i$.

(c) $\sum_{i=1}^{4} x_i^2$;

3.117 Given $x_{11} = 2$, $x_{12} = 3$, $x_{13} = -1$, $x_{21} = 1$, $x_{22} = 2$, $x_{23} = 2$, $x_{31} = 2$, $x_{32} = -2$, $x_{33} = 2$, $x_{41} = 3$, $x_{42} = -4$, and $x_{43} = -3$, find

(a) $\sum_{i=1}^{4} x_{ij}$ for $j = 1, 2$, and 3;

(c) $\sum_{i=1}^{4} \sum_{j=1}^{3} x_{ij}$.

(b) $\sum_{j=1}^{3} x_{ij}$ for $i = 1, 2, 3$, and 4;

3.118 Show that $\sum_{i=1}^{n} (x_i - \bar{x}) = 0$ for any set of x's whose mean is $\bar{x}$.

3.119 Is it true in general that $\left(\sum_{i=1}^{n} x_i \right)^2 = \sum_{i=1}^{n} x_i^2$? (*Hint:* Check whether the equation holds for $n = 2$.)

SOLUTIONS OF PRACTICE EXERCISES

3.1 (a) The data would constitute a population if the producer of the commercials had to determine the tax that is due on the profit.

(b) The data would constitute a sample if the producer wanted to estimate (predict) the production cost of other commercials.

3.4 $\sum x = 48 + 58 + 33 + 42 + 57 + 31 + 52 + 25 + 46 + 60 + 61 + 49 + 38 + 53 + 30 + 47 + 52 + 63 + 41 + 34 = 920$, $n = 20$, so that

$$\bar{x} = \frac{920}{20} = 46$$

3.10 The total weight of the women is $8(123) = 984$ and the total weight of the men is $5(174) = 870$, so that their combined weight is $984 + 870 = 1,854$. Since this does not exceed 2,000, the elevator is not overloaded.

3.17 Substituting $w_1 = 1,000$, $w_2 = 3,000$, $w_3 = 16,000$, $x_1 = 7$, $x_2 = 7.5$, and $x_3 = 8$ into the formula for the weighted mean, we get

$$\bar{x}_w = \frac{1,000(7) + 3,000(7.5) + 16,000(8)}{1,000 + 3,000 + 16,000} = \frac{157,500}{20,000} = 7.875\%$$

3.18 Substituting $\bar{x}_1 = 13,300$, $\bar{x}_2 = 14,500$, $\bar{x}_3 = 21,000$, $n_1 = 13,000$, $n_2 = 17,200$, and $n_3 = 2,400$ into the formula for the grand mean of combined data, we get

$$\bar{\bar{x}} = \frac{13,000(13,300) + 17,200(14,500) + 2,400(21,000)}{13,000 + 17,200 + 2,400}$$

$$= \frac{472,700,000}{32,600}$$

$$= \$14,500$$

3.25 (a) Since $\dfrac{n+1}{2} = \dfrac{15+1}{2} = 8$, the median is the 8th largest value.

(b) Since $\dfrac{n+1}{2} = \dfrac{40+1}{2} = 20.5$, the median is the mean of the 20th and 21st largest values.

3.28 The median position is $\dfrac{n+1}{2} = \dfrac{12+1}{2} = 6.5$; arranged according to size, the sales figures are 42, 44, 46, 58, 58, 64, 66, 70, 79, 85, 89, and 93, so that the median is $\dfrac{64+66}{2} = 65$.

3.37 (a) The medians are 3, 6, 5, 4, 8, 4, 7, 1, 5, 8, 5, 5, 6, 8, and 6; the means are $4\frac{1}{3}$, $6\frac{1}{3}$, $5\frac{2}{3}$, 6, $6\frac{1}{3}$, 5, 7, $1\frac{2}{3}$, 5, 7, 6, $5\frac{1}{3}$, $5\frac{1}{3}$, $5\frac{2}{3}$, and 7.

(b) Grouping the medians and the means into the indicated classes, we get

Medians	Frequency		Means	Frequency
0.5–2.5	1		0.5–2.5	1
2.5–4.5	3		2.5–4.5	1
4.5–6.5	7		4.5–6.5	10
6.5–8.5	4		6.5–8.5	3

(c) The histograms of these distributions are shown in Figure 3.9, and it can be seen that there is less variability among the means than among the

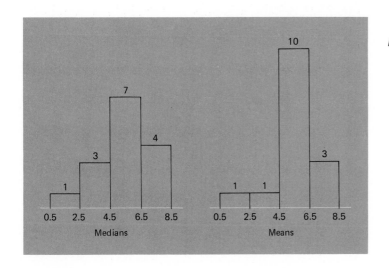

3.9

Histograms for part (c) of Exercise 3.37.

medians. In other words, the means are concentrated more closely about $\mu = 5.5$, the actual mean of the numbers on all the cards.

3.42 The median position is $\dfrac{21 + 1}{2} = 11$, and Q_1 position is $\dfrac{10 + 1}{2} = 5.5$, and the Q_3 position is the 5.5th from the other end, namely, 16.5. Thus, there are five values to the left of the Q_1 position, five values between the Q_1 and median positions, five values between the median and Q_3 positions, and five values to the right of the Q_3 position.

3.45 The median position is $\dfrac{13 + 1}{2} = 7$, the Q_1 position is $\dfrac{6 + 1}{2} = 3.5$, and the Q_3 position is the 3.5th from the other end, namely, 10.5. Arranged according to size, the data are 0, 1, 1, 3, 5, 5, 5, 6, 6, 8, 10, 12, and 15, so that $Q_1 = \dfrac{1 + 3}{2} = 2$ and $Q_3 = \dfrac{8 + 10}{2} = 9$.

3.50 Since 9 occurs three times, 1 and 8 each occurs twice, and 0, 2, 4, 5, 7, 11, 12, 14, and 16 each occurs once, the mode is 9.

3.57 Since the smallest and largest values are \$130 and \$169, the range is \$169 − \$130 = \$39.

3.62 (a) The mean is $\bar{x} = \dfrac{11 + 15 + 12 + 8 + 14}{5} = \dfrac{60}{5} = 12$, so that

$$s = \sqrt{\dfrac{(11 - 12)^2 + (15 - 12)^2 + (12 - 12)^2 + (8 - 12)^2 + (14 - 12)^2}{5 - 1}}$$

$$= \sqrt{\dfrac{1 + 9 + 0 + 16 + 4}{4}}$$

$$= \sqrt{7.5}$$

$$= 2.74$$

(b) Substituting $n = 5$, $\sum x = 60$, and $\sum x^2 = 11^2 + 15^2 + 12^2 + 8^2 + 14^2 = 750$, into the computing formula for s, we get

$$s = \sqrt{\dfrac{5(750) - (60)^2}{5 \cdot 4}}$$

$$= \sqrt{\dfrac{150}{20}}$$

$$= \sqrt{7.5}$$

$$= 2.74$$

3.75 (a) At least $1 - \dfrac{1}{4^2} = \dfrac{15}{16} = 0.9375$, or 93.75%, of the data must lie within $k = 4$ standard deviations on either side of the mean.

(b) At least $1 - \dfrac{1}{12^2} = \dfrac{143}{144} = 0.993$, or 99.3%, of the data must lie within $k = 12$ standard deviations on either side of the mean.

3.76 (a) $\dfrac{8.2 - 5.4}{1.4} = 2$; at least $1 - \dfrac{1}{2^2} = 0.75$, or 75%, of the planes arrive anywhere between 2.6 minutes late and 8.2 minutes late.

(b) $\dfrac{12.4 - 5.4}{1.4} = 5$; at least $1 - \dfrac{1}{5^2} = 0.96$, or 96%, of the planes arrive anywhere between 1.6 minutes early and 12.4 minutes late.

3.82 In standard units, the price of the head of lettuce is $\dfrac{0.78 - 0.71}{0.05} = 1.40$, the price of the pound of tomatoes is $\dfrac{0.45 - 0.40}{0.03} = 1.67$, and the price of the cucumber is $\dfrac{0.21 - 0.19}{0.02} = 1.00$. Therefore, the pound of tomatoes is relatively most overpriced.

3.86 For the first measuring instrument, the coefficient of variation is $\dfrac{0.018}{4.92} \cdot 100 = 0.37\%$, and for the second measuring instrument it is $\dfrac{0.012}{2.54} \cdot 100 = 0.47\%$. Since the first measuring instrument has the smaller coefficient of variation, it is relatively more precise.

3.89

u	f	$u \cdot f$	$u^2 \cdot f$
-2	9	-18	36
-1	20	-20	20
0	36	0	0
1	30	30	30
2	15	30	60
3	11	33	99
4	4	16	64
	125	71	309

Since $x_0 = \dfrac{140.00 + 149.99}{2} = 144.995$, the mean is

$$\bar{x} = 144.995 + \dfrac{71}{125} \cdot 10 = \$150.68$$

to the nearest cent; also

$$s = 10 \sqrt{\frac{125(309) - (71)^2}{125 \cdot 124}} = 10 \sqrt{\frac{33,584}{15,500}} = 10\sqrt{2.17}$$

$$= \$14.73$$

to the nearest cent.

3.93 The median is $\tilde{x} = 139.95 + \dfrac{33.5}{36} \cdot 10 = \$149.30$.

3.97 For Q_1 we must count $\dfrac{70}{4} = 17.5$ values from the bottom of the distribution, and since this falls into the open class "less than 30," Q_1 cannot be found; for Q_3 we must count 17.5 values from the top of the distribution, and we get

$$Q_3 = 59.5 - \frac{4.5}{10} \cdot 15 = 52.75$$

3.102 P_{10} cannot be found as it falls into the "less than 30" open class, but $P_{40} = 29.5 + \dfrac{0}{19} \cdot 15 = 29.5$ and $P_{90} = 74.5 - \dfrac{2}{8} \cdot 15 = 70.75$.

3.105 The Pearsonian coefficient of skewness is $\dfrac{3(150.68 - 149.30)}{14.73} = 0.28$.

3.113 (a) $1 + 3 + 5 + 7 + 9 = 25$.
(b) $1 + 5 + 10 + 3 + 2 = 21$.
(c) $1 \cdot 1 + 3 \cdot 5 + 5 \cdot 10 + 7 \cdot 3 + 9 \cdot 2 = 105$.
(d) $1^2 \cdot 1 + 3^2 \cdot 5 + 5^2 \cdot 10 + 7^2 \cdot 3 + 9^2 \cdot 2 = 605$.

Achievements

Having read and studied these chapters, and having worked a good portion of the exercises, you should be able to:

1. Explain the difference between descriptive statistics and statistical inference.

2. Construct frequency distributions of both numerical and categorical data.

3. Determine the class limits, the class boundaries, the class marks, and the class interval of a distribution of numerical data.

4. Convert frequency distributions into percentage distributions.

5. Convert frequency distributions into cumulative distributions, and percentage distributions into cumulative percentage distributions.

6. Present frequency distributions in the form of histograms, bar charts, frequency polygons, ogives, or pie charts.

7. Construct stem-and-leaf plots.

★ 8. Construct two-way tables.

9. Explain the difference between samples and populations, and the difference between statistics and parameters.

10. Determine the mean of a set of data.

11. Determine the median of a set of data.

12. Determine the mode of a set of data.

13. List some of the desirable and undesirable features of the various measures of location.

14. Calculate weighted means.

15. Determine the grand mean of combined data.

16. Determine the quartiles of a set of data.

17. Draw a box-and-whisker plot.

18. Determine the range of a set of data.

19. Determine the standard deviation (or the variance) of a set of data.

20. List some of the desirable and undesirable features of the varous measures of variation.

21. Explain Chebyshev's theorem.

22. Convert measurements into standard units and explain the advantage of using such units.

23. Calculate the coefficient of variation and explain its significance.

★ 24. Use coding to determine the mean and the standard deviation of a distribution.

★ 25. Determine the median of a distribution.

★ 26. Determine the quartiles of a distribution.

★ 27. Determine the percentiles of a distribution.

★ 28. Describe the shape of a distribution as symmetrical or skewed, as bell-shaped, as U-shaped, and so forth.

★ 29. Determine the Pearsonian coefficient of skewness.

30. Work with the $\sum$ (summation) notation.

Checklist of Key Terms (with page references to their definitions)

94

REVIEW EXERCISES

R.1 The following are the television audience ratings of ten professional football games: 14.6, 16.2, 15.5, 18.3, 13.9, 17.6, 15.9, 19.2, 20.3, and 14.5. Find
   (a) the mean;
   (b) the median.

R.2 According to Chebyshev's theorem, what can we assert about the percentage of any set of data that must lie within k standard deviations on either side of the mean when
   (a) $k = 6$;
   (b) $k = 15$.

R.3 On three consecutive days, a traffic policeman issued 9, 14, and 10 speeding tickets, and 5, 10, and 12 tickets for going through red lights. Which of the following conclusions can be obtained from these data by purely descriptive methods and which require generalizations? Explain your answers.

(a) Altogether on these three days, the policeman issued more speeding tickets than tickets for going through red lights.

(b) On two of the three days, the policeman issued more speeding tickets than tickets for going through red lights.

(c) The policeman issued the smallest number of tickets on the first day because he was new on the job.

(d) This policeman will seldom give more than 15 speeding tickets on any one day.

R.4 Convert the distribution of Exercise 3.91 on page 83 into a
    (a) cumulative "or less" distribution;
    (b) cumulative "or more" distribution.

R.5 The dean of a college has complete records on how many failing grades each faculty member gave to his or her students during the academic year 1984–1985. Give one example each of a situation in which the dean would look upon these data as
    (a) a population;
    (b) a sample.

R.6 For the past nine days, a student has worked after school for 2, 5, 1, 3, 2, 6, 3, 2, and 4 hours. Find the mean number of hours that the student worked after school on these nine days.

R.7 In one year a large employer gave major bonuses averaging $1,200 to 140 employees and minor bonuses averaging $450 to 260 employees. What is the average size of the bonuses given to these 400 employees?

★ R.8 The following is the distribution of the number of mistakes 150 students made in translating a certain passage from German to English:

Number of mistakes	Number of students
17–19	5
20–22	63
23–25	39
26–28	24
29–31	17
32–34	2

Using coding, find
    (a) the mean;
    (b) the standard deviation.

R.9 Draw a histogram of the distribution of the preceding exercise.

R.10 Draw a bar chart of the distribution of Exercise R.8.

R.11 The following are the systolic blood pressures of twenty hospital patients: 165, 135, 151, 153, 155, 182, 142, 158, 146, 149, 124, 162, 173, 204, 159, 130, 177, 162, 141, and 156. Construct a stem-and-leaf plot with the stem labels 12, 13, ..., and 20.

R.12 The ages of a company's employees are to be grouped into the following classes: under 19, 20–24, 25–29, 30–34, 34–39, and over 39. Explain where difficulties might arise.

R.13 An environmental engineer obtained the following data on the concentration of mercury (in parts per million) at eight locations along a stream: 0.064, 0.071, 0.066, 0.062, 0.073, 0.065, 0.061, and 0.066.
   (a) Find $\bar{x}$ and s.
   (b) Find the median and the range.

R.14 In the United States there were 340 art museums in a recent year, 683 history museums, 284 science museums, 186 art and history museums, and 328 other kinds of museums. Construct a pie chart of this categorical distribution.

R.15 Given $x_1 = 6$, $x_2 = 8$, $x_3 = 5$, $x_4 = -1$, $x_5 = 3$, and $x_6 = 4$, find
   (a) $\sum_{i=1}^{6} x_i$;
   (b) $\sum_{i=1}^{6} x_i^2$.

R.16 Explain why each of the following samples may well yield misleading information:
   (a) To ascertain facts about tooth-brushing habits, a sample of the residents of a community are asked how many times they brush their teeth each day.
   (b) To study the spending patterns of families in a certain income group, a sample survey is conducted during the first three weeks of December.

★ R.17 If a distribution has the mean 112.8, the median 96.5, and the standard deviation 24.7, find the Pearsonian coefficient of skewness.

R.18 Having kept records for many years, Mrs. Jones knows that it takes her on the average 48 minutes to get to work; the standard deviation is 2.2 minutes. If she leaves her home each morning at 2 minutes after 8, at least what percentage of the time does she arrive at work between 21 minutes before 9 and 1 minute after 9?

R.19 On five runs, bus A carried 15, 24, 19, 12, and 20 passengers, and bus B carried 18, 21, 16, 14, and 16 passengers.
   (a) Calculate the respective means to decide which bus averaged more passengers on the five runs.
   (b) Use the formula which defines the standard deviation to calculate s for each of the two sets of data.
   (c) Calculate the two coefficients of variation to determine whether the number of passengers is relatively more variable for bus A or bus B.

R.20 The following are the numbers of accidents that occurred in July of 1983 at eighteen intersections without left-turn arrows: 18, 39, 41, 24, 45, 38, 22, 28, 34, 23, 16, 42, 9, 20, 36, 32, 42, and 35. Find
   (a) the median;
   (b) Q_1 and Q_3.

R.21 Use the results of the preceding exercise to draw a box-and-whisker plot for the accident data.

R.22 The following are the 1981 prices of potatoes (in dollars per cwt) and the corresponding quantities produced (in millions of cwt) in three major potato-producing states:

	Price	Quantity
Idaho	4.55	80
Maine	4.80	27
Washington	3.80	52

Find the mean price of the potatoes produced in 1981 in these three states.

R.23 The daily ticket sales of a movie theater are grouped into a distribution having the classes 50–99, 100–149, 150–199, 200–249, 250–299, 300–349, and 350–399. Determine
   (a) the lower class limits;
   (b) the upper class limits;
   (c) the class boundaries;
   (d) the class marks;
   (e) the class interval of the distribution.

R.24 List the data which correspond to the following stems of stem-and-leaf plots:
   (a) 125 | 3 0 4 8 7 6 6 5
   (b) 34 | 67, 05, 19, 48

R.25 Ten athletes ran a 440-yard race in 46.2, 46.9, 48.3, 46.7, 46.3, 46.5, 47.2, 46.1, 46.9, and 46.3 seconds. Calculate the standard deviation.

R.26 The following are data on the weekly loss of labor-hours due to accidents at thirty industrial plants "before and after" certain safety regulations were put into operation: 37 and 28, 72 and 59, 26 and 24, 125 and 120, 45 and 46, 54 and 43, 13 and 15, 79 and 75, 42 and 23, 39 and 35, 73 and 48, 111 and 83, 41 and 38, 77 and 72, 73 and 28, 103 and 69, 32 and 18, 68 and 73, 82 and 65, 47 and 42, 27 and 25, 49 and 18, 66 and 61, 83 and 42, 63 and 57, 86 and 50, 45 and 23, 103 and 67, 55 and 57, and 41 and 28. Tally these data into the following two-way classification and show the frequencies associated with the sixteen cells.

98

| | Before | | | |
	10–39	40–69	70–99	100–129
After 10–39				
40–69				
70–99				
100–129				

R.27 On four Saturdays, a person jogged for 46, 50, 52, and 60 minutes.
   (a) Find the mean, the range, and the standard deviation of these four sample values.
   (b) Subtract 50 minutes from each of the times, recalculate the mean, the range, and the standard deviation, and compare the results with those obtained in part (a).
   (c) Divide each of the original values by 2, recalculate the mean, the range, and the standard deviation, and compare the results with those obtained in part (a).
   (d) Based on these results, what effect does (1) adding or subtracting a constant, and (2) multiplying or dividing by a constant, have on the mean, the range, and the standard deviation of a set of data?

R.28 Mr. Ames lives in a neighborhood where the average family income is $15,000 with a standard deviation of $2,000, and Mr. Brown lives in a neighborhood where average family income is $18,000 with a standard deviation of $3,000. If Mr. Ames and family make $20,000 and Mr. Brown and family make $24,000, which of the two families is relatively better off with respect to the families in their neighborhoods?

R.29 Find the median position for
   (a) $n = 31$;
   (b) $n = 62$.

R.30 The following is the distribution of the number of meals which certain real estate salespersons ate out in a given week:

Number of meals	Frequency
0–1	18
2–3	25
4–5	11
6–7	5
8–9	1
	60

(a) Draw a histogram of this distribution.

(b) Convert this distribution into a cummulative "less than" distribution and draw an ogive.

★ R.31 With reference to the preceding exercise find

(a) the median;

(b) the quartiles Q_1 and Q_3.

★ R.32 With reference to the distribution of Exercise R.30, suppose that the smallest value is 0 and the largest value is 9. Use the results of Exercise R.31 to draw a box-and-whisker plot.

★ R.33 With reference to the distribution of Exercise R.30, find the two percentiles P_{35} and P_{65}.

R.34 A church bulletin shows that on the five preceding Sundays the attendance was 352, 314, 3,360, 375, and 328.

(a) Calculate the mean and the median of these attendance figures.

(b) Assuming that the third value was printed incorrectly and should have been 360 instead of 3,360, recalculate the mean and the median.

(c) Compare the effects of this printing error on the mean and on the median.

R.35 The 25 teachers of an elementary school are given an intensive course in first aid. Find the range of the following numbers of correct answers they gave in a test administered after the completion of the course: 18, 12, 15, 9, 11, 16, 20, 15, 14, 18, 18, 15, 10, 17, 13, 17, 19, 8, 19, 20, 16, 12, 18, 11, and 14.

R.36 Asked whether they ever accept social invitations from their students, 40 tennis pros replied: occasionally, rarely, rarely, never, rarely, frequently, occasionally, occasionally, never, rarely, rarely, never, occasionally, occasionally, frequently, occasionally, rarely, rarely, occasionally, never, occasionally, occasionally, never, never, rarely, rarely, occasionally, occasionally, frequently, occasionally, never, occasionally, rarely, rarely, never, frequently, occasionally, rarely, rarely, and occasionally.

(a) Construct a categorical distribution and display it in the form of a bar chart.

(b) What is the tennis pros' modal reply?

R.37 The following are the numbers of deer observed in 60 sections of land in a wildlife count:

11	14	21	15	16	4	18	11	17	11	7	12
21	0	16	12	20	17	13	10	16	5	10	14
5	19	10	6	15	1	26	8	18	19	2	14
17	6	15	14	22	7	7	13	19	0	15	17
2	16	11	18	10	28	15	4	32	6	20	7

(a) Group these figures into a distribution having the classes 0–4, 5–9, 10–14, 15–19, 20–24, 25–29, and 30–34, and draw a bar chart.

(b) Convert the distribution obtained in part (a) into a cumulative "less than" distribution and draw an ogive.

R.38 The following are the May 15th average low temperatures in 40 selected cities in the United States: 55, 62, 69, 34, 70, 43, 48, 57, 53, 64, 38, 45, 46, 57, 61, 61, 72, 41, 48, 54, 67, 31, 28, 58, 63, 52, 36, 75, 60, 69, 52, 40, 53, 57, 26, 43, 64, 39, 71, and 51. Construct a stem-and-leaf plot with one-digit leaves.

R.39 Use the stem-and-leaf plot obtained in the preceding exercise to find the median of the 40 average low temperatures.

R.40 The following are the numbers of hours that 23 students watched television in a particular week: 3, 6, 14, 21, 4, 15, 20, 5, 15, 20, 28, 45, 20, 5, 4, 4, 35, 10, 60, 11, 12, 9, and 10. Construct a box-and-whisker plot for these data.

101

Possibilities and Probabilities

The study of statistics, with the emphasis on inference, has been referred to as a lesson in "how to live with uncertainty." The key word here is "uncertainty," but our dictionary tells us merely that this is "the state of being uncertain," which does not help. So, we look up "uncertain" and get "dependent on chance," and this in turn leads to

> **CHANCE** ('chan(t)s) The supposed impersonal purposeless determiner of unaccountable happenings.

Does this help? Probably not, but it shows that when it comes to terms such as "uncertainty," "chance," "likelihood," "probability," and "luck," it is very difficult to be precise.

In this chapter we shall see how uncertainties can actually be measured, namely, how they can be assigned numbers, and how these numbers, called **probabilities,** can be used to "live" with the uncertainties. In other words, we shall see how probabilities can, or may be, interpreted, and how they can be used to make choices (among different courses of action) which promise to be the most profitable, or otherwise most desirable.

Sections 4.1, 4.2, and 4.3 are devoted to the problem of determining what is possible in given situations, and in Section 4.4 we learn how to judge also what is probable. In Sections 4.5 and 4.6 we present the concept of a mathematical expectation and its application to problems of decision making.

103

4.1

Counting

In the study of "what is possible," there are essentially two kinds of problems. There is the problem of listing everything that can happen in a given situation, and then there is the problem of determining how many different things can happen (without actually constructing a complete list). In connection with our work in this book, the second kind of problem is especially important; in most cases we will not need a complete list and, hence, can save ourselves a great deal of work.

Although the first kind of problem, that of listing everything that can happen in a given situation, may seem straightforward and easy, this is not always the case.

EXAMPLE   To meet a graduation requirement, each of two students must study a foreign language, including, among others, French, German, and Spanish. List the different ways in which thay can make their choice, if we care only how many of them, not which ones, will study French, how many of them will study German, and how many of them will study Spanish.

Solution   Clearly, there are many possibilities: both students may decide to study German; one of them may decide to study French while the other decides to study Spanish; one of them may decide to study German while the other decides to study a foreign language other than French, German, or Spanish; both may decide to study a foreign language other than French, German, or Spanish; and so forth. Continuing this way carefully, we may be able to complete the list, but the chances are that we will omit at least one of the possibilities.

To handle this kind of problem systematically, it is helpful to draw a **tree diagram** such as that of Figure 4.1. This diagram shows that first there are three possibilities (three branches) corresponding to 0, 1, or 2 of the students deciding to study French. Then, for German there are three branches emanating from the top branch, two from the middle branch, and one from the bottom

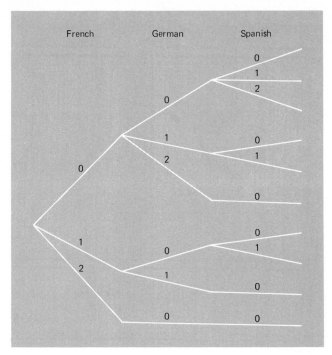

French   German   Spanish

0
1
2

0

1
2

0
1

0

0

1
2

0

0
1

1

0

2

0

0

0

4.1

Tree diagram showing in each case how many of the students choose the different languages.

branch. Evidently, there are again three possibilities, 0, 1, or 2, when neither student is going to study French; two possibilities, 0 or 1, when one of the two students is going to study French; and one possibility, 0, when both students are going to study French. For Spanish the reasoning is the same, and we find that (going from left to right) there are altogether ten different paths along the "branches" of the tree. In other words, there are altogether ten possibilities.

EXAMPLE   In a medical study, patients are classified according to whether they have blood type A, B, AB, or O, and also according to whether their blood pressure is low, normal, or high. In how many different ways can a patient thus be classified according to blood type and blood pressure?

Solution   As is apparent from the tree diagram of Figure 4.2, the answer is 12. Starting at the top, the first path along the "branches" corresponds to a patient having blood type A and low blood pressure, the second path corresponds to a patient having blood type A and normal blood pressure, . . . , and the twelfth path corresponds to a patient having blood type O and high blood pressure.

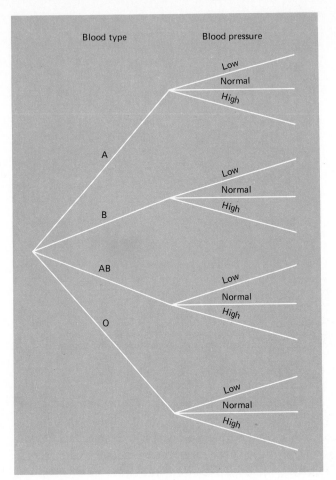

Tree diagram showing combinations of blood types and blood pressure levels.

The answer we got in the second example is $4 \cdot 3 = 12$, namely, the product of the number of blood types and the number of blood pressure levels. Generalizing from this example, let us state the following rule:

Multiplication of choices

> *If a choice consists of two steps, of which the first can be made in m ways and for each of these the second can be made in n ways, then the whole choice can be made in m · n ways.*

To prove this, we have only to draw a tree diagram similar to that of Figure 4.2. First there are *m* branches corresponding to the possibilities in the first step, and then there are n branches emanating from each of these branches to represent the possibilities in the second step. This leads to $m \cdot n$ paths along the branches of the tree diagram, and hence to $m \cdot n$ possibilities.

EXAMPLE   If a restaurant offers nine different desserts, which it serves with coffee, decaffeinated coffee, tea, milk, or hot chocolate, in how many different ways can one order a dessert and a drink?

Solution   Since $m = 9$ and $n = 5$, the answer is $9 \cdot 5 = 45$.

EXAMPLE   If an English department schedules four lecture sections and twelve discussion groups for a course in modern literature, in how many different ways can a student choose a lecture section and a discussion group?

Solution   Since $m = 4$ and $n = 12$, the answer is $4 \cdot 12 = 48$.

By using appropriate tree diagrams, we can easily generalize the foregoing rule for the "multiplication of choices" so that it will apply to choices involving more than two steps. For k steps, where k is a positive integer, we get the following rule:

Multiplication of choices (generalized)

> *If a choice consists of k steps, of which the first can be made in n_1 ways, for each of these the second can be made in n_2 ways, . . . , and for each of these the kth can be made in n_k ways, then the whole choice can be made in $n_1 \cdot n_2 \cdot \ldots \cdot n_k$ ways.*

We simply keep multiplying the numbers of ways in which the different steps can be made.

EXAMPLE   A new-car buyer has the choice of four body styles, three different engines, and ten colors.
  (a) In how many different ways can a person order one of these cars?
  (b) If a person also has the options of ordering the car with or without air conditioning, with or without an automatic transmission, and with or without bucket seats, in how many different ways can he or she order one of these cars?

107

Solution   (a) Since $n_1 = 4$, $n_2 = 3$, and $n_3 = 10$, there are $4 \cdot 3 \cdot 10 = 120$ different ways in which a person can order one of the cars.

(b) Since $n_1 = 4$, $n_2 = 3$, $n_3 = 10$, $n_4 = 2$, $n_5 = 2$, and $n_6 = 2$, there are $4 \cdot 3 \cdot 10 \cdot 2 \cdot 2 \cdot 2 = 960$ different ways in which a person can order one of the cars.

EXAMPLE   A test consists of twelve multiple-choice questions, with each question having four possible answers. In how many different ways can a student check off one answer to each question?

Solution   Since $n_1 = n_2 = \cdots = n_{12} = 4$, there are

$$4 \cdot 4 \cdot 4 \cdot 4 \cdot 4 \cdot 4 \cdot 4 \cdot 4 \cdot 4 \cdot 4 \cdot 4 \cdot 4 = 16{,}777{,}216$$

ways in which a student can check off one answer to each question. Only in one of these cases will all the answers be correct, and in

$$3 \cdot 3 \cdot 3 \cdot 3 \cdot 3 \cdot 3 \cdot 3 \cdot 3 \cdot 3 \cdot 3 \cdot 3 \cdot 3 = 531{,}441$$

cases will all the answers be wrong.

EXERCISES   (Exercises 4.1, 4.7, and 4.8 are practice exercises; their complete solutions are given on pages 134 and 135.)

4.1 Suppose that in a baseball World Series (in which the winner is the first team to win four games) the National League champion leads the American League champion by three games to one. Construct a tree diagram to show the different ways in which these teams can win or lose the remaining game or games.

4.2 There are four routes, A, B, C, and D, between a businessman's home and his office, but route B is one-way so that he cannot take it on the way to work, and route D is one-way so that he cannot take it on the way home.
   (a) Draw a tree diagram showing the various ways he can go to and from work.
   (b) Draw a tree diagram showing the various ways he can go to and from work, given that he never goes by the same route both ways.

4.3 A person with $1 in his pocket bets $1, even money, on the flip of a coin, and he continues to bet $1 so long as he has any money. Draw a tree diagram to show the various things that can happen during the first three flips of the coin. In how many of the cases will he be exactly $1 ahead?

4.4 A student can study 0, 1, or 2 hours for a history test on any given night. Draw a tree diagram to show that there are seven different ways in which she can study altogether 3 hours for the test on three consecutive nights.

4.5 In a union election, Mr. Brown, Ms. Green, and Ms. Jones are running for president, while Mr. Adams, Ms. Roberts, and Mr. Smith are running for vice-president. Construct a tree diagram showing the nine possible outcomes, and use it to determine the number of ways in which these union officials will not both be of the same sex.

4.6 A purchasing agent places his orders by phone, by telegram, or by mail, requesting in each case that his order be confirmed by telegram or by mail. Draw a tree diagram to show the various ways in which one of his orders can be placed and confirmed.

4.7 There are three trails to the top of a mountain. In how many different ways can a person hike up and down the mountain if
  (a) she can take the same trail both ways;
  (b) she can, but need not, take the same trail both ways;
  (c) she does not want to take the same trail both ways?

4.8 If a baseball squad has nine pitchers, 13 infielders, and ten outfielders, in how many different ways can a most-valuable-player award be given to one of the pitchers, one of the infielders, and one of the outfielders?

4.9 A chain of department stores has three warehouses and eight retail outlets in the Dallas–Fort Worth area. In how many different ways can an item be shipped from one of the warehouses to one of the retail outlets?

4.10 Among its attractions for the upcoming season, an auditorium lists 12 concerts and 8 plays. In how many different ways can a person buy tickets for one of the concerts and one of the plays?

4.11 In a certain restaurant, a customer can order a hamburger rare, medium rare, medium, medium well, and well done, and also with or without cheese. In how many ways can a customer order a hamburger in this restaurant?

4.12 In a doctor's office there are six recent issues of *Newsweek*, eight issues of the *New Yorker* magazine, and five issues of the *Reader's Digest*. In how many different ways can a patient waiting to see the doctor glance at one of each kind, if the order in which she looks at these magazines does not matter?

4.13 The menu of a restaurant lists three soups, four salads, fifteen entrées, and six desserts. In how many different ways can one choose a soup, a salad, an entrée, and a dessert?

4.14 A test consists of eight multiple-choice questions, with each question having three possible answers. In how many different ways can a student check off one answer to each question?

4.15 If a test consists of 15 true-false questions, in how many different ways can a student answer all the questions?

4.16 In a census survey, families are classified into six categories according to income, five categories according to family size, three categories according to the education of the head of the household, three categories with regard to home ownership, and eight categories with regard to the ownership of major appliances. In how many different ways can a family thus be classified?

4.17 A tool kit includes six screwdrivers, four wrenches, two hammers, two saws, and three pliers. In how many different ways can a person select one of each kind?

4.18 Trailer license plates in one state consist of three digits, the first of which cannot be 0, followed by two letters of the alphabet, the first of which cannot be I, O, or Q. How many different plates are possible with this scheme?

4.2
Permutations

The rule for the multiplication of choices, and its generalization, are often used when several selections are made from one set of objects, items, or persons, and the order in which they are selected is important.

EXAMPLE   If 16 entries are submitted to an essay contest, in how many different ways can the judges award a first prize and a second prize?

Solution   Since the first prize can be awarded in 16 ways and the second prize must go to one of the other 15 entries, there are altogether $16 \cdot 15 = 240$ possibilities.

EXAMPLE   In how many different ways can the 42 members of a union elect a president, a vice-president, a secretary, and a treasurer?

Solution   Assuming that the treasurer is voted on first, then the secretary, then the vice-president, and finally the president, we find that there are $n_1 = 42$ ways in which they can elect the treasurer, $n_2 = 41$ ways in which they can elect the secretary, $n_3 = 40$ ways in which they can elect the vice-president, and $n_4 = 39$ ways in which they can elect the president. Thus, there are altogether $42 \cdot 41 \cdot 40 \cdot 39 = 2,686,320$ different possibilities.

In general, if r objects are selected from a set of n distinct objects, any particular arrangement (order) of these objects is called a **permutation.** For instance, Maine, Vermont, and Connecticut, in that order, constitute a permutation (a particular ordered arrangement) of three of the six New England states. Also,

110

Purdue, Illinois, Michigan State, and Wisconsin, in that order, constitute a permutation of four of the universities in the Big 10 Conference.

EXAMPLE Determine the number of different permutations of two of the five vowels a, e, i, o, and u, and list them all.

Solution Since $n_1 = 5$ and $n_2 = 4$, there are $5 \cdot 4 = 20$ different permutations, and they are

ae	ai	ao	au	ei	eo	eu	io	iu	ou
ea	ia	oa	ua	ie	oe	ue	oi	ui	uo

To find a formula for the total number of permutations of r objects selected from a set of n distinct objects, observe that the first selection is made from the whole set of n objects, the second selection is made from the $n - 1$ objects which remain after the first selection has been made, the third selection is made from the $n - 2$ objects which remain after the first two selections have been made, . . . , and the rth and final selection is made from the $n - (r - 1) = n - r + 1$ objects which remain after the first $r - 1$ selections have been made. Therefore, direct application of the generalized rule for the multiplication of choices yields the result that the total number of permutations of r objects selected from a set of n distinct objects, which we shall denote $_nP_r$, is

$$n(n - 1)(n - 2) \cdot \ldots \cdot (n - r + 1)$$

Since products of consecutive integers arise in many problems relating to permutations and other kinds of special arrangements or selections, it is convenient to introduce here what is called the **factorial notation.** In this notation, the product of all positive integers less than or equal to the positive integer n is called "n factorial" and denote by $n!$. Thus,

$$1! = 1$$
$$2! = 2 \cdot 1 = 2$$
$$3! = 3 \cdot 2 \cdot 1 = 6$$
$$4! = 4 \cdot 3 \cdot 2 \cdot 1 = 24$$
$$5! = 5 \cdot 4 \cdot 3 \cdot 2 \cdot 1 = 120$$
$$6! = 6 \cdot 5 \cdot 4 \cdot 3 \cdot 2 \cdot 1 = 720$$

$$\cdot \quad \cdot \quad \cdot \quad \cdot \quad \cdot$$

and in general $n! = n(n - 1)(n - 2) \cdot \ldots \cdot 3 \cdot 2 \cdot 1$. Also, to make various formulas more generally applicable, we let $0! = 1$ by definition.

To express the formula for $_nP_r$ in terms of factorials, observe that $12 \cdot 11 \cdot 10! = 12!$, $7 \cdot 6 \cdot 5 \cdot 4 \cdot 3! = 7!$, $35 \cdot 34 \cdot 33 \cdot 32! = 35!, \ldots$, and similarly

$$_nP_r \cdot (n - r)! = n(n - 1)(n - 2) \cdot \ldots \cdot (n - r + 1) \cdot (n - r)!$$
$$= n!$$

so that $_nP_r = \dfrac{n!}{(n - r)!}$. To summarize

Number of permutations of n objects taken r at a time

> *The number of permutations of r objects selected from a set of n distinct objects is*
>
> $$_nP_r = n(n - 1)(n - 2) \cdot \ldots \cdot (n - r + 1)$$
>
> *or, in factorial notation,*
>
> $$_nP_r = \frac{n!}{(n - r)!}$$

The first formula is generally easier to use because it requires fewer steps, but many students find the one in factorial notation easier to remember.

EXAMPLE   Find the number of permutations of four objects selected from a set of 12 distinct objects (say, the number of ways in which four of 12 basketball teams can be ranked first, second, third, and fourth by a panel of coaches).

Solution   For $n = 12$ and $r = 4$, the first formula yields

$$_{12}P_4 = 12 \cdot 11 \cdot 10 \cdot 9 = 11{,}880$$

and the second formula yields

$$_{12}P_4 = \frac{12!}{(12 - 4)!} = \frac{12!}{8!} = \frac{12 \cdot 11 \cdot 10 \cdot 9 \cdot 8!}{8!} = 11{,}880$$

Essentially, the work is the same, but the second formula requires a few extra steps.

To find the formula for the number of permutations of n distinct objects taken all together, we substitute $r = n$ into either formula for $_nP_r$ and get

Number of permutations of n objects taken all together

$$_nP_n = n!$$

EXAMPLE   In how many ways can ten instructors be assigned to ten sections of a course in economics?

Solution   Substituting $n = 10$, we get

$$_{10}P_{10} = 10! = 3,628,800$$

Throughout this discussion it has been assumed that the n objects are all distinct. When this is not the case, the formula for $_nP_n$ can easily be modified, but we shall not go into that in this book.

EXERCISES

(Exercises 4.19, 4.22, and 4.23 are practice exercises; their complete solutions are given on page 135.)

4.19 Check each of the following to see whether it is true or false:
   (a) $8! = 8 \cdot 7 \cdot 6!$;
   (b) $3! \cdot 2! = 6!$;
   (c) $\dfrac{12!}{12 \cdot 11 \cdot 10} = 9!$.

4.20 Check each of the following to see whether it is true or false:
   (a) $3! + 2! = 5!$;
   (b) $14! = \dfrac{15!}{15}$;
   (c) $3! + 0! = 7$.

4.21 Check each of the following to see whether it is true or false:
   (a) $\dfrac{1}{2!} + \dfrac{1}{2!} = 1$;

(b) $\dfrac{14!}{9!} = 14 \cdot 13$;

(c) $\dfrac{9!}{7!2!} = 72$.

4.22 How many different signals can be made by arranging three of six differently colored flags on a vertical pole?

4.23 In how many different ways can a television director schedule five sponsors' commercials during the five time slots allocated to commercials during the telecast of the first quarter of a basketball game?

4.24 If the drama club of a college has to choose four of nine half-hour skits to present on one evening from 8:00 to 10:00, in how many different ways can they arrange their schedule?

4.25 How many four-letter sequences (not necessarily meaningful words) can be formed using the letters in the word "harmony,"
    (a) with no repetition of letters;
    (b) with repetition of letters allowed?

4.26 On a vacation, a person wants to visit three of the nation's 22 historical parks. If the order of the visits matters, in how many ways can this person plan the trip?

4.27 If there are 12 cars in a race, in how many different ways can they place first, second, and third?

4.28 In how many different ways can a person arrange four paintings next to each other horizontally on a wall?

4.29 In how many ways can five new accounts be distributed among ten advertising executives, if none is to receive more than one of the new accounts?

4.30 In how many different ways can the manager of a baseball team arrange the batting order of the nine players in his starting lineup?

4.31 Three married couples have bought six seats in a row for a performance of a musical comedy.
    (a) In how many ways can they be seated?
    (b) In how many ways can they be seated if each couple is to sit together with the husband to the left of his wife?
    (c) In how many ways can they be seated if each couple is to sit together?
    (d) In how many ways can they be seated if all the men are to sit together and all the women are to sit together?

★ **4.32** The number of ways in which n distinct objects can be arranged in a circle is $(n - 1)!$.
    (a) Present an argument to justify this formula.
    (b) In how many ways can six persons be seated at a round table (if it matters only who sits on whose left and right)?
    (c) In how many ways can a window dresser display four shirts in a circular arrangement?

4.3
Combinations

There are many problems in which we must know the number of ways in which *r* objects can be selected from a set of *n* distinct objects, but we do not care about the order in which the selection is made. For instance, we may want to know in how many ways a committee of four can be selected from among the 72 staff members of a hospital, or the number of ways in which the IRS can choose five of 33 tax returns for a special audit.

To obtain a formula which applies to problems like these, let us first examine the following 24 permutations of three of the first four letters of the alphabet.

abc	acb	bac	bca	cab	cba
abd	adb	bad	bda	dab	dba
acd	adc	cad	cda	dac	dca
bcd	bdc	cbd	cdb	dbc	dcb

If we are not concerned with the order in which the three letters are chosen from the four letters a, b, c, and d, there are only four ways in which the selection can be made: abc, abd, acd, and bcd, the values shown in the first column. Note that each row contains the $3! = 6$ different permutations of the corresponding letters shown in the first column.

In general, there are $r!$ permutations of any *r* objects selected from a set of *n* distinct objects, so that the $_nP_r$ permutations of *r* objects selected from a set of *n* distinct objects contain each set of *r* objects $r!$ times. Therefore, to write a formula for the number of ways in which *r* objects can be selected from a set of *n* distinct objects, also called the number of **combinations** of *n* objects taken *r* at a time and denoted $\binom{n}{r}$, we must divide $_nP_r$ by $r!$, and we get

Number of combinations of n objects taken r at a time

> The number of ways in which r objects can be selected from a set of n distinct objects is
>
> $$\binom{n}{r} = \frac{n(n-1)(n-2)\cdot\ldots\cdot(n-r+1)}{r!}$$
>
> or, in factorial notation,
>
> $$\binom{n}{r} = \frac{n!}{r!(n-r)!}$$

Like the two formulas for $_nP_r$, the first of these formulas is generally easier to use, and many students find the one in factorial notation easier to remember.

For $n = 0$ to $n = 20$, the values of $\binom{n}{r}$ may be read from Table X at the end of the book, where they are referred to as **binomial coefficients.** The reason for this is explained in Exercise 4.44 on page 118.

EXAMPLE   In how many different ways can a person invite three of her eight closest friends to a party?

Solution   For $n = 8$ and $r = 3$, the first formula yields

$$\binom{8}{3} = \frac{8 \cdot 7 \cdot 6}{3!} = 8 \cdot 7 = 56$$

and the second formula yields

$$\binom{8}{3} = \frac{8!}{3!5!} = \frac{8 \cdot 7 \cdot 6 \cdot \cancel{5!}}{3!\cancel{5!}} = \frac{8 \cdot 7 \cdot 6}{3 \cdot 2 \cdot 1} = 56$$

Basically, the work is the same, but the first formula required fewer steps.

EXAMPLE   In how many different ways can a committee of four be selected from among the 72 staff members of a hospital?

Solution   For $n = 72$ and $r = 4$, we get

$$\binom{72}{4} = \frac{72 \cdot 71 \cdot 70 \cdot 69}{4!} = \frac{72 \cdot 71 \cdot 70 \cdot 69}{24} = 1{,}028{,}790$$

Observe that the result of the first example, but not that of the second example, can be checked in Table X.

EXAMPLE   In how many different ways can a student select two of six mathematics courses together with three of seven English courses?

Solution   The student can select the two mathematics courses in $\binom{6}{2}$ ways, the three English courses in $\binom{7}{3}$ ways, and, hence, all five courses in

$$\binom{6}{2} \cdot \binom{7}{3} = 15 \cdot 35 = 525 \text{ ways}$$

The values of the binomial coefficients were obtained from Table X.

When we take 7 objects from a set of 10 distinct objects, then $10 - 7 = 3$ of the objects are left. Thus, there are as many ways of leaving (or selecting) 3 objects from a set of 10 distinct objects as there are ways of selecting 7 objects, and we can write $\binom{10}{7} = \binom{10}{3}$. In general, when r objects are selected from a set of n distinct objects, $n - r$ of the objects are left, and consequently, there are as many ways of leaving (or selecting) $n - r$ objects from a set of n distinct objects as there are ways of selecting r objects. Symbolically, we write

Rule for binomial coefficients

$$\binom{n}{r} = \binom{n}{n-r} \qquad \text{for } r = 0, 1, 2, \ldots, n$$

Sometimes this rule serves to simplify calculations and sometimes it is needed in connection with the use of Table X.

EXAMPLE   Determine the value of $\binom{75}{72}$.

Solution   To avoid having to write down the product $75 \cdot 74 \cdot 73 \cdot \ldots \cdot 4$ and cancel $72 \cdot 71 \cdot 70 \cdot \ldots \cdot 4$, we write directly

$$\binom{75}{72} = \binom{75}{3} = \frac{75 \cdot 74 \cdot 73}{3!} = 67{,}525$$

EXAMPLE   Find the value of $\binom{18}{13}$.

Solution $\binom{18}{13}$ cannot be looked up directly in Table X, but making use of the fact that $\binom{18}{13} = \binom{18}{18-13} = \binom{18}{5}$, we look up $\binom{18}{5}$ and get 8,568.

EXERCISES   (Exercises 4.33 and 4.34 are practice exercises; their complete solutions are given on page 135.)

4.33 A gourmet food shop carries 14 kinds of cheese. Calculate the number of ways in which a person can buy a pound each of three kinds of the cheese. Check your answer in Table X.

4.34 Among the 16 candidates for four positions on a city council, nine are Democrats, five are Republicans, and two are Independents. In how many ways can the four councilmen be chosen so that
    (a) three are Democrats and one is a Republican;
    (b) two are Democrats, one is a Republican, and one is an Independent?

4.35 Calculate the number of ways in which a motel chain can select four of 11 sites for the construction of new motels, and check your answer in Table X.

4.36 Draw poker is a card game played with an ordinary deck of 52 cards in which each player is dealt five cards. How many different five-card hands are there?

4.37 A student is required to report on five of twelve books on a reading list. Calculate the number of ways in which she can select the five books, and check your answer in Table X.

4.38 Determine the number of ways in which the IRS can select five of 33 tax returns for a special audit.

4.39 A carton of 12 transistor batteries contains two that are defective. In how many different ways can one choose three of these batteries so that
    (a) none of the defective batteries is included;
    (b) exactly one of the defective batteries is included?

4.40 If a realtor has listings for eight one-family homes and five condominiums, in how many different ways can he choose two of the one-family homes and two of the condominiums to show to a customer?

4.41 A student committee must consist of three juniors and four seniors. If seven juniors and eight seniors are willing to serve on the committee, in how many different ways can it be selected?

4.42 To fill a number of vacancies, the personnel manager of a company has to choose three secretaries from among ten applicants and two bookkeepers from among four applicants. In how many different ways can she fill the five vacancies?

4.43 Among the members of a tennis club there are 38 in the under-35 age group, 16 in the 35–45 age group, and eight in the over-45 age group. Altogether, in how many different ways can the club choose two members in each age group to represent it at a tournament?

4.44 The quantity $\binom{n}{r}$ is called a binomial coefficient because it is, in fact, the coefficient of $a^{n-r}b^r$ in the binomial expansion of $(a + b)^n$. Verify this for $n = 2, 3,$ and 4 by expanding $(a + b)^2$, $(a + b)^3$, and $(a + b)^4$, and comparing the coefficients with the corresponding entries for $n = 2$, $n = 3$, and $n = 4$ in Table X.

4.45 A table of binomial coefficients is easy to construct by following the pattern shown below, which is called **Pascal's triangle.**

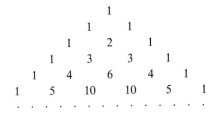

In this arrangement, each row begins with a 1, ends with a 1, and each other entry is the sum of the nearest two values in the row immediately above. Construct the next three rows of Pascal's triangle and verify from Table X that they are, respectively, the binomial coefficients corresponding to $n = 6$, $n = 7$, and $n = 8$.

4.4
Probability

Historically, the oldest way of measuring uncertainty is the **classical concept of probability.** It was developed originally in connection with games of chance, and it lends itself most readily to bridging the gap between possibilities and probabilities. This concept applies only when all possible outcomes are equally likely, in which case we can say that

The classical probability concept

> *If there are n equally likely possibilities, of which one must occur and s are regarded as favorable, or as a "success," then the probability of a "success" is* $\frac{s}{n}$.

In the application of this rule, the terms "favorable" and "success" are used rather loosely—what is favorable to one player is unfavorable to his opponent, and what is a success from one point of view is a failure from another. Thus, the terms "favorable" and "success" can be applied to any kind of outcome, even if "favorable" means that a house gets struck by lightning, or "success" means that a person catches pneumonia. This usage dates back to the days when probabilities were quoted only in connection with games of chance.

EXAMPLE   What is the probability of drawing an ace from a well-shuffled deck of 52 playing cards?

Solution   There are $s = 4$ aces among the $n = 52$ cards, so we get

$$\frac{s}{n} = \frac{4}{52} = \frac{1}{13}$$

EXAMPLE   What is the probability of rolling a 6 with a well-balanced die?

Solution   In this case, $s = 1$ and $n = 6$, so that the probability is

$$\frac{s}{n} = \frac{1}{6}$$

119

EXAMPLE   Find the probability that two cards drawn from an ordinary deck of 52 playing cards will both be black.

Solution   According to what we learned in Section 4.3, the total number of possibilities is $n = \binom{52}{2} = \frac{52 \cdot 51}{2} = 1{,}326$, and the number of favorable possibilities is $s = \binom{26}{2} = \frac{26 \cdot 25}{2} = 325$, since half of the 52 playing cards are black and the others are red. It follows that the probability of drawing two black cards is

$$\frac{s}{n} = \frac{325}{1{,}326} = \frac{25}{102}$$

or approximately 0.245.

Although equally likely possibilities are found mostly in games of chance, the classical probability concept applies also to a great variety of situations where gambling devices are used to make **random selections**—say, when offices are assigned to research assistants by lot, when laboratory animals are chosen for an experiment (perhaps, by the method which is described in Section 8.1) so that each one has the same chance of being selected, when each family in a large apartment complex has the same chance of being included in a sample survey, or when machine parts are chosen for inspection so that each part has the same chance of being selected.

EXAMPLE   A tire manufacturer requires that four of the 20 tires in each production lot must be inspected before they are shipped. If the tires are all satisfactory the whole lot is shipped, but if they are not all satisfactory the remaining 16 tires in the lot are also inspected. What is the probability that such a production lot will pass the inspection when actually two of the tires are defective?

Solution   There are $n = \binom{20}{4} = 4{,}845$ ways of choosing four of the 20 tires, and it will be assumed that they are all equally likely. The number of favorable outcomes is the number of ways in which four of the 18 good tires can be selected, namely, $s = \binom{18}{4} = 3{,}060$, and it follows that the desired probability is

$$\frac{s}{n} = \frac{3,060}{4,845} = \frac{12}{19}$$

or approximately 0.632. The values of the two binomial coefficients, $\binom{20}{4}$ and $\binom{18}{4}$, were obtained directly from Table X.

A major shortcoming of the classical probability concept is its limited applicability, for there are many situations in which the various possibilities cannot all be regarded as equally likely. This would be the case, for instance, if we are concerned with the question of whether it will rain on the next day. Surely, it would be nonsensical to say that either it will rain or it will not rain, and hence the probability for rain is $\frac{1}{2}$; or that there will be no precipitation, rain, hail, or snow, and hence the probability for rain is $\frac{1}{4}$. Also, the various possibilities cannot be regarded as equally likely when we wonder whether a person will get a raise, when we want to predict the outcome of an election or the score of a football game, or when we want to judge whether food prices will go up, go down, or remain the same.

Among the various probability concepts, most widely held is the **frequency interpretation,** according to which

The frequency interpretation of probability

> *The probability of an event (happening or outcome) is the proportion of the time that events of the same kind will occur in the long run.*

If we say that the probability is 0.88 that a jet from Denver to Seattle will arrive on time, we mean that such flights arrive on time 88% of the time. Also, if the Weather Service predicts that there is a 40% chance for rain (namely, that the probability is 0.40 that it will rain), this is meant to imply that under the same weather conditions it will rain 40% of the time. More generally, we say that an event has a probability of, say, 0.90, in the same sense in which we might say that in cold weather our car will start 90% of the time. We cannot guarantee what will happen on any particular occasion—the car may start and then it may not—but if we kept records over a long period of time, we should find that the proportion of "successes" is very close to 0.90.

In accordance with the frequency interpretation of probability, we estimate the probability of an event by observing what fraction of the time similar events have occurred in the past.

EXAMPLE   If data kept by a government agency show that (over a period of time) 528 of 600 jets from Denver to Seattle arrived on time, what is the probability that any one jet from Denver to Seattle will arrive on time?

Solution   Since in the past $\frac{528}{600} = 0.88$ of the flights arrived on time, we use this figure as an estimate of the desired probability; or we say that there is an 88% chance that such a flight will arrive on time.

EXAMPLE   If records show that 506 of 814 automatic dishwashers sold by a large retailer required repairs within the warranty year, what is the probability that an automatic dishwasher sold by this retailer will not require repairs within the warranty year?

Solution   Since $814 - 506 = 308$ of the dishwashers did not require repairs, we estimate the desired probability as $\frac{308}{814} = 0.38$ (rounded to two decimals).

When probabilities are thus estimated, it is only reasonable to ask whether the estimates are any good. Later we shall answer this question in some detail, but for now let us refer to an important theorem called the **Law of Large Numbers.** Informally, this theorem may be stated as follows:

The Law of Large Numbers

> *If a situation, trial, or experiment is repeated again and again, the proportion of successes will tend to approach the probability that any one outcome will be a success.*

In previous editions we illustrated this law by repeatedly flipping a balanced coin and recording the accumulated proportion of heads after each fifth flip. In this edition we shall use instead the computer simulation of Figure 4.3, where the 1's and 0's denote heads and tails. (The computer instruction "BRANDOM 100 N = 1 P = .5 CI" is explained on page 204.)

Reading across successive rows, we find that among the first 5 simulated flips there are 3 heads, among the first 10 there are 6 heads, among the first 15 there are 8 heads, among the first 20 there are 12 heads, among the first 25 there are 14 heads, ..., and among all 100 there are 51 heads. The corresponding proportions, plotted in Figure 4.4, are $\frac{3}{5} = 0.60$, $\frac{6}{10} = 0.60$, $\frac{8}{15} = 0.53$, $\frac{12}{20} = 0.60$,

```
MTB > BRANDOM 100 N=1 P=.5 C1
   100 BINOMIAL EXPERIMENTS WITH N =   1  AND P =  .5000
    0.     0.     1.     1.     1.     1.     1.     0.     0.     1.
    1.     0.     0.     1.     0.     1.     1.     1.     0.     1.
    0.     0.     1.     0.     1.     1.     0.     1.     0.     0.
    1.     1.     0.     1.     0.     0.     1.     1.     1.     0.
    1.     0.     1.     0.     0.     0.     0.     1.     0.     0.
    1.     1.     0.     0.     0.     0.     0.     1.     0.     0.
    1.     1.     0.     0.     1.     1.     1.     0.     1.     1.
    1.     0.     1.     1.     0.     1.     1.     0.     0.     0.
    0.     0.     0.     1.     0.     0.     1.     0.     1.     1.
    1.     0.     1.     1.     1.     1.     0.     1.     0.     1.

   SUMMARY

   VALUE      FREQUENCY
     0          49
     1          51
```

4.3

Computer simulation of 100 flips of a balanced coin.

4.4

Graph illustrating the Law of Large Numbers.

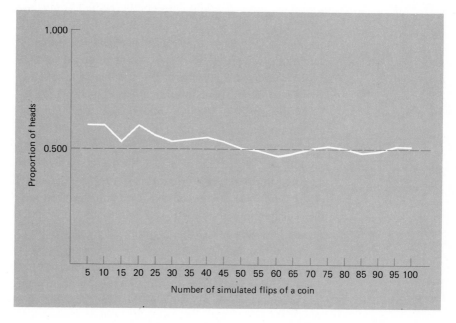

$\frac{14}{25} = 0.56, \ldots,$ and $\frac{51}{100} = 0.51$. Observe that the proportion of heads fluctuates but comes closer and closer to 0.50, the probability of heads for each flip of a coin. Theoretical support for the Law of Large Numbers will be given in Chapter 6.

In the frequency interpretation, the probability of an event is defined in terms of what happens to similar events in the long run, so let us examine briefly whether it is at all meaningful to talk about the probability of an event which can occur only once. For instance, can we assign a probability to the event that Ms. Bertha Jones will be able to leave the hospital within four days after having an appendectomy, or to the event that a certain major-party candidate will win an upcoming gubernatorial election? If we put ourselves in the position of Ms. Jones' doctor, we might check medical records, discover that patients left the hospital within four days after an appendectomy in, say, 34 percent of hundreds of cases, and apply this figure to Ms. Jones. This may not be of much comfort to Ms. Jones, but it does provide a meaning for a probability statement about her leaving the hospital within four days—the probability is 0.34.

This illustrates that when we make a probability statement about a specific (nonrepeatable) event, the frequency interpretation of probability leaves us no choice but to refer to a set of similar events. As can well be imagined, however, this can easily lead to complications, since the choice of "similar" events is generally neither obvious nor straightforward. With reference to Ms. Jones' appendectomy, we might consider as "similar" only cases in which the patients were of the same sex, only cases in which the patients were also of the same age as Ms. Jones, or only cases in which the patients were also of the same height and weight as Ms. Jones. Ultimately, the choice of "similar" events is a matter of personal judgment, and it is by no means contradictory that we can arrive at different probabilities, all valid, concerning the same event.

With regard to the question whether a certain major-party candidate will win an upcoming gubernatorial election, suppose that we ask the persons who have conducted a poll "how sure" they are that the candidate will win. If they say they are "95% sure" (that is, if they assign a probability of 0.95 to the candidate's winning the election), this is not meant to imply that he would win 95% of the time if he ran for office a great number of times. Rather, it means that the pollsters' prediction is based on a method which "works" 95% of the time. It is in this way that we must interpret many of the probabilities attached to statistical results.

Finally, let us mention an alternative concept of probability which is currently gaining in favor. According to this point of view, probabilities are interpreted as **personal** or **subjective evaluations;** they measure one's belief with regard to the uncertainties that are involved. Such probabilities apply especially when there is little or no direct evidence, so that there is really no choice but to consider collateral (indirect) information, "educated guesses," and perhaps intuition and other subjective factors. Subjective probabilities can sometimes be determined by putting the issue on a "put up or shut up" basis, as will be explained in Sections 4.5 and 5.4.

EXERCISES   (Exercises 4.46, 4.47, and 4.58 are practice exercises; their complete solutions are given on page 136.)

4.46 What is the probability of rolling a total of 9 with a pair of well-balanced dice?

4.47 A car rental agency has 20 intermediate-size cars and 10 compact cars. If four of these cars are randomly selected for a safety check, what is the probability that two of them will be intermediate-size cars and two will be compact cars?

4.48 If one card is drawn from a well-shuffled deck of 52 playing cards, what are the probabilities of getting
   (a) a red king;
   (b) a black card;
   (c) a diamond;
   (d) a 3, 4, 5, or 6?

4.49 If H stands for heads and T for tails, the four possible outcomes for two flips of a coin are HH, HT, TH, and TT. If it can be assumed that these four possibilities are equally likely, what are the probabilities of getting 0, 1, or 2 heads?

4.50 If H stands for heads and T for tails, the eight possible outcomes for three flips of a coin are HHH, HHT, HTH, THH, HTT, THT, TTH, and TTT. If it can be assumed that these eight possibilities are equally likely, what are the probabilities of getting 0, 1, 2, or 3 heads?

4.51 A bowl contains 18 red beads, 12 white beads, 14 blue beads, and 6 black beads. If one bead is drawn at random, what are the probabilities of getting
   (a) a red bead;
   (b) a bead that is white or black;
   (c) a bead that is neither red nor black?

4.52 If we roll a well-balanced die, what are the probabilities of getting
   (a) a 1 or a 6;
   (b) an even number?

4.53 If two cards are drawn from a well-shuffled deck of 52 playing cards, what are the probabilities of getting
   (a) two spades;
   (b) two aces;
   (c) a king and a queen?

4.54 Among the 16 applicants for three identical positions at a bank, 12 have college degrees. What are the probabilities that the three positions will be randomly filled with

(a) three of the applicants with college degrees;

(b) two of the applicants with college degrees and one applicant without a college degree?

4.55 A carton of 24 light bulbs includes three that are defective. If two of the bulbs are chosen at random, what are the probabilities that

(a) neither bulb will be defective;

(b) exactly one of the bulbs will be defective;

(c) both bulbs will be defective?

4.56 An apartment building has 24 two-bedroom apartments and 12 three-bedroom apartments. If three of the apartments are chosen at random to be redecorated, what are the probabilities that

(a) all three will be two-bedroom apartments;

(b) two will be two-bedroom apartments and one will be a three-bedroom apartment?

4.57 On a tray there are six pieces of chocolate cake and four pieces of walnut cake. If a waiter randomly picks two pieces of cake from the tray and gives them to customers who ordered chocolate cake, what is the probability that he will be making a mistake?

4.58 Data compiled by the manager of a department store show that 1,018 of 1,956 persons who entered the store on a weekday morning made at least one purchase. Estimate the probability that a person who enters the store on a weekday morning will make at least one purchase.

4.59 If 385 of 493 television viewers in a certain area expressed the opinion that local news coverage is inadequate, estimate the probability that a television viewer in that area will share this opinion?

4.60 In a sample of 1,338 cars stopped at a roadblock, only 201 of the drivers had their seat belts fastened. Estimate the probability that a driver (at least, on that road) will not have his or her seat belt fastened.

4.61 In a sample of 250 students attending a certain large university, 65 expressed the opinion that there was too much emphasis on athletics. Estimate the probability that a student attending this university will share this opinion.

4.62 Among 842 armed robberies in a certain city, 143 were never solved. Estimate the probability that an armed robbery in this city will be solved.

 4.63 Repeat the illustration on page 123; that is, use a computer to simulate 100 flips of a balanced coin and plot the accumulated proportion of heads after each fifth flip (as in Figure 4.4) to illustrate the Law of Large Numbers. If a suitable computer package is not available, actually flip a coin 100 times.

4.64 Use a computer to simulate 200 flips of a balanced coin and plot the accumulated proportion of heads after each fifth flip (as in Figure 4.4) to illustrate the Law of Large Numbers. If a suitable computer package is not available, actually flip a coin 200 times.

 4.65 Use a computer to simulate 150 flips of a pair of coins, recording a 1 or a 0 depending on whether or not both coins come up heads. The probability for this is 0.25 (see Exercise 4.49). Then, to illustrate the Law of Large Numbers, plot the accumulated proportion of successes (both coins come up heads) after each tenth flip. If a suitable computer package is not available, actually flip a pair of coins 150 times.

4.5

Mathematical Expectation

If an insurance agent tells us that in the United States a 45-year-old woman can expect to live 33 more years, this does not mean that anyone really expects a 45-year-old woman to live until her 78th birthday and then pass away the next day. Similarly, if we read that a person living in the United States can expect to eat 276 eggs per year and 128.9 pounds of beef, or that a child in the age group from 6 to 16 can expect to visit a dentist 1.9 times a year, it must be obvious that the word "expect" is not being used in its colloquial sense. A child cannot go to the dentist 1.9 times, and it would be surprising, indeed, if we found somebody who has actually eaten 276 eggs and 128.9 pounds of beef in a given year. So far as 45-year-old women are concerned, some will live another 15 years, some will live another 25 years, some will live another 40 years, . . . , and the life expectancy of "33 more years" will have to be interpreted as an average, or as we shall call it here, a **mathematical expectation.**

Originally, the concept of a mathematical expectation arose in connection with games of chance, and in its simplest form it is the product of the amount a player stands to win and the probability that he or she will win.

EXAMPLE   What is our mathematical expectation if we will receive $20 if and only if a coin comes up heads?

Solution   If we assume that the coin is balanced and randomly tossed, namely, that the probability of heads is $\frac{1}{2}$, our mathematical expectation is $20 \cdot \frac{1}{2} = \$10$.

EXAMPLE   What is our mathematical expectation if we buy one of 2,000 raffle tickets issued for a television set worth $540?

Solution   The probability that we will win is $\dfrac{1}{2,000}$, so our mathematical expecta-

tion is $540 \cdot \dfrac{1}{2,000} = \$0.27$, or 27 cents. Thus, it would be unwise to spend more than 27 cents for the ticket, unless, of course, the proceeds of the raffle go to a worthy cause (or the difference can be credited to whatever pleasure a person might derive from placing a bet).

In both of these examples there is a single prize, but in each case there are two possible payoffs—$20 and $0 in the first example and $540 and $0 in the other. Indeed, in the second example we can argue that 1,999 of the tickets will pay $0 and one of the tickets will pay $540 (or the equivalent in merchandise). Altogether, the 2,000 tickets will thus pay $540, or on the average $\dfrac{540}{2,000} = \$0.27$ per ticket, and this is the mathematical expectation.

To generalize the concept of a mathematical expectation, let us consider the following change in the raffle of the preceding example:

EXAMPLE   What is our mathematical expectation if we buy one of 2,000 raffle tickets issued for a first prize of a television set worth $540, a second prize of a tape recorder worth $180, and a third prize of a pocket radio worth $40?

Solution   Now we can argue that 1,997 of the raffle tickets will not pay anything at all, one ticket will pay $540 (in merchandise), another will pay $180 (in merchandise), and a third will pay $40 (in merchandise). Altogether, the 2,000 tickets will thus pay $540 + 180 + 40 = \$760$ (in merchandise), or on the average $\dfrac{760}{2,000} = \$0.38$ per ticket. As before, this is the mathe-

matical expectation for each ticket. Looking at the problem in a different way we could argue that if the raffle were repeated many times, we would lose $\dfrac{1,997}{2,000} \cdot 100 = 99.85\%$ of the time (or with probability 0.9985) and win each of the prizes $\dfrac{1}{2,000} \cdot 100 = 0.05\%$ of the time (or with probability 0.0005). On the average we would thus win

$$0(0.9985) + 540(0.0005) + 180(0.0005) + 40(0.0005) = \$0.38$$

which is the sum of the products obtained by multiplying each amount by the corresponding proportion or probability.

Generalizing from this example, let us now give the following general definition:

Mathematical expectation

> *If the probabilities of obtaining the amounts $a_1, a_2, \ldots,$ or a_k are, respectively, $p_1, p_2, \ldots,$ and p_k, then the mathematical expectation is*
>
> $$E = a_1p_1 + a_2p_2 + \cdots + a_kp_k$$

Each amount is multipled by the corresponding probability, and the mathematical expectation, E, is the sum of all these products. In the $\sum$ notation,

$$E = \sum a \cdot p$$

So far as the a's are concerned, it is important to keep in mind that they are positive when they represent profits, winnings, or gains (namely, amounts which we receive), and that they are negative when they represent losses, penalties, or deficits (namely, amounts which we have to pay).

EXAMPLE   What is our mathematical expectation if we win \$6 when a die comes up 1 or 2, and lose \$3 when the die comes up 3, 4, 5, or 6?

Solution   The amounts are $a_1 = 6$ and $a_2 = -3$, and if we assume that the die is balanced and randomly tossed, $p_1 = \frac{2}{6}$ and $p_2 = \frac{4}{6}$. Thus, the mathematical expectation is

$$E = 6 \cdot \frac{2}{6} + (-3) \cdot \frac{4}{6} = 0$$

This example illustrates what we call an **equitable** or **fair game.** It is a game which does not favor either player; that is, each player's mathematical expectation is zero.

The probabilities are 0.22, 0.36, 0.28, and 0.14 that an investor will be able to sell a piece of property at a profit of $5,000, that he will be able to sell it at a profit of $2,000, that he will break even, or that he will sell it at a loss of $3,000. What is his expected profit?

Solution Substituting $a_1 = 5,000$, $a_2 = 2,000$, $a_3 = 0$, $a_4 = -3,000$, $p_1 = 0.22$, $p_2 = 0.36$, $p_3 = 0.28$, and $p_4 = 0.14$ into the formula for E, we get

$$E = 5,000(0.22) + 2,000(0.36) + 0(0.28) - 3,000(0.14)$$
$$= \$1,400$$

Although we referred to the quantities $a_1, a_2, \ldots,$ and a_k as "amounts," they need not be cash winnings, losses, penalties, or rewards. When we said on page 127 that a child in the age group from 6 to 16 can expect to visit a dentist 1.9 times a year, we referred to the result which was obtained by multiplying 0, 1, 2, 3, $\ldots$, by the corresponding probabilities that a child in this age group will visit a dentist that many times a year (see Exercise 4.73 on page 133).

EXAMPLE The police chief of a city knows that the probabilities for 0, 1, 2, 3, 4, or 5 car thefts on any given day are, respectively, 0.21, 0.37, 0.25, 0.13, 0.03, and 0.01. How many car thefts can he expect per day?

Solution Substituting into the formula for a mathematical expectation, we get

$$E = 0(0.21) + 1(0.37) + 2(0.25) + 3(0.13) + 4(0.03) + 5(0.01)$$
$$= 1.43$$

In all of the examples in this section we were given the values of a and p (or the values of the a's and p's) and calculated E. Now let us consider an example in which we are given the values of a and E to arrive at some result about p.

EXAMPLE To handle a liability suit, a lawyer has to decide whether to charge a straight fee of $2,400 or a contingent fee of $9,600 which she will get only if her client wins. How does she "feel" about her client's chances, if she prefers the straight fee of $2,400?

Solution   If she "feels" that the probability is p that her client will win and she decides on the contingent fee, her mathematical expectation is $9,600p$. Since she "feels" that the certainty of $2,400 is preferable to the mathematical expectation of $9,600p$, we can write

$$2,400 > 9,600p$$

which yields $p < \dfrac{2,400}{9,600}$ and, hence, $p < 0.25$.

Note that this is one of the two methods of determining subjective probabilities referred to on page 125. To pin it down further, we might ask the lawyer if she would still prefer the straight fee of $2,400 if the contingent fee were, say, $19,200 (see Exercise 4.76 on page 133).

4.6
A Decision Problem ★

When we are faced with uncertainties, mathematical expectations can often be used to great advantage in making decisions. In general, if we have to choose between two or more alternatives, it is considered "rational" to select the one with the "most promising" mathematical expectation: the one which maximizes expected profits, minimizes expected costs, maximizes expected tax advantages, minimizes expected losses, and so on.

EXAMPLE   A clothing manufacturer must decide whether to spend a considerable sum of money to build a new factory. He knows that if the new factory is built and the clothing business has a good sales year, there will be a $451,000 profit; if the new factory is built and the clothing business has a poor sales year, there will be a deficit of $110,000; if the new factory is not built and the clothing business has a good sales year, there will be a $220,000 profit; and if the new factory is not built and the clothing business has a poor year, there will be a $22,000 profit (mostly because of lower overhead cost). If the clothing manufacturer feels that the probabilities for a good sales year or a poor sales year are, respectively, 0.40 and 0.60, would building the new factory maximize his expected profit?

Solution   In problems like this, it usually helps to present the information about profits and deficits in the following kind of table:

	New factory built	New factory not built
Good sales year	451,000	220,000
Poor sales year	−110,000	22,000

As can be seen from this table, it will be advantageous to build the new factory only if the clothing business is going to have a good sales year, and the decision whether to build the new factory will, therefore, have to depend on the chances that this will be the case. Using the manufacturer's probabilities of 0.40 and 0.60 for a good sales year and a poor sales year, we find that if the new factory is built, his expected profit is

$$451,000(0.40) - 110,000(0.60) = \$114,400$$

and if the new factory is not built, his expected profit is

$$220,000(0.40) + 22,000(0.60) = \$101,200$$

Since the first of these two figures exceeds the second, it follows that building the new factory maximizes the manufacturer's expected profit.

The way in which we have studied this problem is called a **Bayesian analysis.** In this kind of analysis, probabilities are assigned to the alternatives about which uncertainties exist (the **states of nature,** which in our example were a good sales year and a poor sales year); then we choose whichever alternative promises the greatest expected profit or the smallest expected loss.

This approach to decision making has great intuitive appeal, but it is not without complications. If mathematical expectations are to be used for making decisions, it is essential that our appraisals of all relevant probabilities are close if not correct, and that we know the exact values of the "payoffs" associated with the various possibilities (see Exercises 4.83 and 4.84).

EXERCISES

(Exercises 4.66, 4.67, 4.75, and 4.79 are practice exercises; their complete solutions are given on pages 136 and 137.)

4.66 At a bazaar held to raise money for a charity, it costs 50 cents to try one's luck in drawing an ace from an ordinary deck of 52 playing cards. What is the expected profit per customer, if they pay $4 if and only if a person draws an ace?

4.67 A union wage negotiator feels that the probabilities are 0.25, 0.50, 0.20, and 0.05, that the members of the union will get a $1.20 raise in their hourly wage, an 80-cent raise, a 40-cent raise, or no raise at all. What is the corresponding expected raise?

4.68 If a service club sells 500 raffle tickets for a cash prize of $100, what is the mathematical expectation of a person who buys one of the tickets?

4.69 In a friendly game, if we receive 10 cents each time we roll a 6 with a balanced die, how much should we pay when we roll a 1, 2, 3, 4, or 5 so as to make the game equitable?

4.70 In the finals of a tennis tournament, the winner gets $48,000 and the loser gets $24,000. What are the two finalists' mathematical expectations if
   (a) they are evenly matched;
   (b) their respective probabilities of winning are $\frac{2}{3}$ and $\frac{1}{3}$?

4.71 A stockbroker feels that the probabilities are 0.40, 0.30, 0.20, and 0.10 that the value of a stock will increase by $1.50, $1.00, 50 cents, or not at all. What is the expected value of the stock?

4.72 An importer is offered a shipment of pineapples for $12,000, and the probabilities that he will be able to sell them for $16,000, $15,000, $14,000, or $13,000 are, respectively, 0.15, 0.41, 0.33, and 0.11. What is the importer's expected gross profit?

4.73 If the probabilities are 0.15, 0.28, 0.27, 0.17, 0.08, 0.03, and 0.02 that a child in the age group from 6 to 16 will visit a dentist 0, 1, 2, 3, 4, 5, or 6 times a year, how many times can a child in this age group expect to visit a dentist in any given year?

4.74 The probabilities that a person who enters "The Department Store" will make 0, 1, 2, 3, 4, or 5 purchases are 0.11, 0.33, 0.31, 0.12, 0.09, and 0.04. How many puchases can a person entering this store be expected to make?

4.75 A salesperson has to choose between a straight salary of $24,000 and a salary of $20,000 plus a bonus of $8,000 if her sales exceed a certain quota. How does she assess her chances of exceeding the quota if she chooses the lower salary with the possibility of the bonus?

4.76 With reference to the example on page 130, suppose that the lawyer is asked whether she would still prefer the straight fee of $2,400 if the contingent fee were doubled to $19,200. What new information do we have about her "feelings" regarding her client's chances, if she prefers the contingent fee of $19,200?

4.77 An insurance company agrees to pay the promoter of a drag race $15,000 if the race has to be canceled because of rain. If the company's actuary feels that a fair net premium for this risk is $2,400, what does this tell us about his assessment of the probability that the race will have to be canceled because of rain?

4.78 One contractor offers to do a road repair job for $45,000, while another contractor offers to do the job for $50,000 with a penalty of $12,500 if the job is not finished on time. If the person who lets out the contract for the job prefers the second offer, what does this tell us about her assessment of the probability that the second contractor will not finish the job on time?

★ **4.79** A truck driver has to deliver a load of building materials to one of two construction sites, which are, respectively, 16 and 20 miles from the lumberyard, but he has misplaced the order telling him where the load should go. The two construction sites are 6 miles apart, and to complicate matters, the telephone at the lumberyard is out of order. If the driver feels that the probabilities are, respectively, $\frac{1}{5}$ and $\frac{4}{5}$ that the load should go to the site which is 16 miles from the lumberyard and the one which is 20 miles from the lumberyard, where should he go first so as to minimize the expected distance he will have to drive?

★ **4.80** With reference to the preceding exercise, where should the driver go first so as to minimize the expected distance he will have to drive, if he reassessed the probabilities as $\frac{1}{8}$ and $\frac{7}{8}$ instead of $\frac{1}{5}$ and $\frac{4}{5}$?

★ **4.81** The management of a mining company must decide whether to continue an operation at a certain location. If they continue and are successful, they will make a profit of $4,500,000; if they continue and are not successful, they will lose $2,700,000; if they do not continue but would have been successful if they had continued, they will lose $1,800,000 (for competitive reasons); and if they do not continue and would not have been successful if they had continued, they will make a profit of $450,000 (because funds allocated to the operation remain unspent). What decision would maximize the company's expected profit if it is felt that there is a fifty-fifty chance for success?

★ **4.82** Show that it does not matter what they decide to do in the preceding exercise if it is felt that the probabilities for and against success are, respectively, $\frac{1}{3}$ and $\frac{2}{3}$.

4.83 With reference to the example on page 131, show that building the new factory would not maximize the expected profit if the manufacturer felt that the probabilities for a good sales year and a poor sales year are, respectively, $\frac{1}{6}$ and $\frac{5}{6}$.

★ **4.84** With reference to the example on page 131, would the clothing manufacturer's decision to build the new factory still maximize his expected profit if it is found that the $110,000 figure is in error and should be a loss of $165,000?

SOLUTIONS
OF
PRACTICE
EXERCISES

4.1 The tree diagram is shown in Figure 4.5, where N denotes a win by the National League champion and A denotes a win by the American League champion.

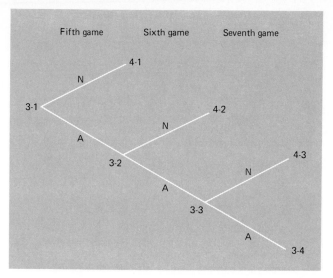

Tree diagram for Exercise 4.1.

4.7 (a) There are three choices corresponding to the three trails.

   (b) $m = 3$, $n = 3$, so that there are $3 \cdot 3 = 9$ possibilities.

   (c) $m = 3$, $n = 2$, so that there are $3 \cdot 2 = 6$ possibilities.

4.8 $n_1 = 9$, $n_2 = 13$, and $n_3 = 10$, and there are $n_1 \cdot n_2 \cdot n_3 = 9 \cdot 13 \cdot 10 = 1{,}170$ possibilities.

4.19 (a) $8! = 8 \cdot 7 \cdot (6 \cdot 5 \cdot 4 \cdot 3 \cdot 2 \cdot 1) = 8 \cdot 7 \cdot 6!$ is true; in general, $n! = n(n - 1)! = n(n - 1)(n - 2)! = n(n - 1)(n - 2)(n - 3)!$, and so forth.

   (b) $3! = 6$, $2! = 2$, $6! = 720$, and $6 \cdot 2$ does not equal 720.

   (c) $\dfrac{12!}{12 \cdot 11 \cdot 10} = \dfrac{12 \cdot 11 \cdot 10 \cdot 9!}{12 \cdot 11 \cdot 10} = 9!$ is true.

4.22 $_6P_3 = 6 \cdot 5 \cdot 4 = 120$.

4.23 $_5P_5 = 5! = 120$.

4.33 $\dbinom{14}{3} = \dfrac{14 \cdot 13 \cdot 12}{3!} = 364$.

4.34 (a) The three Democrats can be chosen in $\binom{9}{3} = 84$ ways, the Republican can be chosen in 5 ways, so that, by the multiplication of choices, there are $84 \cdot 5 = 420$ possibilities.

   (b) The two Democrats can be chosen in $\binom{9}{2} = 36$ ways, the Republican can be chosen in 5 ways, and the Independent can be chosen in 2 ways, so that by the generalized rule for the multiplication of choices there are $36 \cdot 5 \cdot 2 = 360$ possibilities.

4.46 Altogether there are $n = 6 \cdot 6 = 36$ possibilities, and in $s = 4$ of these (3 and 6, 4 and 5, 5 and 4, 6 and 3) the total is 9; therefore, the probability is $\dfrac{s}{n} = \dfrac{4}{36} = \dfrac{1}{9}$.

4.47 Altogether there are $n = \dbinom{30}{4} = \dfrac{30 \cdot 29 \cdot 28 \cdot 27}{4!} = 5 \cdot 29 \cdot 7 \cdot 27$ possibilities;

the two intermediate-size cars can be chosen in $\dbinom{20}{2} = \dfrac{20 \cdot 19}{2} = 10 \cdot 19$ ways,

the two compact cars can be chosen in $\dbinom{10}{2} = \dfrac{10 \cdot 9}{2} = 5 \cdot 9$ ways, so that, by

the multiplication of choices, $s = 10 \cdot 19 \cdot 5 \cdot 9$; it follows that the probability is

$\dfrac{s}{n} = \dfrac{10 \cdot 19 \cdot 5 \cdot 9}{5 \cdot 29 \cdot 7 \cdot 27} = \dfrac{190}{609}$, or approximately 0.31.

4.58 The estimate is $\dfrac{1{,}018}{1{,}956} = 0.52$ rounded to two decimals.

4.66 The mathematical expectation of a person who draws a card is $4 \cdot \frac{1}{13} = \$0.31$ to the nearest cent, so that the expected profit is $50 - 31 = 19$ cents.

4.67 Substituting $a_1 = 120$, $a_2 = 80$, $a_3 = 40$, $a_4 = 0$, $p_1 = 0.25$, $p_2 = 0.50$, $p_3 = 0.20$, and $p_4 = 0.05$ into the formula for a mathematical expectation, we get $120(0.25) + 80(0.50) + 40(0.20) + 0(0.05) = 78$ cents.

4.75 If she feels that the probability of her exceeding the quota is p, the mathematical expectation corresponding to the salary plus bonus alternative is $20{,}000 + 8{,}000p$. Since this exceeds $24{,}000$, we get $20{,}000 + 8{,}000p > 24{,}000$ and, hence, $8{,}000p > 4{,}000$ and $p > \dfrac{4{,}000}{8{,}000} = \dfrac{1}{2}$. Thus, she feels that there is a better than fifty-fifty chance that she will exceed the quota.

4.79 Letting I and II denote the construction sites which are, respectively, 16 and 20 miles from the lumberyard, we can show the distances corresponding to the various alternatives (depending on where the driver goes first and where he should go) in the following table:

		Driver goes first to	
		I	II
Driver should go to	I	16	$20 + 6 = 26$
	II	$16 + 6 = 22$	20

Thus, if he first goes to site I, the expected distance is

$$16 \cdot \frac{1}{5} + 22 \cdot \frac{4}{5} = \frac{16 + 88}{5} = \frac{104}{5} = 20.8 \text{ miles}$$

and if he goes first to site II, the expected distance is

$$26 \cdot \frac{1}{5} + 20 \cdot \frac{4}{5} = \frac{26 + 80}{5} = \frac{106}{5} = 21.2 \text{ miles}$$

Thus, if the driver wants to minimize the expected distance he has to drive, he should go first to site I, that is, the construction site which is 16 miles from the lumberyard.

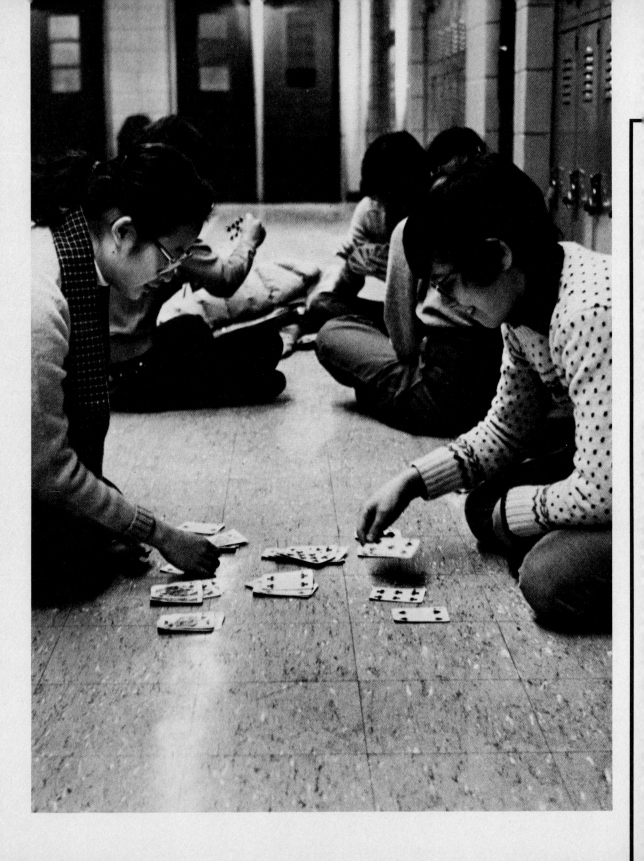

Some Rules of Probability

In the study of probability there are three fundamental kinds of questions: (1) What do we mean when we say that the probability of an event is, say, 0.60, 0.88, or 0.02? (2) How are the numbers we call probabilities determined, or measured, in actual practice? (3) What are the mathematical rules which probabilities must obey?

Since the first two kinds of questions have already been studied in Chapter 4, we shall concentrate here on some of the rules which probabilities must obey, namely, on

> **PROBABILITY THEORY** (ˌpräb-ə-ˈbil-ət-ē ˈthē-ə-rē) The branch of mathematics which concerns the study of probability; the mathematics of chance.

Although probability theory was developed originally in connection with games of chance, it has become an indispensable tool for business managers who must plan inventories without knowing with certainty what products will sell, for military strategists who must commit personnel and equipment to the hazards of battle, for doctors who risk their lives in combatting disease, for anyone who sends a message by mail without an assurance that it will be delivered on time, and so on.

In this chapter, after some preliminaries in Sections 5.1 and 5.2, we shall study some of the basic rules in Section 5.3, the relationship between probabilities and odds in Section 5.4, addition rules in Section 5.5, conditional probabilities and related problems in Sections 5.6, 5.7, and 5.8, and Bayes' theorem in Section 5.8.

5.1

The Sample Space

In statistics, the word "experiment" is used in a very wide, and unconventional, sense. For lack of a better term, it refers to any process of observation or measurement. Thus, an **experiment** may consist of counting how many times a student has been absent; it may consist of the simple process of noting whether a light is on or off, or whether a person is single or married; or it may consist of the very complicated process of obtaining and evaluating data to predict trends in the economy, to find the source of social unrest, or to study the cause of a disease. The results one obtains from an experiment, whether they are instrument readings, counts, "yes or no" answers, or values obtained through extensive calculations, are called the **outcomes** of the experiment.

For each experiment, the set of all possible outcomes is called the **sample space** and it is usually denoted by the letter S. For instance, if a zoologist must choose three of his 24 guinea pigs for an experiment, the sample space consists of the $\binom{24}{3} = 2,024$ ways in which the selection can be made; if the dean of a college has to assign two of her 84 faculty members as advisors to a political science club, the sample space consists of the $\binom{84}{2} = 3,486$ ways in which this can be done. Also, if we are concerned with the number of days that it rains in Chicago during the month of January, the sample space is the set

$$S = \{0, 1, 2, 3, 4, \ldots, 30, 31\}$$

When we study the outcomes of an experiment, we usually identify the various possibilities with numbers, points, or some other kinds of symbols, so that we can treat all questions about them mathematically, without having to go through long verbal descriptions of what has taken place, is taking place, or will take place. For instance, if there are eight candidates for a scholarship and we let a, b, c, d, e, f, g, and h denote that it is awarded to Ms. Adam, Mr. Bean, Miss Clark, and so on, then the sample space for this experiment is the set

$$S = \{a, b, c, d, e, f, g, h\}$$

The use of points rather than letters or numbers has the advantage that it makes it easier to visualize the various possibilities, and perhaps discover some special features which several of the outcomes may have in common.

CHAP. 5: Some Rules of Probability

EXAMPLE   A used-car dealer has two 1984 Chevrolet Camaros on his lot and we
are interested in how many of them each of two salespersons will sell
in a given week.

(a) Using two coordinates so that (0, 1), for example, represents the
outcome that the first salesperson will sell neither of the Camaros
and the second salesperson will sell one, (1, 1) represents the
outcome that each of the two salespersons will sell one of the
Camaros, and (2, 0) represents the outcome that the first salesperson will sell them both, list all possible outcomes of this
experiment.

(b) Draw a figure showing the corresponding points of the sample
space.

Solution   (a) The six possible outcomes are (0, 0), (1, 0), (0, 1), (2, 0), (1, 1), and
(0, 2).

(b) The corresponding points are shown in Figure 5.1, from which it
is apparent, for instance, that they sell equally many 1984 Camaros
in two of the six possibilities, and that they sell both cars in three
of the six possibilities.

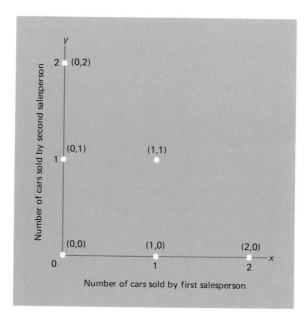

5.1

*Sample space for two-sales-
persons example.*

Generally, we classify sample spaces according to the number of elements,
or points, which they contain. The ones we have studied so far in this section

contained, respectively, 2,024, 3,486, 32, 8, and 6 elements, and we refer to them all as **finite.** In this chapter we shall consider only sample spaces that are finite, but in later chapters we shall consider also sample spaces that are **infinite.** An infinite sample space arises, for example, when we throw a dart at a target and there is a continuum of points we may hit.

5.2
Events

In statistics, any subset of a sample space is called an **event.** By subset we mean any part of a set, including the set as a whole and, trivially, a set called the **empty set** and denoted by $\varnothing$, which has no elements at all. For instance, with reference to the example on page 140 dealing with the number of days that it rains in Chicago during the month of January, the subset

$$M = \{10, 11, 12, \ldots, 19, 20\}$$

is the event that there will be anywhere from 10 to 20 rainy days, and the subset

$$N = \{18, 19, 20, \ldots, 30, 31\}$$

is the event that there will be at least 18 rainy days.

EXAMPLE   With reference to Figure 5.1 on page 141, express in words what events are represented by
(a) $C = \{(0, 0), (1, 1)\}$;
(b) $D = \{(1, 0), (1, 1)\}$;
(c) $E = \{(0, 2)\}$.

Solution   (a). C is the event that the two salespersons will sell equally many of the 1984 Chevrolet Camaros.
(b) D is the event that the first salesperson will sell one and only one of the two cars.
(c) E is the event that the second salesperson will sell both cars.

Note that in both of the examples above we denoted events by capital letters, as is the custom.

In many probability problems we must deal with events which are compounded by forming **unions, intersections,** and **complements.**

The union of two events A and B, denoted by $A \cup B$, is the event which consists of all the elements (outcomes) contained in event A, in event B, or in both.

The intersection of two events A and B, denoted by $A \cap B$, is the event which consists of all the elements (outcomes) contained in both A and B.

The complement of event A, denoted by A', is the event which consists of all the elements (outcomes) of the sample space that are not contained in A.

It is common practice to read $\cup$ as "or," $\cap$ as "and," and A' as "not A."

EXAMPLE   With reference to the illustration dealing with the number of days that it rains in Chicago in January, where we had

$$S = \{0, 1, 2, 3, 4, \ldots, 30, 31\}$$

$$M = \{10, 11, 12, \ldots, 19, 20\}$$

and

$$N = \{18, 19, 20, \ldots, 30, 31\}$$

express each of the following events symbolically and also in words:
(a) $M \cup N$;   (c) M';   (e) $M' \cup N'$;
(b) $M \cap N$;   (d) N';   (f) $M' \cap N'$.

Solution   (a) Since $M \cup N$ contains all the elements that are in M, in N, or in both, we find that

$$M \cup N = \{10, 11, 12, \ldots, 30, 31\}$$

and this is the event that there will be at least 10 rainy days.
(b) Since $M \cap N$ contains all the elements that are in both M and N, we find that

$$M \cap N = \{18, 19, 20\}$$

and this is the event that there will be from 18 to 20 rainy days.
(c) Since M' contains all the elements of the sample space that are not in M, we find that

$$M' = \{0, 1, 2, \ldots, 8, 9, 21, 22, \ldots, 30, 31\}$$

and this is the event that there will be fewer than 10 or more than 20 rainy days.

(d) Since N' contains all the elements of the sample space that are not in N, we find that

$$N' = \{0, 1, 2, \ldots, 16, 17\}$$

and this is the event that there will be at most 17 rainy days.

(e) Since $M' \cup N'$ contains all the elements that are not in M, not in N, or in neither, we find that

$$M' \cup N' = \{0, 1, 2, \ldots, 16, 17, 21, 22, \ldots, 30, 31\}$$

and this is the event that there will be fewer than 18 or more than 20 rainy days (or the event that there will not be 18, 19, or 20 rainy days).

(f) Since $M' \cap N'$ contains all the elements that are in both M' and N', we find that

$$M' \cap N' = \{0, 1, 2, \ldots, 8, 9\}$$

and this is the event that there will be at most 9 rainy days.

Observe also that in the two-salespersons example events D and E have no elements in common. Such events are called **mutually exclusive.**

> Two events A and B are mutually exclusive if they cannot both occur at the same time, namely, when $A \cap B = \varnothing$.

EXAMPLE With reference to the two-salespersons example, which of the following pairs of events are mutually exclusive:

(a) C and D;
(b) C and E?

Solution (a) Since events C and D both contain $(1, 1)$, they are not mutually exclusive.

(b) Since $C \cap E = \varnothing$, events C and E are mutually exclusive.

Sample spaces and events, particularly relationships among events, are often pictured by means of **Venn diagrams** such as those of Figures 5.2 and 5.3.

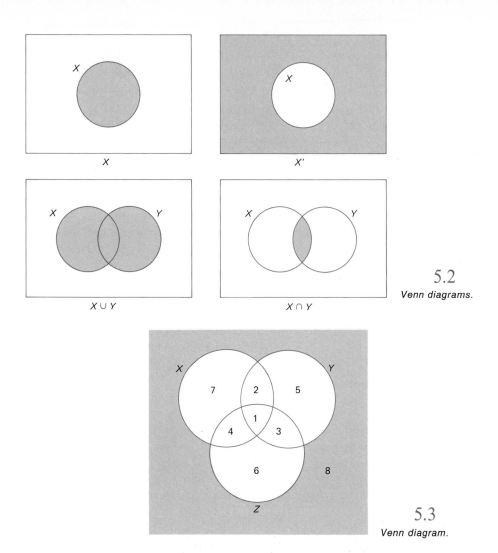

X

X'

X ∪ Y

X ∩ Y

5.2
Venn diagrams.

5.3
Venn diagram.

In each case, the sample space is represented by a rectangle, while events are represented by regions within the rectangle, usually by circles or parts of circles. The tinted regions of the four Venn diagrams of Figure 5.2 represent the event X, the complement of event X, the union of two events X and Y, and the intersection of two events X and Y.

EXAMPLE   If X is the event that a given high school student will be accepted at Duke University and Y is the event that he will be accepted at the University of Virginia, what events are represented by the tinted regions of the four Venn diagrams of Figure 5.2?

145

Solution   The tinted region of the first diagram represents the event that the student will be accepted at Duke University; the tinted region of the second diagram represents the event that he will not be accepted at Duke University; the tinted region of the third diagram represents the event that he will be accepted at Duke University, the University of Virginia, or both; and the tinted region of the fourth diagram represents the event that he will be accepted at Duke University as well as the University of Virginia.

When we deal with three events, we draw the circles as in Figure 5.3. In this diagram, the circles divide the sample space into eight regions, which we numbered 1 through 8, and it is easy to determine whether each of the corresponding events is contained in X or in X', in Y or in Y', and in Z or in Z'.

EXAMPLE   Suppose that the employees of a company are planning a picnic and that X is the event that the weather will be good, Y is the event that there will be enough food, and Z is the event that they will have a good time. With reference to the Venn diagram of Figure 5.3, express in words what events are represented by the following regions:
   (a) region 4;
   (b) regions 1 and 3 together;
   (c) regions 3, 5, 6, and 8 together.

Solution   (a) Since this region is contained in X and in Z but not in Y, it represents the event that the weather will be good and they will have a good time, but there will not be enough food.
   (b) Since this is the region common to Y and Z, it represents the event that there will be enough food and that they will have a good time.
   (c) Since this is the entire region outside X, it represents the event that the weather will not be good.

EXERCISES   (Exercises 5.1, 5.2, and 5.12 are practice exercises; their complete solutions are given on pages 182 and 183.)

5.1 With reference to the illustration on page 140, suppose that $a, b, c, d, e, f, g,$ and h denote, respectively, that Ms. Adam, Mr. Bean, Miss Clark, Mrs. Daly, Mr. Earl, Ms. Fuentes, Ms. Garner, or Mr. Hall will be awarded the scholarship, and that $A = \{a, c, f, g\}$, $B = \{d, f, h\}$, and $C = \{a, b, e, g\}$. List the outcomes which comprise each of the following events and also express the events in words:
   (a) A';   (c) $A \cap B$;   (e) $B \cap C$;
   (b) $A \cup B$;   (d) $A \cap C$;   (f) $A' \cup B$.

5.2 A small restaurant has two chefs and five waiters, and at least one chef and two waiters have to be present at all times.

    (a) Using two coordinates so that (1, 3), for example, represents the event that one chef and three waiters are present, and (2, 4) represents the event that two chefs and four waiters are present, draw a diagram similar to that of Figure 5.1, showing the eight points of the sample space.

    (b) Describe in words what events are represented by $Q = \{(1, 4), (2, 4)\}$, $R = \{(1, 2), (2, 3)\}$, and $T = \{(1, 2), (1, 3), (1, 4), (1, 5)\}$.

    (c) List the outcomes which comprise each of the following events and also express the events in words: $Q \cup T$, $R \cap T$, and T'.

    (d) With reference to part (b), which of the pairs of events, Q and R, Q and T, and R and T, are mutually exclusive?

5.3 In an experiment, we roll a pair of dice and observe the total number of points.

    (a) List the eleven elements of the sample space.

    (b) If $A = \{2, 3, 4, 7, 8, 9, 10\}$ and $B = \{4, 5, 6, 7, 8\}$, list the outcomes which comprise each of the following events and also express the events in words: A', $A \cup B$, and $A \cap B$.

5.4 In an experiment, persons are asked to pick a number from 1 to 10, so that for each person the sample space is the set $S = \{1, 2, \ldots, 9, 10\}$. If $F = \{1, 2, 3, 4, 5, 6\}$, $G = \{5, 6, 7, 8, 9, 10\}$, and $H = \{4, 5, 6, 7\}$, list the outcomes which comprise each of the following events and also express the events in words:

(a) $F \cup G$;	(c) $G \cap H'$;
(b) $F \cap G$;	(d) $G' \cap H'$.

5.5 If one card is drawn from an ordinary deck of 52 playing cards, the sample space may be written as

$$S = \{A_c, 2_c, \ldots, K_c, A_d, 2_d, \ldots, K_d, A_h, 2_h, \ldots, K_h, A_s, 2_s, \ldots, K_s\}$$

where c, d, h, and s denote the four suits: clubs, diamonds, hearts, and spades. If

$$M = \{Q_c, K_c, Q_d, K_d, Q_h, K_h, Q_s, K_s\}$$

and

$$N = \{10_s, J_s, Q_s, K_s\}$$

list the outcomes which comprise each of the following events and also express the events in words:

(a) $M \cup N$;	(c) M';	(e) $M' \cup N'$;
(b) $M \cap N$;	(d) N';	(f) $M' \cap N'$.

5.6 To construct sample spaces for experiments in which we deal with categorical data, we often code the various alternatives by assigning them numbers. For instance, if persons are asked whether their favorite color is red, yellow, blue, green, brown, white, purple, or some other color, we might assign these alternatives the codes 1, 2, 3, 4, 5, 6, 7, and 8. If $A = \{3, 4\}$, $B = \{1, 2, 3, 4, 5, 6, 7\}$, and $C = \{6, 7, 8\}$, list the outcomes which comprise each of the following events:

(a) B'; (c) $A \cap B$; (e) $B \cup C$;
(b) $A \cup B$; (d) C'; (f) $B \cap C'$.

5.7 With reference to the preceding exercise, which of the pairs of events, A and B, A and C, and B and C, are mutually exclusive?

5.8 A movie critic has two days in which to view some of the pictures that have recently been released. She wants to see at least three of the movies but not more than three on either day.

(a) Using two coordinates so that (3, 1), for example, represents the event that she will see three of the movies on the first day and one on the second day, draw a diagram similar to that of Figure 5.1 showing the ten points of the corresponding sample space.

(b) If T is the event that altogether she will see three of the movies, U is the event that she will see more of the movies on the second day than on the first, V is the event that she will see three of the movies on the first day, and W is the event that she will see equally many movies on both days, list the outcomes which comprise each of these four events.

5.9 With reference to the preceding exercise, which of the six pairs of events, T and U, T and V, T and W, U and V, U and W, and V and W, are mutually exclusive?

5.10 A small marina has three fishing boats which are sometimes in dry dock for repairs.

(a) Using two coordinates so that (2, 1), for example, represents the event that two of the fishing boats are in dry dock and one is rented out for the day, and (0, 2) represents the event that none of the boats is in dry dock and two are rented out for the day, draw a diagram similar to that of Figure 5.1 showing the ten points of the corresponding sample space.

(b) If K is the event that at least two of the boats are rented out for the day, L is the event that more boats are in dry dock than are rented out for the day, and M is the event that all the boats that are not in dry dock are rented out for the day, list the outcomes which comprise each of these three events.

(c) With reference to part (b), list the outcomes which comprise K' and $L \cap M$, and also express these events in words.

(d) With reference to part (b), which of the three pairs of events, K and L, K and M, and L and M, are mutually exclusive?

5.11 Which of the following pairs of events are mutually exclusive? Explain your answers.

(a) A driver getting a ticket for speeding and a ticket for going through a red light.

(b) Being foreign-born and being President of the United States.

(c) A baseball player getting a walk and hitting a home run in the same at bat.

(d) A baseball player getting a walk and hitting a home run in the same game.

(e) Having rain and sunshine on July 4, 1986.

5.12 In Figure 5.4, *D* is the event that a person vacationing in Southern California visits Disneyland and *U* is the event that he visits Universal Studios. Explain in words what events are represented by regions 1, 2, 3, and 4.

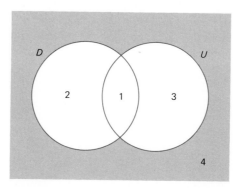

5.4

Venn diagram for Exercise 5.12.

5.13 With reference to the preceding exercise, what events are represented by
(a) regions 1 and 2 together;
(b) regions 2 and 3 together;
(c) regions 2 and 4 together?

5.14 In Figure 5.5, *C* is the event that a burglary suspect is caught and *G* is the event that she is found guilty. Explain in words what events are represented by regions 1, 2, 3, and 4.

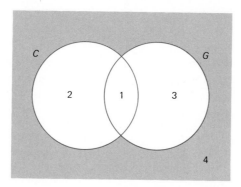

5.5

Venn diagram for Exercise 5.14.

5.15 With reference to the preceding exercise, what events are represented by
(a) regions 1 and 3 together;
(b) regions 1 and 4 together;
(c) regions 3 and 4 together?

5.16 With reference to the example on page 146 and Figure 5.3, what regions or combinations of regions represent the following events?

(a) There will be enough food.

(b) The weather will be good or they will have a good time, or both, but there will not be enough food.

(c) The weather will be good and they will have enough food.

(d) The weather will be good, but they will have neither enough food nor a good time.

5.17 In Figure 5.6, B is the event that someone will bring beer to a party, P is the event that someone will bring pretzels, and C is the event that someone will bring cheese. Explain in words what events are represented by

(a) region 1;

(b) region 3;

(c) region 6;

(d) region 8;

(e) regions 1 and 4 together;

(f) regions 3 and 5 together;

(g) regions 1, 3, 4, and 6 together;

(h) regions 2, 5, 7, and 8 together.

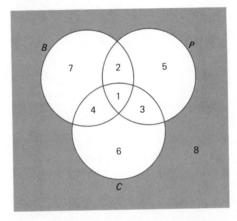

5.6

Venn diagram for Exercise 5.17.

5.18 With reference to the preceding exercise and the Venn diagram of Figure 5.6, list the regions or combinations of regions which represent the following events:

(a) Someone will bring pretzels, but no one will bring beer or cheese.

(b) Someone will bring beer, but no one will bring pretzels.

(c) Someone will bring beer, and someone will bring pretzels or someone will bring cheese.

5.3

Some Basic Rules of Probability

Probabilities always pertain to the occurrence or nonoccurrence of events (whether a coin will come up heads, whether a letter will arrive on time, whether it will not rain, whether a project will succeed, . . .), and now that we have learned how to deal with events, let us turn to the rules according to which probabilities must "behave," namely, the mathematical rules which they must obey. To express these rules symbolically, we shall denote events with capital letters as in Section 5.2, and write the probability of event A as $P(A)$, the probability of event B as $P(B)$, and so forth.

The following are some of the most basic rules of probability, and we shall justify them here with reference to the frequency interpretation; in Exercise 5.29, the reader will be asked to justify them with reference to the classical probability concept, and in Section 5.4 we shall see to what extent they are compatible with subjective probabilities.

Basic rules of probability

1. *Probabilities are real numbers on the interval from 0 to 1.*
2. *If an event is certain to occur, its probability is 1, and if an event is certain not to occur, its probability is 0.*
3. *If two events are mutually exclusive, the probability that one or the other will occur equals the sum of their probabilities.*
4. *The sum of the probabilities that an event will occur and that it will not occur is equal to 1.*

With reference to the frequency interpretation, the first of these rules simply expresses the fact that an event cannot occur less than 0% or more than 100% of the time, namely, that the proportion of successes, or the proportion of successes in the long run, cannot be negative or exceed 1. Symbolically, we write $0 \le P(A) \le 1$ for any event A. By the same token, the second rule expresses the fact that an event which is certain to occur will occur 100% of the time, and an event which is certain not to occur will occur 0% of the time. Symbolically, we write $P(S) = 1$ for any sample space S, which expresses the certainty that one of the outcomes in any sample space S must occur, and $P(\emptyset) = 0$, which expresses the fact that an event which cannot occur (which does not include any of the outcomes) has zero probability.

To show that the third rule is satisfied by the frequency interpretation, we have only to observe that if one event occurs, say, 15% of the time, another event occurs, say, 28% of the time, and the two events cannot both occur at the same time (that is, they are mutually exclusive), then one or the other occurs $15 + 28 = 43\%$ of the time. Of course, the same argument applies also to proportions. If the probabilities are, respectively, 0.45 and 0.17 that the weather will improve or that it will remain the same, then the probability is $0.45 + 0.17 = 0.62$ that it will improve or remain the same. Symbolically, we write $P(A \cup B) = P(A) + P(B)$ for any two mutually exclusive events A and B.

151

The fourth rule expresses the certainty that an event either will or will not occur. If someone is late for work 22% of the time, he is not late 78% of the time, the corresponding probabilities are 0.22 and 0.78, and their sum is $0.22 + 0.78 = 1$. Symbolically, we write $P(A) + P(A') = 1$ for any event A, or $P(A') = 1 - P(A)$, if we want to calculate the probability that an event will not occur in terms of the probability that it will occur.

The examples which follow illustrate how these rules are put to use in actual practice.

EXAMPLE   If A is the event that a student will stay home to study and B is the event that she will instead go to a movie, $P(A) = 0.64$, and $P(B) = 0.21$, find
(a) $P(A')$;   (b) $P(A \cup B)$;   (c) $P(A \cap B)$.

Solution   (a) Using the fourth rule, we find that the probability of A', the event that the student will not stay home to study is $1 - P(A) = 1 - 0.64 = 0.36$.
(b) Since A and B are mutually exclusive, we can use the third rule and write $P(A \cup B) = P(A) + P(B) = 0.64 + 0.21 = 0.85$ for the probability that the student will either stay home to study or go to a movie.
(c) Since A and B are mutually exclusive, they cannot possibly both occur and, hence, $P(A \cap B) = P(\varnothing) = 0$.

In problems like this, it often helps to draw a Venn diagram, fill in the probabilities associated with the various regions, and then read the answers directly off the diagram.

EXAMPLE   If C is the event that at 9:30 A.M. a certain doctor is in his office and D is the event that he is in the hospital, $P(C) = 0.48$, and $P(D) = 0.27$, find $P(C' \cap D')$, which is the probability that he is neither in his office nor in the hospital.

Solution   Drawing the Venn diagram as in Figure 5.7, we first put a 0 probability into region 1 because the events C and D are mutually exclusive. It follows that the 0.48 probability of event C must go into region 2, the 0.27 probability of event D must go into region 3, and since the probability for the entire sample space must total 1, we put $1 - (0.48 + 0.27) = 0.25$ into region 4. Since the event $C' \cap D'$ is represented by

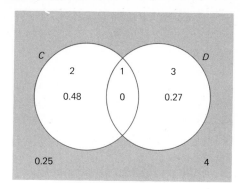

Venn diagram.

the region outside both circles, namely, region 4, we find that the answer
is $P(C' \cap D') = 0.25$.

5.4

Probabilities and Odds

If an event is twice as likely to occur than not to occur, we say that the **odds** are
2 to 1 that it will occur; if an event is three times as likely to occur than not to
occur, we say that the odds are 3 to 1; if an event is ten times as likely to occur
than not to occur, we say that the odds are 10 to 1; and so forth. In general,

> **The odds that an event will occur are given by the ratio of the
> probability that it will occur to the probability that it will not
> occur.**

Symbolically, if the probability of an event is p, the odds for its occurrence are
a to b, where a and b are positive values such that

$$\frac{a}{b} = \frac{p}{1 - p}$$

It is customary to express odds in terms of positive integers having no common
factors.

EXAMPLE   What are the odds for the occurrence of an event if its probability is
(a) $\frac{5}{9}$;
(b) 0.85?

Solution   (a) By definition, the odds are $\frac{5}{9}$ to $1 - \frac{5}{9} = \frac{4}{9}$, or 5 to 4.

(b) By definition, the odds are 0.85 to $1 - 0.85 = 0.15$, 85 to 15, or better, 17 to 3.

If an event is more likely not to occur than to occur, it is customary to quote the odds that it will not occur rather than the odds that it will occur.

EXAMPLE   What are the odds, if the probability of an event is 0.20?

Solution   The odds for the occurrence of the event are 0.20 to $1 - 0.20 = 0.80$, or 1 to 4, but it is customary to say instead that the odds against the occurrence of the event are 4 to 1.

In betting, the word "odds" is also used to denote the ratio of the wager of one party to that of another. For instance, if a gambler says that he will give 3 to 1 odds on the occurrence of an event, he means that he is willing to bet $3 against $1 (or perhaps $30 against $10 or $1,500 against $500) that the event will occur. If such **betting odds** actually equal the odds that the event will occur, we say that the betting odds are **fair.**

EXAMPLE   Records show that $\frac{1}{12}$ of the trucks weighed at a certain check point in Nevada carry too heavy a load. If someone offers to bet $40 against $4 that the next truck weighed at this check point will not carry too heavy a load, are these betting odds fair?

Solution   Since the probability is $1 - \frac{1}{12} = \frac{11}{12}$ that the truck will not carry too heavy a load, the odds are 11 to 1, and the bet would be fair if the person offered to bet $44 against $4 that the next truck weighed at the check point will not carry too heavy a load. Thus, the $40 against $4 bet is not fair; it favors the person offering the bet.

This discussion of odds and betting odds provides the groundwork for a way of measuring subjective probabilities. If a businessman "feels" that the odds for the success of a new clothing store are 3 to 2, this means that he is willing to bet (or considers it fair to bet) $300 against $200, or perhaps $3,000 against $2,000, that the new store will be a success. In this way he is expressing his belief regarding the uncertainties connected with the success of the store, and to convert it into a probability we take the equation $\frac{a}{b} = \frac{p}{1 - p}$ and solve it for *p.* Leaving the details to the reader in Exercise 5.37, let us merely state the result that

---

Formula relating probabilities to odds

If the odds are a to b that an event will occur, the probability of its occurrence is

$$p = \frac{a}{a+b}$$

EXAMPLE   Convert the businessman's 3 to 2 odds for the success of the new clothing store into a probability.

Solution   Substituting $a = 3$ and $b = 2$ into the formula for *p*, we get

$$p = \frac{3}{3+2} = \frac{3}{5}$$

EXAMPLE   If an applicant for a managerial position feels that the odds are 7 to 4 that she will get the job, what probability is she thus assigning to her getting the job?

Solution   Substituting $a = 7$ and $b = 4$ into the formula for *p*, we get

$$p = \frac{7}{7+4} = \frac{7}{11}$$

or approximately 0.64.

Let us now see whether subjective probabilities, determined in this way, "behave" in accordance with the four rules on page 151. So far as the first and fourth rules are concerned, this is easy to see. Since *a* and *b* are positive quantities, $\frac{a}{a+b}$ is a fraction between 0 and 1; since the probabilities that an event will occur and that it will not occur are, respectively, $\frac{a}{a+b}$ and $\frac{b}{b+a}$, their sum is $\frac{a}{a+b} + \frac{b}{b+a} = \frac{a+b}{a+b} = 1$. So far as the second rule is concerned, observe that the surer we are that an event will occur, the "better" odds we should be willing to give—say, 100 to 1, 1,000 to 1, or perhaps even 1,000,000 to 1. The corresponding probabilities are $\frac{100}{100 + 1} = 0.99$, $\frac{1,000}{1,000 + 1} = 0.999$, and $\frac{1,000,000}{1,000,000 + 1} = 0.999999$, and it can be seen that the surer we are that an event will occur, the closer its probability will be to 1. By the same token, it can be shown that the surer we are that an event will not occur, the closer its probability will be to 0.

This leaves only the third rule—$P(A \cup B) = P(A) + P(B)$ for any two mutually exclusive events A and B—and this rule is not necessarily satisfied when it comes to subjective probabilities. Indeed, proponents of the subjectivist point of view impose it as a **consistency criterion,** and this provides a means of "policing" a person's subjective probabilities.

EXAMPLE An economist feels that the odds are 2 to 1 that the price of beef will go up during the next month, 1 to 5 that it will remain unchanged, and 8 to 3 that it will go up or remain unchanged. Are the corresponding probabilities consistent?

Solution The corresponding probabilities that the price of beef will go up during the next month, that it will remain unchanged, and that it will go up or remain unchanged are, respectively, $\frac{2}{2 + 1} = \frac{2}{3}$, $\frac{1}{1 + 5} = \frac{1}{6}$, and $\frac{8}{8 + 3} = \frac{8}{11}$, and since $\frac{2}{3} + \frac{1}{6} = \frac{5}{6}$ and not $\frac{8}{11}$, the probabilities are not consistent. Hence, the economist's judgment must be questioned.

EXERCISES (Exercises 5.19, 5.22, 5.30, and 5.31 are practice exercises; their complete solutions are given on page 183.)

5.19 In a study of the future needs of a community, D stands for the event that there will be enough doctors and H stands for the event that there will be enough hospital beds. State in words what probabilities are expressed by
(a) $P(D')$;
(b) $P(H')$;
(c) $P(D \cup H)$;
(d) $P(D \cap H)$;
(e) $P(D' \cap H')$;
(f) $P(D \cap H')$.

5.20 In a study of the adequacy of fuel supplies, C stands for the event that a power plant will use coal and E is the event that it will be able to provide enough electricity. State in words what probabilities are expressed by
(a) $P(C')$;
(d) $P(C \cap E)$;

(b) $P(E')$;

(e) $P(C' \cup E)$;

(c) $P(C \cup E)$;

(f) $P(C' \cap E')$.

5.21 If J is the event that Harry will find a job and M is the event that he will manage financially, express symbolically the probabilities that he will
  (a) not find a job;
  (b) find a job and manage financially;
  (c) not find a job and not manage financially;
  (d) find a job, manage financially, or both.

5.22 Explain why there must be a mistake in each of the following statements:
  (a) Since there is not a cloud in the sky, the probability that it will rain later on in the day is -0.99.
  (b) The probability that a mineral specimen will contain copper is 0.38 and the probability that it will not contain copper is 0.52.
  (c) The probability that Mary will get an A in a certain test is 0.22, the probability that she will get a B is 0.37, and the probability that she will get an A or a B is 0.61.

5.23 Explain why there must be a mistake in each of the following statements:
  (a) Since Nancy studied very hard for the test, the probability that she will pass is at least 2.
  (b) The probability that an experiment will succeed is 0.73 and the probability that it will fail is 0.42.
  (c) In a primary election, the probability that a person will vote for candidate X is 0.43, the probability that he will vote for candidate Y is 0.29, and the probability that he will vote for either of these candidates is 0.74.
  (d) The probability that Jim will be hired for a job is 0.47, and the probability that Joan will be hired for the same job is 0.60.

5.24 Draw Venn diagrams to determine why there must be a mistake in each of the following statements:
  (a) The probability that a student will pass an English test is 0.63 and the probability that she will pass the English test as well as a French test is 0.82.
  (b) The manager of a restaurant figures that the probability is 0.33 that a person will have chocolate cake for dessert, and 0.27 that a person will have chocolate cake or apple pie for dessert.

5.25 Given the mutually exclusive events U and V for which $P(U) = 0.41$ and $P(V) = 0.36$, find
  (a) $P(U')$;

  (c) $P(U \cup V)$;

  (b) $P(V')$;

  (d) $P(U \cap V)$.

5.26 With reference to the preceding exercise, draw a Venn diagram, fill in the probabilities associated with the different regions, and thus determine $P(U' \cap V')$.

5.27 The probabilities that critics' reactions to a new movie will be favorable or indifferent are, respectively, 0.18 and 0.55. Find the probabilities that their reactions

    (a) will not be favorable;

    (b) will be favorable or indifferent;

    (c) will be neither favorable nor indifferent.

5.28 The probabilities that a student will make less than five mistakes in a mathematics assignment or anywhere from five to ten mistakes are, respectively, 0.66 and 0.21. Find the probabilities that the student will make

    (a) five or more mistakes;

    (b) at most ten mistakes;

    (c) more than ten mistakes.

5.29 Show that the four basic rules on page 151 are satisfied if we interpret probabilities in accordance with the classical probability concept.

5.30 Convert each of the following probabilities to odds, or odds to probabilities:

    (a) The probability of getting three heads and three tails in six flips of a balanced coin is $\frac{5}{16}$.

    (b) The odds against rolling "7 or 11" with a pair of balanced dice are 7 to 2.

    (c) If a pollster randomly selects five of 24 households to be included in a survey, the probability is $\frac{5}{24}$ that any particular household will be included.

    (d) If three eggs are randomly chosen from a carton of 12 eggs of which three are cracked, the odds are 34 to 21 that at least one of them will be cracked.

5.31 A government statistician claims that the odds are 2 to 1 that unemployment will go up and 5 to 1 that it will go down. Can these odds be right? Explain.

5.32 Convert each of the following probabilities to odds, or odds to probabilities:

    (a) The probability of drawing a black Jack from an ordinary deck of 52 playing cards is $\frac{1}{26}$.

    (b) The odds are 27 to 5 against getting four heads and one tail in five flips of a balanced coin.

    (c) If a secretary randomly puts special delivery stamps on three of six letters, of which three are supposed to go by special delivery, the probability is $\frac{1}{20}$ that she will put all these stamps on the right letters.

    (d) If we arbitrarily arrange the letters in the word "nest," the odds are 5 to 1 that we will not get a meaningful word in the English language.

5.33 A football coach claims that the odds are 2 to 1 that his team will win an upcoming game, and that the odds against his team's losing or tieing are, respectively, 4 to 1 and 9 to 1. Can these odds be right? Explain.

5.34 If a student is anxious to bet \$25 against \$5 that she will pass a certain course, what does this tell us about the probability she assigns to her passing the course? (*Hint:* The answer should read "greater than . . . .")

5.35 Asked about his political future, a party official replies that the odds are 2 to 1 that he will not run for the House of Representatives and 5 to 1 that he will not run for the Senate. Furthermore, he feels that it is an even-money bet that he will run for one or the other. Are the corresponding probabilities consistent?

5.36 A real estate salesperson feels that the odds are 7 to 1 against her selling a certain house and 3 to 1 against her being able to rent it. To be consistent, what should she consider fair odds that she will either sell the house or rent it?

5.37 Verify algebraically that the formula $\dfrac{a}{b} = \dfrac{p}{1-p}$, solved for p, yields $p = \dfrac{a}{a+b}$.

5.5
Addition Rules

The third rule on page 151 is sometimes referred to as the **special addition rule**—it is special in that it applies only to mutually exclusive events, and it is an addition rule because we add the probabilities of the two events. As we shall see in this section, it can be generalized so that it will apply to more than two mutually exclusive events, and also to events which need not be mutually exclusive. Repeatedly applying the third rule on page 151, we can show that

Special addition rule for two or more events

> *If k events are mutually exclusive, the probability that one of them will occur equals the sum of their respective probabilities; symbolically*
>
> $$P(A_1 \cup A_2 \cup \cdots \cup A_k) = P(A_1) + P(A_2) + \cdots + P(A_k)$$
>
> *for any mutually exclusive events $A_1, A_2, \ldots,$ and A_k.*

where, again, $\cup$ is usually read "or."

EXAMPLE   The probabilities that a woman will buy a new dress for a party at Bullock's, the Broadway Southwest department store, or the May Co., are 0.22, 0.18, and 0.35. What is the probability that she will buy the new dress at one of these stores?

Solution   Since the three possibilities are mutually exclusive, the answer is

$$0.22 + 0.18 + 0.35 = 0.75$$

EXAMPLE   The probabilities that a consumer testing service will rate a new anti-pollution device for cars very poor, poor, fair, good, very good, or excellent are 0.07, 0.12, 0.17, 0.32, 0.21, and 0.11. What is the probability that it will rate the device poor, fair, good, or very good?

Solution   Since the four possibilities are mutually exclusive, we get

$$0.12 + 0.17 + 0.32 + 0.21 = 0.82$$

The job of assigning probabilities to all possible events connected with a given situation can be very tedious, to say the least. If a sample space has ten elements (outcomes) we can form more than a thousand different events, and if a sample space has twenty elements (outcomes) we can form more than a million different events.[†] Fortunately, it is seldom, if ever, necessary to find the probabilities of all possible events relating to a given situation; the following rule, which is a direct application of the special addition rule for two or more events, makes it easy to determine the probability of any event on the basis of the probabilities assigned to the individual elements (outcomes) which form the sample space.

Rule for calculating probability of an event

> *The probability of any event A is given by the sum of the probabilities of the individual outcomes comprising A.*

In the special case where the outcomes are all equiprobable, this rule leads to the formula $P(A) = \dfrac{s}{n}$, which we studied earlier in connection with the classical probability concept. Here, n is the total number of outcomes and s is the number of "successes," namely, the number of outcomes in event A.

EXAMPLE   Referring again to the example on page 141, which dealt with the two salespersons trying to sell two 1984 Chevrolet Camaros, suppose that the six points of the sample space have the probabilities shown in Figure 5.8. Find the probabilities that
(a) the first salesperson will not sell either of the two cars;
(b) both cars will be sold;
(c) the second salesperson will sell at least one of the two cars.

[†] In general, if a sample space has n elements, we can form 2^n different events. Each element is either included or excluded for a given event, so by the multiplication of choices there are $2 \cdot 2 \cdot 2 \cdot \ldots \cdot 2 = 2^n$ possibilities.

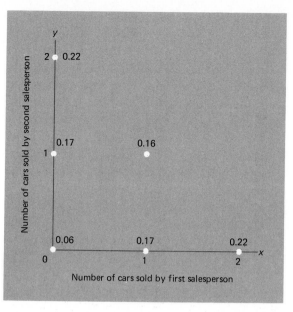

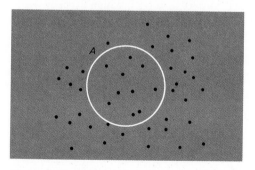

5.8

Sample space for two-sales-persons example with proba-bilities assigned to the outcomes.

Solution (a) Adding the probabilities associated with the points (0, 0), (0, 1), and (0, 2) we get $0.06 + 0.17 + 0.22 = 0.45$.

(b) Adding the probabilities associated with the points (2, 0), (1, 1), and (0, 2), we get $0.22 + 0.16 + 0.22 = 0.60$.

(c) Adding the probabilities associated with the points (0, 1), (1, 1), and (0, 2), we get $0.17 + 0.16 + 0.22 = 0.55$.

EXAMPLE Assuming that the 44 points (outcomes) of the sample space of Figure 5.9 are all equiprobable, find $P(A)$.

5.9

Sample space with 44 outcomes.

Solution Since there are $s = 10$ outcomes in A and the $n = 44$ outcomes are all equiprobable, it follows that $P(A) = \frac{10}{44} = \frac{5}{22}$.

Since the addition rules we have studied so far apply only to mutually exclusive events, they cannot be used, for example, to find the probability that at least one of two friends will pass a final examination; the probability that in an automobile accident the driver will break an arm, a rib, or a leg; or the probability that a customer will buy a shirt or a sweater at a given department store. Both friends can pass the final examination; the driver of the car can break an arm and a rib, and also a leg; and the customer of the department store can buy a sweater as well as a shirt.

To find a formula for $P(A \cup B)$ which holds regardless of whether the events A and B are mutually exclusive, let us consider the Venn diagram of Figure 5.10, which concerns the election of a mayor. The letter G stands for the election of a person who is a college graduate, and the letter M stands for the election of a member of a certain minority group. It follows from the figures in the Venn diagram that

$$P(G) = 0.44 + 0.15 = 0.59$$

$$P(M) = 0.15 + 0.07 = 0.22$$

and

$$P(G \cup M) = 0.44 + 0.15 + 0.07 = 0.66$$

where we were able to add the respective probabilities because they pertain to mutually exclusive events (namely, to regions of the Venn diagram which do not overlap).

Had we erroneously used the special addition rule, the third of the four basic rules on page 151, to calculate $P(G \cup M)$, we would have obtained $P(G) + P(M) = 0.59 + 0.22 = 0.81$, which exceeds the correct value by 0.15. This error results from including $P(G \cap M) = 0.15$ twice, once in $P(G) = 0.59$ and once in

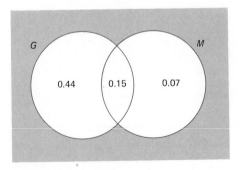

5.10
Venn diagram.

CHAP. 5: Some Rules of Probability

$P(M) = 0.22$, and we could correct for it by subtracting 0.15 from 0.81. Symbolically, we could thus write

$$P(G \cup M) = P(G) + P(M) - P(G \cap M)$$
$$= 0.59 + 0.22 - 0.15$$
$$= 0.66$$

and this agrees, as it should, with the result obtained before.

Since the argument which we used in this example holds for any two events A and B, we can now state the following **general addition rule,** which applies regardless of whether A and B are mutually exclusive events:

General addition rule

$$P(A \cup B) = P(A) + P(B) - P(A \cap B)$$

Note that when A and B are mutually exclusive, $P(A \cap B) = 0$ and the general addition rule reduces to the special addition rule, the third of the basic rules which we gave on page 151.

EXAMPLE   If one card is drawn from an ordinary deck of 52 playing cards, what is the probability that it will be either a club or a face card (king, queen, or jack)?

Solution   If C denotes drawing a club and F denotes drawing a face card, then $P(C) = \frac{13}{52}$, $P(F) = \frac{12}{52}$, and $P(C \cap F) = \frac{3}{52}$, so that

$$P(C \cup F) = \frac{13}{52} + \frac{12}{52} - \frac{3}{52} = \frac{22}{52}$$

EXAMPLE   If the probabilities are, respectively, 0.92, 0.33, and 0.29 that a person vacationing in Washington, D.C., will visit the Capitol building, the Smithsonian Institution, or both, what is the probability that a person vacationing there will visit at least one of these buildings?

Solution   Substituting into the formula, we get

$$0.92 + 0.33 - 0.29 = 0.96$$

Note that if we had incorrectly used the special addition rule (for mutually exclusive events), we would have obtained the impossible answer $0.92 + 0.33 = 1.25$.

The general addition rule can be generalized further so that it will apply to more than two events, but we shall not go into that in this book.

EXERCISES

(Exercises 5.38, 5.43, 5.47, and 5.50 are practice exercises; their complete solutions are given on page 184.)

5.38 A police department needs new tires for its patrol cars and the probabilities are, respectively, 0.22, 0.17, 0.21, and 0.09 that it will buy Uniroyal, Goodyear, Goodrich, or Michelin tires. What is the probability that it will buy one of these four kinds of tires?

5.39 The probabilities that a review board will rate a given movie X, R, or PG are, respectively, 0.41, 0.30, and 0.12. What is the probability that the movie will get one of these three ratings?

5.40 The probabilities that a student will get an A, a B, or a C in a course are, respectively, 0.05, 0.14, and 0.47. What is the probability that the student will get a grade lower than a C?

5.41 The probabilities that a TV station will receive 0, 1, 2, 3, ..., 7, or at least 8 complaints after showing a controversial program are, respectively, 0.02, 0.04, 0.07, 0.12, 0.15, 0.19, 0.18, 0.14, and 0.09. What are the probabilities that after showing such a program the station will receive
  (a) at least 5 complaints;
  (b) at most 3 complaints;
  (c) anywhere from 2 to 4 complaints?

5.42 The probabilities that the serviceability of a new typewriter will be rated very difficult, difficult, average, easy, or very easy are, respectively, 0.11, 0.16, 0.35, 0.28, and 0.10. Find the probabilities that the serviceability of the new typewriter will be rated
  (a) difficult or very difficult;
  (b) difficult, average, or easy;
  (c) average or better.

5.43 Figure 5.11 pertains to the number of persons who are invited to a conference and the number of persons who attend. If each of the 35 points of the sample space has the probability $\frac{1}{35}$, what are the probabilities that
  (a) all the persons who are invited will attend;
  (b) at most two persons will attend;
  (c) at least seven persons will be invited;
  (d) one or two of the persons who are invited will not attend?

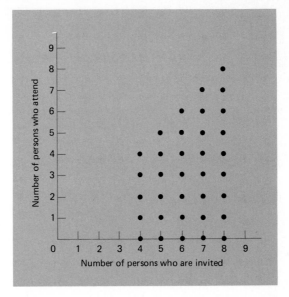

5.11

*Sample space for
Exercise 5.43.*

5.44 If each point of the sample space of Figure 5.12 represents an outcome having the probability $\frac{1}{32}$, find

(a) $P(A)$;

(b) $P(B)$;

(c) $P(A \cap B)$;

(d) $P(A \cup B)$;

(e) $P(A' \cap B)$;

(f) $P(A' \cap B')$.

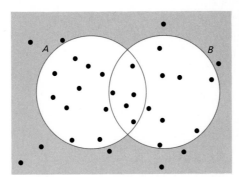

5.12

Sample space for Exercise 5.44.

5.45 With reference to the example on page 141 and Figure 5.8, find the probabilities that

(a) only one of the two cars will be sold;

(b) either salesperson will sell both cars;

(c) the two salespersons will sell equally many cars.

5.46 If H stands for heads and T for tails, the 16 possible outcomes for four flips of a coin are HHHH, HHHT, HHTH, HTHH, THHH, HHTT, HTHT, HTTH,

THHT, THTH, TTHH, HTTT, THTT, TTHT, TTTH, and TTTT. Assuming that these 16 possibilities are all equally likely, what are the probabilities of getting 0, 1, 2, 3, or 4 heads in four flips of a balanced coin?

5.47 An artist, who has entered a large oil painting and a small oil painting in a show, feels that the probabilities are, respectively, 0.15, 0.18, and 0.11 that she will sell the large oil painting, the small one, or both. What is the probability that she will sell either of the two paintings?

5.48 The probabilities that a reckless driver will be fined, get his license revoked, or both are, respectively, 0.88, 0.60, and 0.55. What is the probability that he will be fined or get his license revoked?

5.49 The probabilities that a dentist's receptionist, his assistant, or both will be sick on a given day are, respectively, 0.04, 0.07, and 0.02. What is the probability that at least one of the two will be sick on that day?

5.50 Given $P(K) = 0.45$, $P(L) = 0.27$, and $P(K \cap L) = 0.13$, draw a Venn diagram, fill in the probabilities associated with the various regions, and thus determine
    (a) $P(K \cap L')$;   (d) $P(K' \cup L)$;
    (b) $P(K' \cap L)$;   (e) $P(K' \cap L')$;
    (c) $P(K \cup L)$;   (f) $P(K' \cup L')$.

5.51 Among the 64 doctors on the staff of a hospital, 58 carry malpractice insurance, 33 are surgeons, and 31 of the surgeons carry malpractice insurance. If one of these doctors is chosen by lot to represent the hospital staff at an AMA convention (that is, each of the doctors has a probability of $\frac{1}{64}$ of being selected), what is the probability that the one chosen is not a surgeon and does not carry malpractice insurance?

5.52 Given $P(M) = 0.72$, $P(N) = 0.45$, and $P(M \cap N) = 0.36$, draw a Venn diagram, fill in the probabilities associated with the various regions, and thus determine
    (a) $P(M \cap N')$;   (c) $P(M \cup N)$;
    (b) $P(M' \cap N)$;   (d) $P(M' \cap N')$.

5.6
Conditional Probability

Difficulties can easily arise when we ask for the probability of an event without specifying the sample space. For instance, the probability that a person will find a bargain depends on where she shops, the chances that a person will get a traffic ticket depends on how well and how carefully he drives, and the odds that a person will get promoted depends on how well she is trained, or how hard she is willing to work.

Since the choice of the sample space (namely, the set of all possibilities under consideration) is rarely self-evident, it helps to use the symbol $P(A|S)$ to denote the **conditional probability** of event A relative to the sample space S, or as we often say "the probability of A given S." The symbol $P(A|S)$ makes it explicit that we are referring to a particular sample space S, and it is preferable to the abbreviated notation $P(A)$ unless the tacit choice of S is clearly understood.

To elaborate on the idea of a conditional probability, suppose that a consumer research organization has studied the service provided by the 150 appliance repairpersons in a certain city, and that their findings are summarized in the following table:

	Good service	Poor service	Total
Factory-trained	48	16	64
Not factory-trained	24	62	86
Total	72	78	150

If one of these repairpersons is randomly selected (that is, each has the probability $\frac{1}{150}$ of being selected), we find that the probabilities of choosing one who provides good service, one who is factory-trained, or one who provides good service and is factory-trained are, respectively,

$$P(G) = \frac{72}{150} = 0.48$$

$$P(F) = \frac{64}{150} = 0.43$$

rounded to two decimals, and

$$P(G \cap F) = \frac{48}{150} = 0.32$$

All these probabilities were calculated by means of the formula $\frac{s}{n}$ for equally likely possibilities.

Since the first of these probabilities is particularly disconcerting—there is less than a fifty-fifty chance of choosing a repairperson who provides good service—let us see what will happen if we limit the choice to those who are factory-trained. Looking at the reduced sample space represented by the first row of the table, we get

$$P(G|F) = \frac{48}{64} = 0.75$$

and this is quite an improvement over $P(G) = 0.48$, as might have been expected. Note that this conditional probability, 0.75, can also be written as

$$P(G|F) = \frac{\dfrac{48}{150}}{\dfrac{64}{150}} = \frac{P(G \cap F)}{P(F)}$$

which is the ratio of the probability of choosing a factory-trained repairperson who provides good service to the probability of choosing a repairperson who is factory-trained.

Generalizing from this example, let us now make the following definition of conditional probability, which applies to any two events A and B belonging to a given sample space S:

*Definition
of conditional
probability*

> *If $P(B)$ is not equal to zero, then the conditional probability of A relative to B, namely, the probability of A given B, is*
>
> $$P(A|B) = \frac{P(A \cap B)}{P(B)}$$

EXAMPLE   With reference to the appliance repairpersons of the example above, what is the probability that one of them who is not factory-trained will provide good service?

Solution   As can be seen from the table, $P(G \cap F') = \frac{24}{150}$ and $P(F') = \frac{86}{150}$, so that substitution into the formula yields

$$P(G|F') = \frac{P(G \cap F')}{P(F')} = \frac{\dfrac{24}{150}}{\dfrac{86}{150}} = \frac{24}{86} = 0.28$$

rounded to two decimals. Of course, the fraction $\frac{24}{86}$ could have been obtained directly by considering only the second row of the table.

Although we introduced the formula for $P(A|B)$ by means of an example in which the possibilities were all equally likely, this is not a requirement for its use. The only restriction is that $P(B)$ must not equal zero.

EXAMPLE Police records show that in a certain city the probability is 0.35 that a burglar will be caught, and 0.14 that a burglar will be caught and convicted. What is the probability that a burglar, if caught, will be convicted?

Solution If A and B are, respectively, the events that a burglar will be convicted and that he will be caught, then $P(B) = 0.35$, $P(A \cap B) = 0.14$, and substitution into the formula for a conditional probability yields

$$P(A|B) = \frac{P(A \cap B)}{P(B)} = \frac{0.14}{0.35} = 0.40$$

EXAMPLE The probability that there will be a shortage of cement is 0.28, and the probability that there will not be a shortage of cement and a construction job will be finished on time is 0.64. What is the probability that the construction job will be finished on time given that there will not be a shortage of cement?

Solution If we let N denote the event that there will not be a shortage of cement and F the event that the construction job will be finished on time, then $P(N) = 1 - 0.28 = 0.72$, $P(F \cap N) = 0.64$, and it follows that

$$P(F|N) = \frac{P(F \cap N)}{P(N)} = \frac{0.64}{0.72} = \frac{8}{9}$$

or approximately 0.89.

5.7
Independent Events

To introduce another concept which is important in the study of probability, let us consider the following problem:

EXAMPLE The probabilities that a student will get passing grades in Algebra, in Literature, or in both are, respectively, $P(A) = 0.75$, $P(L) = 0.84$, and $P(A \cap L) = 0.63$. What is the probability that the student will get a passing grade in Algebra given that he or she gets a passing grade in Literature?

169

Solution　Substituting into the formula which defines conditional probability, we get

$$P(A|L) = \frac{P(A \cap L)}{P(L)} = \frac{0.63}{0.84} = 0.75$$

What is special, and interesting, about this result is that $P(A|L) = P(A) = 0.75$, namely, that the probability of event A is the same regardless of whether event L has occurred (occurs, or will occur).

In general, if $P(A|B) = P(A)$, we say that event A is **independent** of event B; that is,

> Event A is independent of event B if the probability of A is not affected by the occurrence or nonoccurrence of B.

Since it can be shown that event B is independent of event A whenever event A is independent of event B, it is customary to say simply that A **and** B **are independent** whenever one is independent of the other (see Exercise 5.69 on page 176). If two events A and B are not independent, we say that they are **dependent.**

5.8
Multiplication Rules

So far we have used the formula $P(A|B) = \dfrac{P(A \cap B)}{P(B)}$ only to calculate conditional probabilities, but if we multiply on both sides of the equation by $P(B)$, we get the following formula, called the **general multiplication rule,** which enables us to calculate the probability that two events will both occur.

General multiplication rule

$$P(A \cap B) = P(B) \cdot P(A|B)$$

In words, the probability that two events will both occur is the product of the probability that one of the events will occur and the conditional probability that the other event will occur given that the first event has occurred (occurs, or will occur). As it does not matter which event is denoted by A and which is denoted by B, the formula above can also be written as

General multiplication rule (alternative form)

$$P(A \cap B) = P(A) \cdot P(B|A)$$

EXAMPLE A jury consists of nine persons who are native-born and three persons who are foreign-born. If two of the jurors are randomly picked for an interview, what is the probability that they will both be foreign-born?

Solution Let A be the event that the first juror is foreign-born and B the event that the second juror is foreign-born. If we assume equal probabilities for each choice (which is, in fact, what we mean by the selection being random), the probability that the first juror picked will be foreign-born is $P(A) = \frac{3}{12}$. Then, if the first juror picked is foreign-born, the probability that the second juror will also be foreign-born is $P(B|A) = \frac{2}{11}$. Hence, the probability of getting two jurors who are both foreign-born is

$$P(A \cap B) = P(A) \cdot P(B|A) = \frac{3}{12} \cdot \frac{2}{11} = \frac{1}{22}$$

When A and B are independent, we can substitute $P(A)$ for $P(A|B)$ in the first of the two formulas for $P(A \cap B)$, or $P(B)$ for $P(B|A)$ in the second, and we obtain

Special multiplication rule (independent events)

$$P(A \cap B) = P(A) \cdot P(B)$$

In words, the probability that two independent events will both occur is simply the product of their respective probabilities. This rule is sometimes used as the definition of independence; in any case, it may be used to check whether two given events are independent.

EXAMPLE What is the probability of getting two heads in two flips of a balanced coin?

Solution Since the probability of heads is $\frac{1}{2}$ for each flip of the coin, the answer is $\frac{1}{2} \cdot \frac{1}{2} = \frac{1}{4}$.

EXAMPLE If $P(C) = 0.65$, $P(D) = 0.40$, and $P(C \cap D) = 0.24$, are the events C and D independent?

Solution Since $P(C) \cdot P(D) = (0.65)(0.40) = 0.26$ and not 0.24, the two events are not independent.

EXAMPLE   What is the probability of getting two aces in a row when two cards are drawn from an ordinary deck of 52 playing cards, if
(a) the first card is replaced before the second card is drawn;
(b) the first card is not replaced before the second card is drawn?

Solution   (a) Since there are four aces among the 52 cards, we get $\frac{4}{52} \cdot \frac{4}{52} = \frac{1}{169}$.
(b) Since there are only three aces among the 51 cards which remain after one ace has been removed from the deck, the answer is $\frac{4}{52} \cdot \frac{3}{51} = \frac{1}{221}$.

Observe the difference in the calculations and the results of the two parts of the preceding exercise. They serve to illustrate the distinction between **sampling with replacement** and **sampling without replacement.**

The special multiplication rule can easily be generalized so that it applies to three or more independent events—again, we simply multiply all their probabilities.

EXAMPLE   If the probability is 0.25 that a person will name red as his or her favorite color, what is the probability that three totally unrelated persons will all name red as their favorite color?

Solution   Assuming independence, we get $(0.25)(0.25)(0.25) = 0.0156$ rounded to four decimals.

When three or more events are not independent, the multiplication rule becomes more complicated—we form the product of the probability that one of the events will occur, the conditional probability that a second event will occur given that the first event has occurred, the conditional probability that a third event will occur given that the first two events have occurred, and so on.

EXAMPLE   A department store which bills its charge-account customers once a month has found that if a customer pays promptly one month, the probability is 0.90 that he will also pay promptly the next month; however, if a customer does not pay promptly one month, the probability that he will pay promptly the next month is only 0.40. What is the probability that a customer who pays promptly one month will not pay promptly the next three months?

Solution For the three months, the probabilities that the customer will not pay promptly are $1 - 0.90 = 0.10$, $1 - 0.40 = 0.60$, and again $1 - 0.40 = 0.60$. Thus, the answer is

$$(0.10)(0.60)(0.60) = 0.036$$

EXERCISES

(Exercises 5.53, 5.54, 5.59, 5.60, and 5.76 are practice exercises; their complete solutions are given on pages 184 and 185.)

5.53 If F is the event that a student will get financial aid, J is the event that he will find a part-time job, and G is the event that he will graduate, express symbolically the probabilities that
   (a) a student who gets financial aid will graduate;
   (b) a student who gets no financial aid will find a part-time job;
   (c) a student who gets no financial aid will neither find a part-time job nor graduate;
   (d) a student who gets financial aid and finds a part-time job will not graduate.

5.54 If E is the event that a female applicant for a sales position has had prior experience, C is the event that she owns a car, and G is the event that she is a college graduate, state in words what probabilities are expressed by
   (a) $P(C|G)$; (d) $P(G'|C')$;
   (b) $P(E|C')$; (e) $P(C|E \cup G)$;
   (c) $P(C'|E)$; (f) $P(E \cap C'|G)$.

5.55 If W is the event that a worker is well-trained and Q is the event that he or she meets the production quota, express symbolically the probabilities that
   (a) a worker who is well-trained will meet the production quota;
   (b) a worker who meets the production quota is not well-trained;
   (c) a worker who is not well-trained will not meet the production quota.

5.56 With reference to the preceding exercise, state in words what probabilities are expressed by
   (a) $P(W|Q)$;
   (b) $P(Q'|W)$;
   (c) $P(W'|Q')$.

5.57 If H is the event that a probation officer is honest, E is the event that he or she is easy-going, and W is the event that he or she is well-liked, express symbolically the probabilities that
   (a) a probation officer who is easy-going will be well-liked;
   (b) a probation officer who is dishonest will be easy-going;
   (c) a probation officer who is honest and easy-going will be well-liked.

173

5.58 With reference to the preceding exercise, state in words what probabilities are expressed by
- (a) $P(H|W')$;
- (b) $P(W'|E')$;
- (c) $P(W \cap E|H)$.

5.59 Among the 400 inmates of a prison, some are first offenders, some are hardened criminals, some serve terms of less than five years, and some serve longer terms, with the exact breakdown being

	Terms of less than five years	Longer terms
First offenders	120	40
Hardened criminals	80	160

If one of the inmates is to be selected at random to be interviewed about prison conditions, H is the event that he is a hardened criminal, and L is the event that he is serving a longer term, determine each of the following probabilities directly from the entries and the row and column totals of the table:
- (a) $P(H)$;
- (b) $P(L)$;
- (c) $P(L \cap H)$;
- (d) $P(H' \cap L)$;
- (e) $P(L|H)$;
- (f) $P(H'|L)$.

5.60 Use the results of the preceding exercise to verify that

(a) $P(L|H) = \dfrac{P(L \cap H)}{P(H)}$;

(b) $P(H'|L) = \dfrac{P(H' \cap L)}{P(L)}$.

5.61 In some of the delinquent charge accounts of a store the amount owed is less than $100, in some it is $100 or more, some have been delinquent for less than a month, and some have been delinquent for a month or more, with the exact breakdown being

	Less than a month	A month or more
Less than $100	132	48
$100 or more	33	27

If one of these delinquent accounts is to be selected at random for a new credit check, L is the event that the amount owed is less than $100, and M is the event that the account has been delinquent for a month or more, determine each of the following probabilities directly from the entries and the row and column totals of the table:

(a) $P(L)$;

(b) $P(M)$;

(c) $P(L \cap M')$;

(d) $P(M' \cap L')$;

(e) $P(L|M')$;

(f) $P(M'|L')$.

5.62 Use the results of the preceding exercise to verify that

(a) $P(L|M') = \dfrac{P(L \cap M')}{P(M')}$;

(b) $P(M'|L') = \dfrac{P(M' \cap L')}{P(L')}$.

5.63 In the table shown below, 60 college students are classified according to their class standing and also according to their favorite pizza topping:

	Anchovies	Onions	Mushrooms	Hamburger
Freshman	7	6	7	3
Sophomore	1	9	0	9
Junior	3	2	5	8

If one of these students is selected at random, F, S, and J denote the three classes, and A, O, M, and H denote the four kinds of pizza topping, find

(a) $P(M \cup J)$;

(b) $P(H|F)$;

(c) $P(O \cap S)$;

(d) $P(F'|A)$;

(e) $P(M \cup H|J')$;

(f) $P(J|A \cup M)$.

5.64 With reference to the preceding exercise, find the probabilities that the student chosen will be

(a) a freshman whose favorite pizza topping is mushrooms;

(b) an anchovy pizza eater given that he or she is a junior;

(c) a sophomore given that he or she is not a junior.

5.65 The probability that a bus from Buffalo to Rochester will leave on time is 0.70, and the probability that it will leave on time and also arrive on time is 0.56. What is the probability that if such a bus leaves on time it will also arrive on time?

5.66 The probability that a certain concert will be well-advertised is 0.80, and the probability that it will be well-advertised and also a great success is 0.76. What is the probability that if the concert is well-advertised it will be a great success?

5.67 An English professor figures that the probability is 0.60 that a term paper she receives will be well-written. If the probability is 0.51 that such a term paper will be well-written and also receive a good grade, what is the probability that a well-written term paper will receive a good grade?

5.68 If the odds are 5 to 3 that event K will not occur, 2 to 1 that event L will occur, and 3 to 1 that they will not both occur, are these two events independent?

5.69 With reference to the example on page 169, show that event L is also independent of event A, namely, that $P(L|A) = P(L) = 0.84$.

5.70 If $P(A) = 0.80$, $P(B) = 0.35$, and $P(A \cap B) = 0.28$, check whether events A and B are independent.

5.71 For two rolls of a balanced die, find the probabilities of getting
(a) two 6's;
(b) first a 6 and then some other number.

5.72 If two cards are drawn from an ordinary deck of 52 playing cards, what are the probabilities of getting two diamonds if the drawing is
(a) with replacement;
(b) without replacement?

5.73 A fifth-grade class consists of 16 boys and 14 girls. If one pupil is chosen each week by lot to assist the teacher, what are the probabilities that a boy will be chosen two weeks in a row if
(a) the same pupil can serve two weeks in a row;
(b) the same pupil cannot serve two weeks in a row?

5.74 If a zoologist has four male guinea pigs and eight female guinea pigs and randomly chooses two of them for an experiment, what are the probabilities that
(a) both will be males;
(b) both will be females;
(c) there will be one of each sex?

5.75 Find the probabilities of getting
(a) eight heads in a row with a balanced coin;
(b) no fives in four rolls of a balanced die.

5.76 If five of a company's 15 delivery trucks do not meet emission standards, and four of them are randomly chosen for inspection, what is the probability that all of them will meet emission standards?

5.77 A carton contains 12 shirts of which four have blemishes and the rest are good. What is the probability that if three of the shirts are randomly selected from the carton, they will all have blemishes?

5.78 With reference to the example on page 172, what are the probabilities that
(a) a customer who pays promptly one month will pay promptly the next month, not promptly the month after that, and then promptly the month after that;
(b) a customer who does not pay promptly one month will pay promptly the next two months, and then again not promptly the two months after that?

5.79 In a certain city, the probability that it will snow on a December day is 0.40, the probability that a December day on which it snows will be followed by a day on which it snows is 0.70, and the probability that a December day on which it does not snow will be followed by a day on which it snows is 0.20. What are the probabilities that on three consecutive December days it will

(a) snow on each day;

(b) not snow on the first day and then snow on the next two days?

5.80 With reference to the preceding exercise, what is the probability that it will snow, snow, not snow, and snow on four consecutive December days?

5.9

Bayes' Theorem ★

Although the symbols $P(A|B)$ and $P(B|A)$ may look very much alike, there is a great difference between the probabilities which they represent. For instance, on page 167 we calculated the probability $P(G|F)$ that a factory-trained appliance repairperson will provide good service, but what do we mean when we write $P(F|G)$? This is the probability that a repairperson who provides good service will have been factory-trained. Thus, we turned things around—cause becomes effect and effect becomes cause. To give another example, suppose that someone, applying for a loan, has his credit checked, and that R denotes the event that a person is a bad risk while J denotes the event that a person is judged a bad risk. Then, $P(J|R)$ is the probability that a person who is a bad risk will be judged a bad risk, and $P(R|J)$ is the probability that a person who is judged a bad risk actually is a bad risk.

Since there are many problems which involve such pairs of conditional probabilities, let us try to find a formula which expresses $P(B|A)$ in terms of $P(A|B)$ for any two events A and B. Fortunately, we do not have to look very far; all we have to do is equate the expressions for $P(A \cap B)$ in the two forms of the general multiplication rule on page 170, and we get

$$P(A) \cdot P(B|A) = P(B) \cdot P(A|B)$$

and, hence,

$$P(B|A) = \frac{P(B) \cdot P(A|B)}{P(A)}$$

after dividing by $P(A)$.

EXAMPLE   In a state where cars have to be tested for the emission of pollutants, 25% of all cars emit excessive amounts of pollutants. When tested, 99% of all cars that emit excessive amounts of pollutants will fail, but 17%

of the cars that do not emit excessive amounts of pollutants will also fail. What is the probability that a car which fails the test actually emits excessive amounts of pollutants?

Solution Letting A denote the event that a car fails the test and B the event that it emits excessive amounts of pollutants, we can translate the given percentages into probabilities and write $P(B) = 0.25$, $P(A|B) = 0.99$, and $P(A|B') = 0.17$.

Before we can calculate $P(B|A)$ by means of the formula given above, we will first have to determine $P(A)$, and to this end let us look at the tree diagram of Figure 5.13. Here A is reached either along the branch which passes through B or along the branch which passes through B', and the probabilities of this happening are, respectively, $(0.25)(0.99) = 0.2475$ and $(0.75)(0.17) = 0.1275$. Since the alternatives represented by the two branches are mutually exclusive, we find that $P(A) = 0.2475 + 0.1275 = 0.3750$, and substitution into the formula for $P(B|A)$ given above yields

$$P(B|A) = \frac{P(B) \cdot P(A|B)}{P(A)} = \frac{(0.25)(0.99)}{0.3750} = 0.66$$

This is the probability that a car which fails the test actually emits excessive amounts of pollutants.

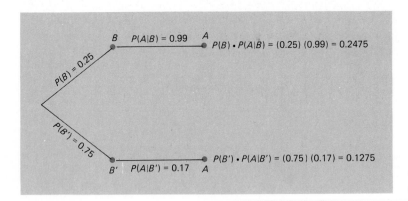

B $P(A|B) = 0.99$ A
$P(B) \cdot P(A|B) = (0.25)\,(0.99) = 0.2475$

$P(B) = 0.25$

$P(B') = 0.75$

$P(B') \cdot P(A|B') = (0.75)\,(0.17) = 0.1275$

B' $P(A|B') = 0.17$ A

5.13
Tree diagram for emission testing example.

With reference to the tree diagram of Figure 5.13 we can say that $P(B|A)$ is the probability that event A was reached via the upper branch of the tree, and we showed that its value is given by the ratio of the probability associated

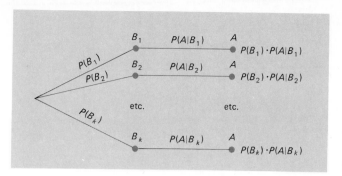

5.14

Tree diagram for Bayes' theorem.

with that branch to the sum of the probabilities associated with both branches of the tree. This argument can be generalized to the case where there are more than two possible "causes," namely, more than two branches leading to an event A. With reference to Figure 5.14 we can say that $P(B_i|A)$ is the probability that event A was reached via the ith branch of the tree (for $i = 1, 2, \ldots,$ or k), and it can be shown that its value is given by the ratio of the probability associated with the ith branch to the sum of the probabilities associated with all the branches. Symbolically, this result, called **Bayes' theorem,** is given by

Bayes' theorem

$$P(B_i|A) = \frac{P(B_i) \cdot P(A|B_i)}{P(B_1) \cdot P(A|B_1) + P(B_2) \cdot P(A|B_2) + \cdots + P(B_k) \cdot P(A|B_k)}$$

for $i = 1, 2, \ldots,$ or k.

EXAMPLE   In a cannery, assembly lines I, II, and III account, respectively, for 37%, 42%, and 21% of the total output. If 0.6% of the cans from assembly line I are improperly sealed, while the corresponding percentages for assembly lines II and III are 0.4% and 1.2%, what is the probability that an improperly sealed can (discovered at the final inspection of outgoing products) came from assembly line III?

Solution   Letting A denote the event that a can is improperly sealed and B_1, B_2, and B_3 denote the events that a can comes from assembly lines I, II, or III, we can translate the given percentages into probabilities and write $P(B_1) = 0.37$, $P(B_2) = 0.42$, $P(B_3) = 0.21$, $P(A|B_1) = 0.006$, $P(A|B_2) = 0.004$, and $P(A|B_3) = 0.012$. Then, with reference to the tree diagram of Figure 5.15 we find that the probabilities associated with the three branches are, respectively, $(0.37)(0.006) = 0.00222$,

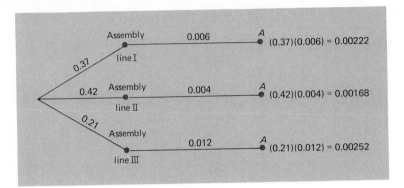

Assembly line I   0.006   A   (0.37)(0.006) = 0.00222

0.37

0.42   Assembly line II   0.004   A   (0.42)(0.004) = 0.00168

0.21   Assembly line III   0.012   A   (0.21)(0.012) = 0.00252

5.15

Tree diagram for cannery example.

$(0.42)(0.004) = 0.00168$, and $(0.21)(0.012) = 0.00252$. Thus, the required probability is

$$P(B_3|A) = \frac{0.00252}{0.00222 + 0.00168 + 0.00252} = 0.39$$

rounded to two decimals.

If we had wanted to work this problem without referring to a tree diagram, direct substitution into the formula for Bayes' theorem would have yielded

$$P(B_3|A) = \frac{(0.21)(0.012)}{(0.37)(0.006) + (0.42)(0.004) + (0.21)(0.012)}$$

$$= 0.39$$

and all the details of the calculations are, of course, the same.

As can be seen from our two examples, Bayes' theorem is a relatively simple mathematical rule. There can be no question about its validity, but criticism has frequently been raised about its applicability. This is because it involves a "backward" or "inverse" sort of reasoning—namely, reasoning from effect to cause. In our examples we used it to determine the probability that a car which fails the emission test actually emits excessive amounts of pollutants, and the probability that an improperly sealed can was "caused" by assembly line III. It is precisely this aspect of Bayes' theroem which makes it play an important role in statistical inference, where our reasoning goes from sample data that are observed to the populations from which they came. Discussion of such inferences, appropriately called **Bayesian inferences,** may be found in more advanced texts.

(Exercise 5.81 is a practice exercise; its complete solution is
given on page 185.)

⋆ **5.81** The probability is 0.70 that a rare tropical disease will be diagnosed correctly. If it is diagnosed correctly, the probability is 0.90 that the patient will be cured; if it is not diagnosed correctly, the probability is 0.40 that the patient will be cured. If a patient having this disease is cured, what is the probability that it was diagnosed correctly?

⋆ **5.82** At a shoe factory, it is known from past experience that the probability is 0.82 that a worker who has attended the factory's training program will meet the production quota, and that the corresponding probability is 0.53 for a worker who has not attended the factory's training program. If 60% of the workers attend the factory's training program, what is the probability that a worker who meets the production quota will have attended the training program?

⋆ **5.83** The probability that a one-car accident is due to faulty brakes is 0.05, the probability that a one-car accident is correctly attributed to faulty brakes is 0.82, and the probability that a one-car accident is incorrectly attributed to faulty brakes is 0.03. What is the probability that a one-car accident attributed to faulty brakes was actually due to faulty brakes?

⋆ **5.84** In a T-maze, a rat is given food if it turns left and an electric shock if it turns right. On the first trial there is a fifty-fifty chance that a rat will turn either way; then if it receives food on the first trial, the probability is 0.72 that it will turn left on the next trial, and if it receives a shock on the first trial, the probability is 0.88 that it will turn left on the next trial. If a rat turns left on the second trial, what is the probability that it turned left also on the first trial?

⋆ **5.85** With reference to the example on page 177, suppose that the state agency doing the testing can reduce to 0.03 the probability that a car which does not emit excessive amounts of pollutants will fail the test. How will this affect the probability that a car which fails the test actually emits excessive amounts of pollutants?

⋆ **5.86** With reference to the example on page 179, what are the probabilities that the improperly sealed can came from
   (a) assembly line I;
   (b) assembly line II?

⋆ **5.87** A hotel gets cars for its guests from three rental agencies, 25% from agency X, 25% from agency Y, and 50% from agency Z. If 8% of the cars from X, 6% of the cars from Y, and 15% of the cars from Z need tune-ups, what is the probability that a car needing a tune-up which is delivered to a guest of the hotel came from rental agency Y?

⋆ **5.88** A retailer of automobile parts has four employees K, L, M, and N, who make mistakes in filling an order one time in 100, four times in 100, two times in 100, and six times in 100. Of all the orders filled, K, L, M, and N fill, respectively, 20, 40, 30, and 10%. If a mistake is found in a particular order, what are the probabilities that it was filled by K, L, M, or N?

5.1 (a) $A' = \{b, d, e, h\}$ is the event that Mr. Bean, Mrs. Daly, Mr. Earl, or Mr. Hall will be awarded the scholarship.

(b) $A \cup B = \{a, c, d, f, g, h\}$ is the event that Ms. Adams, Miss Clark, Mrs. Daly, Ms. Fuentes, Ms. Garner, or Mr. Hall will be awarded the scholarship.

(c) $A \cap B = \{f\}$ is the event that Ms. Fuentes will be awarded the scholarship.

(d) $A \cap C = \{a, g\}$ is the event that Ms. Adam or Ms. Garner will be awarded the scholarship.

(e) $B \cap C = \varnothing$ is the empty set.

(f) $A' \cup B = \{b, d, e, f, h\}$ is the event that Mr. Bean, Mrs. Daly, Mr. Earl, Ms. Fuentes, or Mr. Hall will be awarded the scholarship.

5.2 (a) The sample space is shown in Figure 5.16.

(b) Q is the event that four waiters are present, R is the event that the number of waiters exceeds the number of chefs by one, and T is the event that only one chef is present.

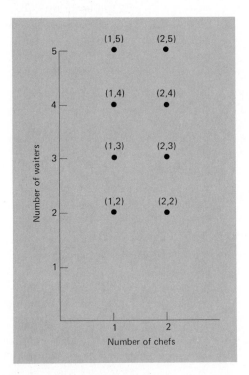

5.16

Diagram for Exercise 5.2.

(c) $Q \cup T = \{(1, 4), (2, 4), (1, 2), (1, 3), (1, 5)\}$ is the event that four waiters are present or one chef is present, $R \cap T = \{(1, 2)\}$ is the event that one chef and two waiters are present, and $T' = \{(2, 2), (2, 3), (2, 4), (2, 5)\}$ is the event that both chefs are present.

(d) Q and R are mutually exclusive, Q and T are not mutually exclusive, and R and T are not mutually exclusive.

5.12 Region 1 represents the event that a person vacationing in Southern California visits Disneyland and Universal Studios, region 2 represents the event that a person vacationing in Southern California visits Disneyland but not Universal Studios, region 3 represents the event that a person vacationing in Southern California visits Universal Studios but not Disneyland, and region 4 represents the event that a person vacationing in Southern California visits neither Disneyland nor Universal Studios.

5.19 (a) The probability that there will not be enough doctors.

(b) The probability that there will not be enough hospital beds.

(c) The probability that there will be enough doctors, enough hospital beds, or both.

(d) The probability that there will be enough doctors and enough hospital beds.

(e) The probability that there will be neither enough doctors nor enough hospital beds.

(f) The probability that there will be enough doctors but not enough hospital beds.

5.22 (a) A probability cannot be a negative number.

(b) The sum of the two probabilities should be 1 and not 0.90.

(c) Since the two events are mutually exclusive, the sum of the first two probabilities must equal the third.

5.30 (a) The odds against getting three heads and three tails in six flips of a balanced coin are 11 to 5.

(b) The probability of rolling "7 or 11" with a pair of balanced dice is $\dfrac{2}{2 + 7} = \dfrac{2}{9}$.

(c) If a pollster randomly selects five of 24 households to be included in a survey, the odds are 19 to 5 that any particular household will not be included.

(d) If three eggs are randomly chosen from a carton of 12 eggs of which three are cracked, the probability is $\dfrac{34}{34 + 21} = \dfrac{34}{55}$ that at least one of them will be cracked.

5.31 The corresponding probabilities are $\dfrac{2}{2 + 1} = \dfrac{2}{3}$ and $\dfrac{5}{5 + 1} = \dfrac{5}{6}$, and since the events are mutually exclusive we can write $\frac{2}{3} + \frac{5}{6} = \frac{3}{2}$ for the probability that unemployment will go up or down. Since a probability cannot exceed 1, this is an impossible value, and the original odds must have been wrong.

5.38 $0.22 + 0.17 + 0.21 + 0.09 = 0.69$.

5.43 (a) The event consists of $(4, 4), (5, 5), (6, 6), (7, 7)$, and $(8, 8)$, so that the probability is $\frac{5}{35} = \frac{1}{7}$.

(b) The event consists of $(4, 0), (4, 1), (4, 2), (5, 0), (5, 1), (5, 2), (6, 0), (6, 1), (6, 2), (7, 0), (7, 1), (7, 2), (8, 0), (8, 1)$, and $(8, 2)$, and the probability is $\frac{15}{35} = \frac{3}{7}$.

(c) The event consists of $(7, 0), (7, 1), (7, 2), (7, 3), (7, 4), (7, 5), (7, 6), (7, 7), (8, 0), (8, 1), (8, 2), (8, 3), (8, 4), (8, 5), (8, 6), (8, 7)$, and $(8, 8)$, and the probability is $\frac{17}{35}$.

(d) The event consists of $(4, 2), (4, 3), (5, 3), (5, 4), (6, 4), (6, 5), (7, 5), (7, 6), (8, 6)$, and $(8, 7)$, and the probability is $\frac{10}{35} = \frac{2}{7}$.

5.47 $0.15 + 0.18 - 0.11 = 0.22$.

5.50 The probabilities associated with the various regions are shown in the Venn diagram of Figure 5.17.

(a) 0.32;

(b) 0.14;

(c) $0.32 + 0.13 + 0.14 = 0.59$;

(d) $0.13 + 0.14 + 0.41 = 0.68$;

(e) 0.41;

(f) $1 - 0.13 = 0.87$.

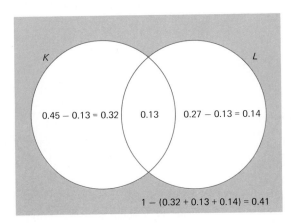

5.17

Venn diagram for Exercise 5.50.

5.53 (a) $P(G|F)$;

(b) $P(J|F')$;

(c) $P(J' \cap G'|F')$;

(d) $P(G'|F \cap J)$.

5.54 (a) The probability that a female applicant who is a college graduate will own a car.

(b) The probability that a female applicant who does not own a car will have had prior experience.

(c) The probability that a female applicant with prior experience will not own a car.

(d) The probability that a female applicant who does not own a car will not be a college graduate.

(e) The probability that a female applicant who has had prior experience or is a college graduate will own a car.

(f) The probability that a female applicant who is a college graduate will have had prior experience but not own a car.

5.59 (a) $\dfrac{80 + 160}{400} = \dfrac{240}{400} = 0.60$;   (d) $\dfrac{40}{400} = 0.10$;

(b) $\dfrac{40 + 160}{400} = \dfrac{200}{400} = 0.50$;   (e) $\dfrac{160}{80 + 160} = \dfrac{2}{3}$;

(c) $\dfrac{160}{400} = 0.40$;   (f) $\dfrac{40}{40 + 160} = 0.20$.

5.60 (a) $\dfrac{2}{3} = \dfrac{0.40}{0.60}$, which checks;   (b) $0.20 = \dfrac{0.10}{0.50}$, which checks.

5.76 The probability that the first one chosen will meet emission standards is $\frac{10}{15}$; the probability that the second one will meet emission standards given that the first one met emission standards is $\frac{9}{14}$; the probability that the third one will meet emission standards given that the first two met emission standards is $\frac{8}{13}$; and the probability that the fourth one will meet emission standards given that the first three met emission standards is $\frac{7}{12}$. Thus, the probability that all of them will meet emission standards is

$$\frac{10}{15} \cdot \frac{9}{14} \cdot \frac{8}{13} \cdot \frac{7}{12} = \frac{2}{13}$$

5.81 Substituting into the formula for Bayes' theorem, we get

$$P(D\,|\,C) = \frac{(0.70)(0.90)}{(0.70)(0.90) + (0.30)(0.40)} = \frac{0.63}{0.63 + 0.12} = \frac{0.63}{0.75} = 0.84.$$

Review: Chapters

4 & 5

Achievements

Having read and studied these chapters, and having worked a good portion of the exercises, you should be able to:

1. Draw tree diagrams to determine all the alternatives that are possible in given situations.

2. Apply the formula for the multiplication of choices and its generalization.

3. Work with the factorial notation.

4. Determine the number of permutations of n distinct objects taken r at a time.

5. Determine the number of combinations of n distinct objects taken r at a time.

6. Use the table of binomial coefficients.

7. Explain the classical probability concept.

8. Explain the frequency interpretation of probability.

9. Calculate mathematical expectations.

★10. Use mathematical expectations as the basis for making rational decisions.

11. Explain what is meant by "experiment," "outcome," "sample space," "event," and "mutually exclusive events."

12. Construct compound events by forming unions, intersections, and complements.

13. Picture events with the use of Venn diagrams.

14. List, justify, and apply the basic rules of probability.

15. Convert probabilities to odds and odds to probabilities.

16. Use betting odds to measure subjective probabilities.

17. State and apply the special addition rule and its generalization.

18. State and apply the general addition rule for events which need not be mutually exclusive.

19. Define "conditional probability" and apply the formula.

20. Explain what is meant by "independent events" and check whether two events are independent or dependent.

21. State and apply the general multiplication rule.

22. State and apply the special multiplication rule.

23. State and apply generalizations of the multiplication rules.

★ 24. Use Bayes' theorem.

Checklist of Key Terms (with page references to their definitions)

Addition rules, 159
★ Bayes' theorem, 179
★ Bayesian analysis, 132
Betting odds, 154
Binomial coefficients, 118
Classical probability concept, 119
Combinations, 115
Complement, 143
Conditional probability, 166, 168
Consistency criterion, 156
Dependent events, 170
Empty set, 142
Equitable game, 129
Event, 142
Experiment, 140
Factorial notation, 111
Fair game, 129
Frequency interpretation of probability, 121
General addition rule, 163
General multiplication rule, 170

Independent events, 170
Intersection, 143
Law of Large Numbers, 122
Mathematical expectation, 127, 129
Multiplication of choices, 106, 107
Multiplication rules, 170
Mutually exclusive events, 144
Odds, 153
Outcome, 140
Pascal's triangle, 118
Permutations, 110
Sample space, 140
Sampling with replacement, 172
Sampling without replacement, 172
Special addition rule, 159
Special multiplication rule, 171
★ States of nature, 132
Subjective probability, 124
Tree diagram, 104
Union, 143
Venn diagram, 144

R.41 On a weekday afternoon, a television station schedules four soap operas, two situation comedies, and two game shows. In how many different ways can a viewer choose two of the soap operas, one of the situation comedies, and one of the game shows.

R.42 With reference to the two salespersons and Figure 5.1, list the points of the sample space which comprise the following events:

   (a) One of the salespersons sells both cars.

   (b) The second salesperson sells one of the cars.

   (c) The first salesperson sells at least one car.

R.43 A small real estate office has four part-time salespersons. Using two coordinates so that (3, 1), for example, represents the event that three of the salespersons are at work and one of them is busy with a customer, and (2, 0) represents the event that two of the salespersons are at work but none of them is busy with a customer, draw a diagram similar to that of Figure 5.1, showing the 15 points of the corresponding sample space.

R.44 With reference to the preceding exercise, if each of the 15 points of the sample space has the probability $\frac{1}{15}$, find the probabilities that

   (a) all the salespersons that are at work are busy with customers;

   (b) at least three of the salespersons are at work;

   (c) at least three salespersons are busy with a customer;

   (d) none of the salespersons who are at work are busy with customers.

R.45 The following table gives the probabilities that a woman who enters "The Dress Shop" will buy 0, 1, 2, 3, or 4 dresses:

Number of dresses	0	1	2	3	4
Probability	0.11	0.37	0.35	0.12	0.05

How many dresses can a woman entering this shop be expected to buy?

R.46 If $P(M) = 0.55$, $P(N) = 0.18$, and $P(M \cap N) = 0.099$, are the events M and N independent or dependent?

R.47 Among six applicants for an executive job, A is a college graduate, foreign born, and single; B is not a college graduate, foreign born, and married; C is a college graduate, native born, and married; D is not a college graduate, native born, and single; E is a college graduate, native born, and married; and F is not a college graduate, native born, and married. One of these applicants is to get the job, and the event that the job is given to a college graduate, for example, is denoted $\{A, C, E\}$. State in a similar manner the event that the job is given to

   (a) a single person;

   (b) a native-born college graduate;

   (c) a married person who is foreign born.

R.48 Certain government employees are classified into six categories according to age and four categories according to marital status. In how many ways can one of these employees be classified?

R.49 A grab bag contains 12 packages worth 80 cents apiece, 15 packages worth 40 cents apiece, and 25 packages worth 30 cents apiece. Is it worthwhile to pay 50 cents for the privilege of picking one of the packages at random?

R.50 If a district attorney feels that the odds are 43 to 17 that she will win her case, what is her personal probability that she will win the case?

R.51 Among the classified ads of a newspaper listing foreign-made cars, there are listings for eight Japanese cars, six German cars, and two Italian cars. In how many different ways can a person choose two of the Japanese cars, three of the German cars, and one of the Italian cars to inspect?

★ **R.52** During a time of national emergency, a country uses lie detectors to uncover security risks. Since lie detectors are not infallible, let us suppose that the probabilities are 0.10 and 0.04 that a lie detector will fail to detect a security risk or incorrectly label a person a security risk. If 2% of the persons who are given lie detector tests are security risks, what is the probability that a person labeled a security risk by a lie detector actually is a security risk?

R.53 On a college faculty there are three instructors named Brown: Jim Brown, George Brown, and Norma Brown. Draw a tree diagram to show the different ways in which the college's payroll department can distribute their paychecks so that each of the Browns receives a check made out to one of the Browns. In how many of the possibilities will
(a) only one of them get the right check;
(b) none of them get the right check?

R.54 A carton contains 15 hair dryers of which 4 have minor blemishes and the rest are in perfect condition. What is the probability that if 3 of the hair dryers are randomly selected from the carton, they will all have blemishes?

R.55 A psychologist preparing three-letter nonsense words for use in a memory test chooses the first letter from among the consonants q, w, x, and z; the second letter from among the vowels e, i, and u; and the third letter from among the consonants c, f, p, and v.
(a) How many different three-letter nonsense words can he construct?
(b) How many of these nonsense words will begin with the letter w?
(c) How many of these nonsense words will end either with the letter f or the letter p?

R.56 Convert each of the following odds to probabilities:
(a) If three eggs are randomly chosen from a carton of twelve eggs of which three are cracked, the odds are 34 to 21 that at least one of them will be cracked.
(b) If a person has eight $1 bills, five $5 bills, and one $20, and randomly selects three of them, the odds are 11 to 2 that they will not all be $1 bills.

189

R.57 If C and W are the events that a customer will order a cocktail or wine before dinner at a certain restaurant, $P(C) = 0.43$ and $P(W) = 0.21$, find the probabilities that a customer will
  (a) not order a cocktail before dinner;
  (b) order either a cocktail or wine before dinner;
  (c) order neither a cocktail nor wine before dinner.

R.58 At a restaurant, the probabilities are, respectively, 0.12, 0.35, and 0.06 that a person will ask for a hamburger with lettuce, with tomatoes, or with both. What is the probability that a person will ask for a hamburger with lettuce or tomatoes?

R.59 If there are eight horses in a race, in how many different ways can they place first, second, and third?

R.60 The probabilities are 0.15, 0.26, and 0.08 that a family driving through a Western city will spend the night at one of its hotels, at one of its motels, or at its camp-ground. What is the probability that a family driving through this city will spend the night at one of these kinds of facilities?

R.61 If 768 of 1,200 letters mailed by a government agency were delivered within 48 hours, estimate the probability that any one letter mailed by the agency will be delivered within 48 hours.

R.62 Among the cars worked on by a garage over a certain period of time, some had been in accidents, some had not, some required minor repairs, and some required major repairs, with the exact breakdown being

	Accident	*No accident*
Minor repairs	30	50
Major repairs	180	20

If one of the cars is randomly selected to check on the quality of the workmanship, A is the event that the car had been in an accident, and M is the event that it required only minor repairs, determine each of the following probabilities directly from the entries and the row and column totals of the table:
  (a) $P(A)$;   (d) $P(A|M')$;
  (b) $P(M')$;   (e) $P(M'|A)$;
  (c) $P(A \cap M')$;   (f) $P(A' \cup M)$.

R.63 With reference to the preceding exercise, verify that

$$P(A|M') = \frac{P(A \cap M')}{P(M')}$$

R.64 If the probability is 0.26 that any one woman will name yellow or orange as her favorite color, what is the probability that four women, selected at random, will all name yellow or orange as their favorite color?

R.65 If the probability is 0.36 that next year's inflation rate will exceed this year's, what are the corresponding odds?

★ **R.66** Ms. Cooper is planning to attend a convention in San Diego, and she must send in her room reservations immediately. The convention is so large that the activities are held partly in hotel A and partly in hotel B, and Ms. Cooper does not know whether the particular session she wants to attend will be held in hotel A or hotel B. She is planning to stay only one day, which would cost her $40.00 at hotel A and $36.40 at hotel B, but it will cost her an extra $6.00 for cab fare if she stays at the wrong hotel. Where should she make her reservation if she feels that the probability is 0.75 that the session she wants to attend will be held at hotel A and she wants to minimize her expected cost?

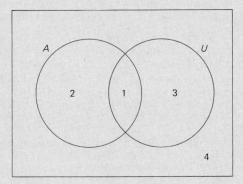

R.1

Venn diagram for Exercise R.67.

R.67 If A is the event that a school's football team is rated among the top twenty by AP and U is the event that it is rated among the top twenty by UPI, what events are represented by the four regions of the Venn diagram of Figure R.1?

R.68 A guidance department gives students various kinds of tests. If I is the event that a student scores high in intelligence, A is the event that a student rates high on a social adjustment scale, and N is the event that a student displays neurotic tendencies, express each of the following probabilities in symbolic form:

(a) The probability that a student who scores high in intelligence will display neurotic tendencies.

(b) The probability that a student who does not rate high on the social adjustment scale will not score high in intelligence.

(c) The probability that a student who displays neurotic tendencies will neither score high in intelligence nor rate high on the social adjustment scale.

(d) The probability that a student who scores high in intelligence and rates high on the social adjustment scale will not display any neurotic tendencies.

R.69 If someone feels that 17 to 8 are fair odds that a paint job will be finished on time, what subjective probability does he assign to this event?

R.70 If the probabilities are respectively, 0.02, 0.07, 0.15, 0.20, 0.20, 0.16, 0.10, 0.06, 0.03, and 0.01 that 0, 1, 2, 3, 4, 5, 6, 7, 8, or 9 students will be absent from class on a given day, find the probabilities that on a given day
    (a) at least 6 students will be absent;
    (b) at most 3 students will be absent;
    (c) anywhere from 5 to 7 students will be absent.

R.71 Among the eight nominees for two vacancies on a school board are four men and four women. In how many ways can these vacancies be filled
    (a) with any two of the eight nominees;
    (b) with any two of the female nominees;
    (c) with one of the male nominees and one of the female nominees?

R.72 In Figure R.2, E, T, and N are the events that a car brought to a garage needs an engine overhaul, transmission repairs, or new tires. Express in words what events are represented by
    (a) region 1;   (d) regions 1 and 4 together;
    (b) region 3;   (e) regions 2 and 5 together;
    (c) region 7;   (f) regions 3, 5, 6, and 8 together.

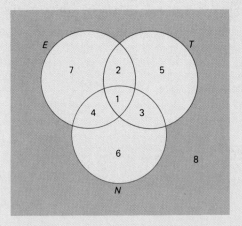

R.2
Venn diagram for Exercise R.72.

R.73 With reference to the preceding exercise and the Venn diagram of Figure R.2, list the regions or combinations of regions which represent the events that a car brought to the garage needs
    (a) transmission repairs, but neither an engine overhaul nor new tires;
    (b) an engine overhaul and transmission repairs;
    (c) transmission repairs or new tires, but not an engine overhaul;
    (d) new tires.

R.74 Explain why there must be a mistake in each of the following statements:
    (a) The probability that a missile will explode on lift-off is -0.0002.
    (b) The probability that a student will get an A in a course is 0.15, and the probability that he will get a C is ten times as large.

(c) The probability that a person is allergic to ragweed is 0.37, the probability that he or she is allergic to Bermuda grass is 0.13, and the probability that he or she is allergic to one or the other is 0.70.

(d) The probability that everyone will show up for the party is 0.69, and the probability that this will not be the case is 0.21.

R.75 Suppose that the numbers 1, 2, 3, 4, 5, and 6 are used to denote that a committee of parents and teachers decides that a certain education program is terrible, poor, fair, good, very good, or excellent. If $L = \{2, 3, 4, 5\}$ and $R = \{4, 5, 6\}$, list the elements of the sample space comprising each of the following events, and also express the events in words:

(a) L';

(b) $L \cup R$;

(c) $L \cap R$;

(d) $L \cap R'$.

R.76 For a given city, police records show that the odds are, respectively, 7 to 5, 1 to 3, and 5 to 1 that a person arrested on charges of robbery will be under 21 years old, from 21 to 24 years old, or under 25 years old. Are the corresponding probabilities consistent?

R.77 A police department needs new tires for its patrol cars and the probabilities are 0.17, 0.22, 0.03, 0.29, 0.21, and 0.08 that it will buy Uniroyal tires, Goodyear tires, Michelin tires, General tires, Goodrich tires, or Armstrong tires. Find the probabilities that it will buy

(a) Goodyear or Goodrich tires;

(b) Uniroyal, General, or Goodrich tires;

(c) Michelin or Armstrong tires;

(d) Goodyear, General, or Armstrong tires.

R.78 If E is the event that an applicant for a home mortgage is employed, G is the event that he has a good credit rating, and A is the event that the application is approved, state in words what probabilities are expressed by

(a) $P(A|E)$;

(b) $P(A|G)$;

(c) $P(A'|E')$;

(d) $P(A|E \cap G)$.

R.79 In the Sonora desert, the probability that a cactus will show damage due to frost is 0.20, that it will show damage due to too much moisture is 0.16, and that it will show both kinds of damage is 0.11. What is the probability that a cactus will show damage due to frost, excessive moisture, or both?

R.80 Mr. Clark is willing to bet Mrs. Clark $24 to her $8, but not $24 to her $6 that they will be late for the theater. What does this tell us about the probability which he assigns to their being late?

R.81 Given $P(A) = 0.65$, $P(B) = 0.20$, and $P(A \cap B) = 0.13$, verify that

(a) $P(A|B) = P(A)$;

(b) $P(A|B') = P(A)$;

(c) $P(B|A) = P(B)$;

(d) $P(B|A') = P(B)$.

193

Probability Distributions

If a teacher records how many students are absent, a social scientist determines what percentage of a panel of jurors are over 60 years old, a geologist measures the hardness of a rock, or an engineer figures out the projected cost of a new engine, they are in each case concerned with a number that is associated with the outcome of a situation involving an element of chance, namely, with the value of a

> **RANDOM VARIABLE** ('ran-dəm 'ver-ē-ə-bəl) A quantity which can take on the values of a given set with specified probabilities.

This definition may not be as rigorous as can be—strictly speaking, a random variable is a function—but most beginners find it easiest to think of random variables simply as quantities which can take on different values depending on chance. In addition to the examples cited above, this includes such things as the number of speeding tickets issued each day on a freeway between two cities, the annual production of coffee in Brazil, the wind velocity at Kennedy airport, the size of the audience at a baseball game, and the number of mistakes a person makes computing his or her income tax.

Since random variables are neither random nor variables, the reader may be curious to know why they are called by that name. This is hard to say, but a mathematician with a good sense of humor likened them to alligator pears, or avocados, which are neither alligators nor pears.

In the study of random variables, we are usually interested mostly in the probabilities associated with all their values, namely, in their **probability distributions.** The general study of probability distributions in Section 6.1 is followed by the discussion of various special probability distributions in Sections 6.2 through 6.5, and the description of their most important properties in Sections 6.6 through 6.8.

195

6.1

Probability Distributions

Random variables are usually classified according to the number of values which they can assume. In this chapter we shall limit our discussion to random variables called **discrete random variables** which can take on only a finite number of values, or a countable infinity of values (as many as there are whole numbers). For instance, the number of points we roll with a pair of dice is a discrete random variable which can take on only the finite, or fixed, number of values 2, 3, 4, 5, 6, 7, 8, 9, 10, 11, or 12. In contrast, the number of the flip on which a coin comes up heads for the first time is a discrete random variable which can take on the countable infinity of values 1, 2, 3, 4, 5, ... It is possible, though highly unlikely, that a coin will come up tails a million times, a billion times, or even more, before it finally comes up heads.

The tables in the two examples which follow serve to illustrate what we mean by the **probability distribution** of a random variable—it is a correspondence which assigns probabilities to its values.

EXAMPLE   Construct a table showing the probabilities of rolling a 1, 2, 3, 4, 5, or 6 with a balanced die.

Solution   Since "balanced" means that the outcomes are all equiprobable, each value has the probability $\frac{1}{6}$, and we get

Number of points rolled with a die	Probability
1	$\frac{1}{6}$
2	$\frac{1}{6}$
3	$\frac{1}{6}$
4	$\frac{1}{6}$
5	$\frac{1}{6}$
6	$\frac{1}{6}$

EXAMPLE   Construct a table showing the probabilities of getting 0, 1, 2, or 3 heads in three flips of a balanced coin.

Solution   The eight equally likely possibilities are HHH, HHT, HTH, THH, HTT, THT, TTH, and TTT, where H stands for heads and T for tails.

Counting the number of heads in each case and using the formula $\frac{s}{n}$ for equiprobable outcomes, we obtain the following probability distribution for the total number of heads:

Number of heads	Probability
0	$\frac{1}{8}$
1	$\frac{3}{8}$
2	$\frac{3}{8}$
3	$\frac{1}{8}$

Whenever possible or necessary, we try to express probability distributions by means of mathematical formulas which enable us to calculate the probabilities associated with the various values of a random variable.[†] For instance, for the number of points we roll with a balanced die we can write

$$f(x) = \tfrac{1}{6} \quad \text{for } x = 1, 2, 3, 4, 5, \text{ and } 6$$

where $f(1)$ represents the probability of rolling a 1, $f(2)$ represents the probability of rolling a 2, and so on, in the usual functional notation. [Most of the time we write the probability that a random variable takes on the value x as $f(x)$, but we could just as well write $g(x)$, $h(x)$, $b(x)$, ...]

EXAMPLE   Verify that the probability distribution of the number of heads obtained in three flips of a balanced coin is given by

$$f(x) = \frac{\binom{3}{x}}{8} \quad \text{for } x = 0, 1, 2, \text{ and } 3$$

[†] It would be necessary, for example, when a random variable can take on infinitely many different values and we cannot possibly write down all the probabilities.

Solution Looking up the binomial coefficients in Table X, we find that $\binom{3}{0} = 1$, $\binom{3}{1} = 3$, $\binom{3}{2} = 3$, and $\binom{3}{3} = 1$. Thus, the probabilities for $x = 0, 1, 2$, and 3 are $\frac{1}{8}$, $\frac{3}{8}$, $\frac{3}{8}$, and $\frac{1}{8}$, and this agrees with the results obtained above.

Notice that each probability in the examples above is between 0 and 1, inclusive, and that the sum of the probabilities is in each case equal to 1. These facts suggest the following two general rules which apply to any probability distribution:

1. **Since the values of a probability distribution are probabilities, they must be numbers in the interval from 0 to 1.**
2. **Since a random variable has to take on one of its values, the sum of all the values of a probability distribution must be equal to 1.**

EXAMPLE Check whether the following function can serve as the probability distribution of an appropriate random variable

$$f(x) = \frac{x + 2}{12} \qquad \text{for } x = 1, 2, \text{ and } 3$$

Solution Substituting $x = 1, 2$, and 3, we get $f(1) = \frac{3}{12}$, $f(2) = \frac{4}{12}$, and $f(3) = \frac{5}{12}$. Since none of these values is negative or greater than 1, and since their sum is $\frac{3}{12} + \frac{4}{12} + \frac{5}{12} = 1$, the given function can serve as the probability distribution of a random variable.

6.2
The Binomial Distribution

There are many applied problems in which we are interested in the probability that an event will occur x times out of n. For instance, we may be interested in the probability of getting 75 responses to 200 mail questionnaires sent out as part of a sociological survey, the probability that 12 of 50 tagged wild turkeys will be recaptured, the probability that 45 of 300 drivers stopped at a road block will be wearing their seatbelts, or the probability that 66 of 200 television viewers (interviewed by a market research organization) will recall what products were advertised on a given program. To borrow from the language of games of chance, we could say that in each of these examples we are interested in the probability of getting "x successes in n trials," or in other words, "x successes and $n - x$ failures in n attempts."

In the problems we study in this section, we shall always make the following assumptions:

1. **There is a fixed number of trials.**
2. **The probability of a success is the same for each trial.**
3. **The trials are all independent.**

This means that the theory we shall develop will not apply, for example, if we are interested in the number of dresses a woman may try on before she buys one (where the number of trials is not fixed), if we check every hour whether traffic is congested at a certain intersection (where the probability of "success" is not constant), or if we are interested in the number of times that a person voted for the Republican candidate in the last five presidential elections (where the trials are not independent).

To give an example of a problem where the three conditions apply, suppose we are interested in the probability that a student will guess right on six of the eight questions of a multiple-choice test, in which each question has five choices with one correct answer. If C denotes a correct answer and I an incorrect answer, one way in which the student can get six correct answers and two incorrect answers is given by the sequence of letters

$$CCCCCCII$$

Here, the first six questions are answered correctly and the last two are answered incorrectly. Since $P(C) = \frac{1}{5}$ and $P(I) = \frac{4}{5}$, and the trials are all independent, the probability of this particular sequence of correct and incorrect answers is

$$\frac{1}{5} \cdot \frac{1}{5} \cdot \frac{1}{5} \cdot \frac{1}{5} \cdot \frac{1}{5} \cdot \frac{1}{5} \cdot \frac{4}{5} \cdot \frac{4}{5} = \left(\frac{1}{5}\right)^6 \left(\frac{4}{5}\right)^2$$

However, there are many other sequences of six C's and two I's—for instance, $CCIICCCC$, $ICICCCC$, and $CCICCCIC$—and by the same argument as before, each one has the probability $(\frac{1}{5})^6(\frac{4}{5})^2$.

To get the probability of six C's and two I's *in any order*, we must count the number of ways of arranging six C's and two I's (the number of combinations of eight objects taken six at a time) and multiply by $(\frac{1}{5})^6(\frac{4}{5})^2$. Thus, the result is given by

$$\binom{8}{6} \cdot \left(\frac{1}{5}\right)^6 \left(\frac{4}{5}\right)^2 = 0.00115$$

rounded to five decimals.

This suggests that, in general, if n is the number of trials, p is the probability of a success for each trial, and the trials are all independent, then the probability of x successes in n trials is

Binomial distribution

$$f(x) = \binom{n}{x} p^x (1 - p)^{n - x} \qquad for\ x = 0, 1, 2, \ldots, or\ n$$

It is customary to say here that the number of successes in n trials is a random variable having the **binomial probability distribution,** or simply the **binomial distribution.** The binomial distribution is called by this name because for $x = 0, 1, 2, \ldots$, and n, the values of the probabilities are the successive terms of the binomial expansion of $[(1 - p) + p]^n$.

EXAMPLE   If the probability is 0.70 that a student with very high grades will get into law school, what is the probability that three of five students with very high grades will get into law school?

Solution   Substituting $x = 3$, $n = 5$, $p = 0.70$, and $\binom{5}{3} = 10$ into the formula for the binomial distribution, we get

$$f(3) = \binom{5}{3}(0.70)^3(1 - 0.70)^{5-3}$$

$$= 10(0.70)^3(0.30)^2$$

$$= 0.3087$$

or approximately 0.31.

The following is an example in which we calculate all the probabilities of a binomial distribution:

EXAMPLE   The probability is 0.60 that a person shopping at a certain market will spend at least $25.
(a) Find the probabilities that among five persons shopping at this market 0, 1, 2, 3, 4, or 5 will spend at least $25.
(b) Use the results of part (a) to draw a histogram of the binomial distribution with $n = 5$ and $p = 0.60$.

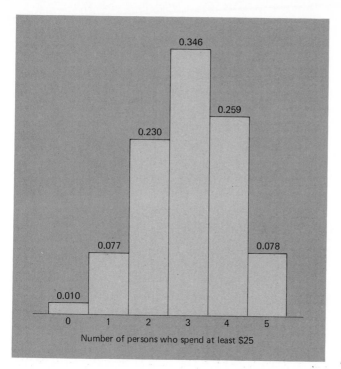

0.346

0.259

0.230

0.077

0.078

0.010

| 0 | 1 | 2 | 3 | 4 | 5 |

Number of persons who spend at least $25

6.1

Histogram of binomial distribution with n = 5 and p = 0.60.

Solution   (a) Substituting $n = 5$, $p = 0.60$, and, respectively, $x = 0, 1, 2, 3, 4$, and 5 into the formula for the binomial distribution, we get

$$f(0) = \binom{5}{0}(0.60)^0(1 - 0.60)^{5-0} = 0.010$$

$$f(1) = \binom{5}{1}(0.60)^1(1 - 0.60)^{5-1} = 0.077$$

$$f(2) = \binom{5}{2}(0.60)^2(1 - 0.60)^{5-2} = 0.230$$

$$f(3) = \binom{5}{3}(0.60)^3(1 - 0.60)^{5-3} = 0.346$$

$$f(4) = \binom{5}{4}(0.60)^4(1 - 0.60)^{5-4} = 0.259$$

$$f(5) = \binom{5}{5}(0.60)^5(1 - 0.60)^{5-5} = 0.078$$

where the probabilities are all rounded to three decimals.
(b) The histogram is shown in Figure 6.1.

201

In actual practice, binomial probabilities are seldom obtained by direct substitution into the formula. Sometimes we use the approximations discussed later in this chapter and in Chapter 7, but more often we refer to special tables such as Table I at the end of this book or the more detailed tables listed in the Bibliography on page 500. Table I gives the binomial probabilities for $n = 2$ to $n = 15$ and $p = 0.05, 0.1, 0.2, 0.3, 0.4, 0.5, 0.6, 0.7, 0.8, 0.9,$ and 0.95, all rounded to three decimals. Values omitted in the table are 0.0005 or less and, therefore, 0.000 rounded to three decimals.

EXAMPLE   If the probability is 0.40 that a divorcee will remarry within three years, find the probabilities that of ten divorcees
(a) at most three will remarry within three years;
(b) at least seven will remarry within three years;
(c) from two to five will remarry within three years;
(d) at least two will remarry within three years.

Solution   (a) From Table I, the probabilities for $n = 10$, $p = 0.40$, and $x = 0, 1,$ 2, and 3 are 0.006, 0.040, 0.121, and 0.215; thus, the probability that at most three of ten divorcees will remarry within three years is

$$0.006 + 0.040 + 0.121 + 0.215 = 0.382$$

(b) From Table I, the probabilities for $n = 10$, $p = 0.40$, and $x = 7, 8,$ 9, and 10 are 0.042, 0.011, 0.002, and 0.000; thus, the probability that at least seven of ten divorcees will remarry within three years is

$$0.042 + 0.011 + 0.002 + 0.000 = 0.055$$

(c) From Table I, the probabilities for $n = 10$, $p = 0.40$, and $x = 2, 3,$ 4, and 5 are 0.121, 0.215, 0.251, and 0.201; thus, the probability that from two to five of ten divorcees will remarry within three years is

$$0.121 + 0.215 + 0.251 + 0.201 = 0.788$$

(d) From Table I, the probabilities for $n = 10$, $p = 0.40$, and $x = 0$ and 1 are 0.006 and 0.040. Thus, the probability that at least two of ten divorcees will remarry within three years is

$$1 - (0.006 + 0.040) = 0.954$$

If the probability that a divorcee will remarry within three years had been 0.42 instead of 0.40, we could not have worked the preceding example by using Table I. In general, if n is greater than 15 and/or p takes on a value other than 0.05, 0.1, 0.2, ..., 0.9, or 0.95, we will have to use a more detailed table (see the Bibliography at the end of the book), refer to the formula for the binomial distribution, or employ a computer as in the following example.

EXAMPLE Use the computer printout of Figure 6.2 to rework the preceding example with $p = 0.42$ instead of $p = 0.40$.

```
MTB > BINOMIAL N=10 P=.42

  BINOMIAL PROBABILITIES FOR N =  10  AND P =  .420000

    K             P( X = K)           P(X LESS OR = K)
    0              .0043                  .0043
    1              .0312                  .0355
    2              .1017                  .1372
    3              .1963                  .3335
    4              .2488                  .5822
    5              .2162                  .7984
    6              .1304                  .9288
    7              .0540                  .9828
    8              .0147                  .9975
    9              .0024                  .9998
   10              .0002                 1.0000
```

6.2

Computer printout of the binomial distribution with $n = 10$ and $p = 0.42$.

Solution (a) Adding the probabilities corresponding to 0, 1, 2, and 3, we get

$$0.0043 + 0.0312 + 0.1017 + 0.1963 = 0.3335$$

Note, however, that since the printout also gives the cumulative "or less" probabilities, the answer is the value in the right-hand column corresponding to $K = 3$.

(b) Adding the probabilities corresponding to 7, 8, 9, and 10, we get

$$0.0540 + 0.0147 + 0.0024 + 0.0002 = 0.0713$$

Since the probability of "at least 7" is 1 minus the probability of "6 or less," we get $1 - 0.9288 = 0.0712$, where 0.9288 is the value in the right-hand column corresponding to $K = 6$. The difference between the two results is due to rounding.

SEC. 6.2: The Binomial Distribution

(c) Adding the probabilities corresponding to 2, 3, 4, and 5, we get

$$0.1017 + 0.1963 + 0.2488 + 0.2162 = 0.7630$$

Since the probability of "2 to 5" equals the difference between the probabilities of "5 or less" and "1 or less," the values in the right-hand column yield $0.7984 - 0.0355 = 0.7629$. Again, the difference between the two results is due to rounding.

(d) Subtracting from 1 the probability of "1 or less," we get

$$1 - (0.0043 + 0.0312) = 0.9645$$

from the column headed $P(X = K)$, and $1 - 0.0355 = 0.9645$ from the column on the right. In this case, the results are the same.

When we observe a value of a random variable having the binomial distribution—for instance, when we observe the number of heads in 50 flips of a coin, the number of seeds (in a package of 24 seeds) that germinate, the number of students (among 200 interviewed) who are opposed to an increase in tuition, or the number of automobile accidents (among 100 investigated) that are due to drunk driving—we say that we are **sampling a binomial population.** This terminology is widely used in statistics.

Computers can also be used to simulate sampling binomial populations; that is, to simulate observations of random variables having given binomial distributions. In fact, that is what we did in the example on page 122, which served to illustrate the Law of Large Numbers. In Figure 4.3 on page 123, the instruction "BRANDOM 1øø N = 1 P = .5 CI" tells the computer to take 100 random observations of a random variable having the binomial distribution with $n = 1$ and $p = 0.50$.

EXERCISES

(Exercises 6.1, 6.5, 6.6, 6.7, and 6.18 are practice exercises; their complete solutions are given on pages 229 and 230.)

6.1 In each case determine whether the given values can be looked upon as the values of a probability distribution of a random variable which can take on the values 1, 2, and 3, and explain your answers:

(a) $f(1) = 0.37$, $f(2) = 0.35$, $f(3) = 0.30$;

(b) $f(1) = \frac{4}{9}$, $f(2) = \frac{4}{9}$, $f(3) = \frac{1}{9}$;

(c) $f(1) = 0.57$, $f(2) = 0.59$, $f(3) = -0.16$.

6.2 Check in each case whether the given function can serve as the probability distribution of an appropriate random variable:

(a) $f(x) = \frac{1}{4}$ for $x = 0, 1, 2, 3, 4$;

(b) $f(x) = \dfrac{x^2}{30}$ for $x = 0, 1, 2, 3, 4$.

6.3 In each case determine whether the given values can be looked upon as the values of a probability distribution of a random variable which can take on the values 1, 2, 3, and 4. Explain your answers.

(a) $f(1) = 0.23$, $f(2) = 0.27$, $f(3) = 0.27$, $f(4) = 0.23$;

(b) $f(1) = 0.01$, $f(2) = 0.02$, $f(3) = 0.03$, $f(4) = 1.04$;

(c) $f(1) = 0.35$, $f(2) = 0.60$, $f(3) = 0.25$, $f(4) = -0.20$.

6.4 Check in each case whether the given function can serve as the probability distribution of an appropriate random variable:

(a) $f(x) = \dfrac{\binom{2}{x}}{4}$ for $x = 0, 1, 2$;

(b) $f(x) = \dfrac{x-2}{9}$ for $x = 1, 2, 3, 4, 5, 6$.

6.5 In a certain town, the probability is 0.80 that a stolen car will be recovered. Use the formula for the binomial distribution to calculate the probability that four of six stolen cars will be recovered.

6.6 Incompatibility is given as the legal reason for 50% of all divorce cases filed in a given county. Find the probability that incompatibility will be the reason given in two of the next four cases

(a) using the formula for the binomial distribution;

(b) referring to Table I.

6.7 A civil service examination is designed so that 90% of all high school graduates can pass. Use Table I to find the probabilities that among 14 high school graduates

(a) at least 12 will pass the test;

(b) at most 10 will pass the test.

6.8 In a given city, medical expenses are given as the reason for 75% of all personal bankruptcies. What is the probability that medical expenses will be given as the reason in three of the next four personal bankruptcies filed in that city?

6.9 If the probability is 0.15 that a set of tennis will go into a tie breaker, what is the probability that two of three sets will go into tie breakers?

6.10 A social scientist claims that only 40% of all high school seniors capable of doing college work actually go to college. If this is so, use the formula for the binomial distribution to calculate the probability that among eight high school seniors capable of doing college work only three will go to college.

6.11 If the probability is 0.65 that a bank's checking-account customer will pay extra for personalized checks, find the probability that only two of five of the bank's checking-account customers, selected at random, will pay extra for personalized checks.

6.12 If it is true that 80% of all industrial accidents can be prevented by paying strict attention to safety regulations, find the probability that five of seven industrial accidents can thus be prevented
   (a) by using the formula for the binomial distribution;
   (b) by referring to Table I.

6.13 It is known from experience that 50% of all persons who get a certain mail-order catalog will order something. Find the probability that only three of nine persons who get this catalog will order something
   (a) by using the formula for the binomial distribution;
   (b) by referring to Table I.

6.14 It is known from experience that 20% of the persons shopping at a certain market regularly buy a particular brand of ice cream. Use Table I to find the probabilities that among 12 randomly selected persons shopping at that market
   (a) at least four regularly buy the ice cream;
   (b) at most two regularly buy the ice cream;
   (c) anywhere from two through five regularly buy the ice cream.

6.15 A study conducted at a certain college shows that 60% of the school's graduates obtain a job in their chosen field within a year after graduation. Use Table I to find the probabilities that, within a year after graduation, among 14 randomly selected graduates of that college
   (a) at least six will find a job in their chosen field;
   (b) at most three will find a job in their chosen field;
   (c) anywhere from five through eight will find a job in their chosen field.

6.16 Research shows that 30% of all women taking a certain medication do not respond favorably. Find the probabilities that among nine randomly selected women taking the medication
   (a) at least four will respond favorably;
   (b) at least four will not respond favorably.

6.17 A study has shown that 50% of the families in a certain large area have at least two cars. Find the probabilities that among 15 families randomly selected in this area
   (a) eight have at least two cars;
   (b) more than ten have at least two cars;
   (c) six have fewer than two cars;
   (d) more than eight have fewer than two cars.

6.18 A quality control engineer wants to check whether (in accordance with specifications) 95% of the products shipped are in perfect condition. To this end, he randomly selects 10 items from each large lot ready to be shipped and passes it only if they are all in perfect condition; otherwise, each item in the lot is checked. Use Table I to find the probabilities that he will commit the error of

 (a) holding a lot for further inspection even though 95% of the items are in perfect condition;

 (b) letting a lot pass through even though only 90% of the items are in perfect condition;

 (c) letting a lot pass through even though only 80% of the items are in perfect condition.

6.19 A food distributor claims that 80% of her 6-ounce cans of mixed nuts contain at least three pecans. To check on this, a consumer testing service decides to examine six of these 6-ounce cans of mixed nuts from a very large production lot, and reject the claim if fewer than four of them contain at least three pecans. Use Table I to find the probabilities that the testing service will commit the error of

 (a) rejecting the claim even though it is true;

 (b) not rejecting the claim when in reality only 60% of the cans of mixed nuts contain at least three pecans;

 (c) not rejecting the claim when in reality only 40% of the cans of mixed nuts contain at least three pecans.

 6.20 With reference to Exercise 6.15, suppose that the study had shown that 61% of the school's graduates obtain a job in their chosen field within a year after graduation. Use a computer printout of the binomial distribution with $n = 14$ and $p = 0.61$ to rework the three parts of that exercise.

 6.21 With reference to Exercise 6.16, suppose that the study had shown that 33% of all women taking the medication do not respond favorably. Use a computer printout of the binomial distribution with $n = 9$ and $p = 0.33$ to rework both parts of that exercise.

 6.22 With reference to Exercise 6.17, suppose that the study had shown that 49% of the families in that area have at least two cars. Use a computer printout of the binomial distribution with $n = 15$ and $p = 0.49$ to rework all four parts of that exercise.

6.3

The Hypergeometric Distribution

To illustrate another important kind of probability distribution, let us consider the following problem. A factory ships certain tape recorders in lots of 16, and when they arrive at their destination, three are randomly selected from each lot for inspection. Now suppose that four of the tape recorders in a lot are defective, and we are interested in the probability that among the three tape recorders selected, two will be all right and one will be defective. In other words, we are interested in the probability of "one success (defective tape recorder) in three trials," and we might be tempted to argue that since four of the 16 tape recorders are defective, the probability of a success is $\frac{4}{16} = \frac{1}{4}$ and, hence, the desired probability is

$$f(1) = \binom{3}{1}\left(\frac{1}{4}\right)^1\left(1 - \frac{1}{4}\right)^{3-1} = 0.422$$

rounded to three decimals. This result, obtained by means of the formula for the binomial distribution, would be correct if sampling is with replacement; namely, if each tape recorder is replaced in the lot before the next one is selected.

In actual practice, we seldom, if ever, sample with replacement. To obtain the correct answer for our problem when sampling is without replacement, we might argue as follows. There are altogether $\binom{16}{3}$ ways of choosing three of the 16 tape recorders, and they are all equiprobable by virtue of the assumption that the selection is random. Since there are $\binom{4}{1}$ ways of choosing one of the defective tape recorders and $\binom{12}{2}$ ways of choosing two that are not defective, there are $\binom{4}{1} \cdot \binom{12}{2}$ ways of getting "one success (defective tape recorder) in three trials." Thus, it follows by the special formula $\dfrac{s}{n}$ for equiprobable outcomes that the desired probability is

$$\frac{\binom{4}{1} \cdot \binom{12}{2}}{\binom{16}{3}} = \frac{4 \cdot 66}{560} = 0.471$$

rounded to three decimals.

This suggests that, in general, if n objects are chosen at random from a set consisting of a objects of one kind (successes) and b objects of another kind (failures), and the selection is without replacement, then the probability of "x successes and $n - x$ failures" is

Hypergeometric distribution

$$f(x) = \frac{\binom{a}{x} \cdot \binom{b}{n-x}}{\binom{a+b}{n}} \qquad \textit{for } x = 0, 1, 2, \ldots, \textit{or } n$$

where x cannot exceed a and $n - x$ cannot exceed b, since we cannot get more successes (or failures) than there are in the whole set. This is the formula for the **hypergeometric distribution.**

EXAMPLE   A secretary is supposed to send six of 15 letters by airmail, but she gets them all mixed up and randomly puts airmail stamps on six of the letters. What is the probability that only three of the letters which should go by airmail get an airmail stamp?

Solution   Since this problem involves sampling without replacement, the situation requires the formula for the hypergeometric distribution. Substituting $a = 6$, $b = 9$, $n = 6$, and $x = 3$ into this formula, we get

$$f(3) = \frac{\binom{6}{3}\binom{9}{3}}{\binom{15}{6}} = \frac{20 \cdot 84}{5,005} = 0.336$$

rounded to three decimals.

The following is an example in which we calculate all the probabilities of a hypergeometric distribution:

EXAMPLE   Among a department store's 16 delivery trucks, five have worn brakes. If three of the trucks are randomly picked for a general overhaul, what are the probabilities that this will include 0, 1, 2, or 3 of the trucks with worn brakes?

Solution   Again, the problem involves sampling without replacement. Thus, substituting $a = 5$, $b = 11$, $n = 3$, and $x = 0, 1, 2$, and 3 into the formula for the hypergeometric distribution, we get

$$f(0) = \frac{\binom{5}{0}\binom{11}{3}}{\binom{16}{3}} = \frac{1 \cdot 165}{560} = 0.295$$

$$f(1) = \frac{\binom{5}{1}\binom{11}{2}}{\binom{16}{3}} = \frac{5 \cdot 55}{560} = 0.491$$

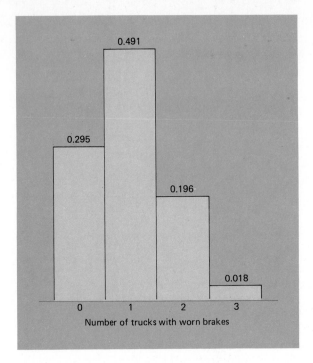

*Histogram of hypergeometric
distribution with a = 5, b = 11,
and n = 3.*

$$f(2) = \frac{\binom{5}{2}\binom{11}{1}}{\binom{16}{3}} = \frac{10 \cdot 11}{560} = 0.196$$

and

$$f(3) = \frac{\binom{5}{3}\binom{11}{0}}{\binom{16}{3}} = \frac{10 \cdot 1}{560} = 0.018$$

where the probabilities are all rounded to three decimals. A histogram
of this probability distribution is shown in Figure 6.3.

As we have pointed out, the binomial distribution does not apply when
sampling is without replacement, but sometimes it can be used as an
approximation.

EXAMPLE   In a federal prison for women, 100 of the 240 inmates have radical political views. If five of them are randomly chosen to appear before a legislative committee, find the probability that only one of them will have radical political views by using

(a) the formula for the hypergeometric distribution;

(b) the formula for the binomial distribution with $n = 5$ and $p = \frac{100}{240} = \frac{5}{12}$ as an approximation.

Solution   (a) Substituting $a = 100$, $b = 140$, $n = 5$, and $x = 1$ into the formula for the hypergeometric distribution, we get

$$f(1) = \frac{\binom{100}{1}\binom{140}{4}}{\binom{240}{5}} = \frac{100 \cdot \dfrac{140 \cdot 139 \cdot 138 \cdot 137}{4!}}{\dfrac{240 \cdot 239 \cdot 238 \cdot 237 \cdot 236}{5!}} = 0.2409$$

(b) Substituting $n = 5$, $p = \frac{5}{12}$, and $x = 1$ into the formula for the binomial distribution, we get

$$f(1) = \binom{5}{1}\left(\frac{5}{12}\right)^1\left(1 - \frac{5}{12}\right)^{5-1} = 5 \cdot \frac{5}{12} \cdot \frac{2{,}401}{20{,}736} = 0.2412$$

and since the difference between the two values is only 0.0003, the approximation is very good.

It is generally agreed that the approximation illustrated by this example may be used so long as the sample does not exceed 5% of the population, namely, so long as

Condition for binomial approximation to hypergeometric distribution

$$n \leq (0.05)(a + b)$$

Note that in our example $n = 5$ is less than $0.05(100 + 140) = 12$, so that the condition is satisfied. The main advantages of the binomial approximation to the hypergeometric distribution are that the binomial distribution has been tabulated much more extensively than the hypergeometric distribution and that the calculations are generally less involved.

211

6.23 Of 18 secretaries in a business office, 12 are graduates of a secretarial school. If five of these secretaries are randomly chosen for a certain task, what is the probability that only two of them will be graduates of the secretarial school?

6.24 Among the 12 houses for sale in a development, nine have air conditioning. If four of the houses are randomly chosen for a full-page newspaper ad, what is the probability that three of them will have air conditioning?

6.25 Find the probability that an IRS auditor will get two income tax returns with unallowable deductions, if she randomly selects five returns from among 20 returns of which 14 contain unallowable deductions.

6.26 Among the 14 applicants for a job, 10 have college degrees. If three of the applicants are randomly chosen to be interviewed, what is the probability that two of them will have college degrees?

6.27 To pass a quality control inspection, two radios are randomly chosen from each lot of 12 car radios, and the lot is passed only if neither radio has any defects. Find the probabilities that a lot will
(a) pass the inspection when two of the 12 radios are defective;
(b) fail the inspection when four of the 12 radios are defective.

6.28 Among a person's 10 pairs of socks, four pairs need mending. If he randomly picks three pairs of these socks to take along on a trip, what are the probabilities that
(a) none of the socks will need mending;
(b) one pair will need mending;
(c) two pairs will need mending;
(d) all three pairs will need mending.

6.29 Check in each case whether the condition for the binomial approximation to the hypergeometric distribution is satisfied:
(a) $a = 40$, $b = 160$, and $n = 15$;
(b) $a = 380$, $b = 120$, and $n = 22$;
(c) $n = 400$, $b = 240$, and $n = 30$.

6.30 A shipment of 200 burglar alarms contains six defectives. If three of these burglar alarms are randomly selected and shipped to a customer, find the probability that she will get one bad unit using
(a) the formula for the hypergeometric distribution;
(b) the binomial distribution as an approximation.

6.31 Check in each case whether the condition for the binomial approximation to the hypergeometric distribution is satisfied:
(a) $a = 200$, $b = 200$, and $n = 32$;
(b) $a = 150$, $b = 180$, and $n = 12$;
(c) $a = 100$, $b = 120$, and $n = 14$.

6.32 Among the 200 employees of a company, 140 are union members and the others are nonunion. If four of the employees are chosen by lot to serve on a grievance committee, find the probability that two of them will be union members and the others nonunion, using

    (a) the formula for the hypergeometric distribution;

    (b) the binomial distribution as an approximation.

6.33 A panel of prospective jurors includes 240 whites and 60 blacks. Use the binomial distribution (and Table I) to approximate the probability that a jury of 12 randomly chosen from this panel will include only one black.

6.4

The Poisson Distribution

When n is large and p is small, binomial probabilities are often approximated by means of the formula

Poisson approximation to binomial distribution

$$f(x) = \frac{(np)^x \cdot e^{-np}}{x!} \qquad for\ x = 0, 1, 2, 3, \ldots$$

which is a special form of the **Poisson distribution;** the more general form is given on page 216. It is difficult to state precisely what we mean by "n is large and p is small," and there are other situations where the approximation may be used, but we shall use it here only when n is at least 100 and np is less than 10; symbolically, when

Conditions for Poisson approximation to binomial distribution

$$n \geq 100 \quad and \quad np < 10$$

To get some idea of the closeness of the Poisson approximation to the binomial distribution, consider the computer printout of Figure 6.4, which gives the binomial distribution with $n = 100$ and $p = 0.015$ and the corresponding Poisson distribution with $np = 100(0.015) = 1.5$.[†] Comparing the probabilities in the columns headed $P(X = K)$, we find that the maximum difference, corresponding to $K = 0$, is $0.2231 - 0.2206 = 0.0025$.

In the formula for the Poisson distribution, e is the number $2.71828 \ldots$ used in connection with natural logarithms, and the values of e^{-np} may be

---

[†] In the computer printout of Figure 6.4, the quantity $np = 1.5$ is referred to as the mean of the Poisson distribution. This will be explained on page 221.

```
MTB > BINOMIAL N=100 P=.015

    BINOMIAL PROBABILITIES FOR N = 100   AND P =   .015000

        K              P( X = K)           P(X LESS OR = K)
        0                 .2206                  .2206
        1                 .3360                  .5566
        2                 .2532                  .8098
        3                 .1260                  .9358
        4                 .0465                  .9823
        5                 .0136                  .9959
        6                 .0033                  .9992
        7                 .0007                  .9999
        8                 .0001                 1.0000
        9                 .0000                 1.0000
       10                 .0000                 1.0000
       11                 .0000                 1.0000



MTB > POISSON MU=1.5

    POISSON PROBABILITIES FOR MEAN =   1.500

        K              P(X = K)            P(X LESS OR = K)
        0                 .2231                  .2231
        1                 .3347                  .5578
        2                 .2510                  .8088
        3                 .1255                  .9344
        4                 .0471                  .9814
        5                 .0141                  .9955
        6                 .0035                  .9991
        7                 .0008                  .9998
        8                 .0001                 1.0000
```

6.4

Computer printout of the binomial distribution with n = 100 and p = 0.015 and the Poisson distribution with np = 1.5.

obtained from Table XI at the end of the book. For the Poisson distribution, x can take on the countably infinite set of values 0, 1, 2, 3, ..., but this poses no problems, since the probabilities usually become negligible after relatively few values of x.

EXAMPLE   A very large shipment of books contains 2% with defective bindings. Use the Poisson approximation to the binomial distribution to find the probability that among 400 books taken at random from this shipment only five will have defective bindings.

Solution   Substituting $x = 5$ and $np = 400(0.02) = 8$ into the formula for the Poisson distribution and getting $e^{-8} = 0.00034$ from Table XI, we obtain

$$f(5) = \frac{8^5 \cdot e^{-8}}{5!} = \frac{(32,768)(0.00034)}{120} = 0.093$$

It would have been possible, though very cumbersome, to use the formula for the binomial distribution in this example.

EXAMPLE Records show that the probability is 0.00005 that a car will have a flat tire while driving through a certain tunnel. Use the Poisson approximation to the binomial distribution to find the probabilities that among 10,000 cars passing through this tunnel
   (a) at least two will have a flat tire;
   (b) at most two will have a flat tire.

Solution (a) For this probability, we shall subtract from 1 the probability that zero or one of the cars will have a flat tire. Substituting $np = 10,000(0.00005) = 0.5$ and, respectively, $x = 0$ and $x = 1$ into the formula for the Poisson distribution, and getting $e^{-0.5} = 0.607$ from Table XI, we obtain

$$f(0) = \frac{(0.5)^0(0.607)}{0!} = 0.607$$

$$f(1) = \frac{(0.5)^1(0.607)}{1!} = 0.304$$

Thus, the answer is $1 - (0.607 + 0.304) = 0.089$.
   (b) This probability is $f(0) + f(1) + f(2)$. Substituting $np = 0.5$, $e^{-0.5} = 0.607$, and $x = 2$ into the formula for the Poisson distribution, we get

$$f(2) = \frac{(0.5)^2(0.607)}{2!} = 0.076$$

Combining this result with the two probabilities calculated in part (a), we find that the answer is

$$0.607 + 0.304 + 0.076 = 0.987$$

215

```
MTB > POISSON MU=.5

    POISSON PROBABILITIES FOR MEAN =       .500

        K              P(X = K)           P(X LESS OR = K)
        0               .6065                .6065
        1               .3033                .9098
        2               .0758                .9856
        3               .0126                .9982
        4               .0016                .9998
        5               .0002               1.0000
```

6.5

Computer printout of the Poisson distribution with np = 0.5.

In actual practice, Poisson probabilities are seldom obtained by direct substitution into the formula. Sometimes we refer to tables of Poisson probabilities, which may be found in handbooks of statistical tables, and sometimes we use a computer. For instance, had we used the computer printout of Figure 6.5 in the preceding example, we would have found that the answer to part (a) is $1 - 0.9098 = 0.0902$, where 0.9098 is the value in the right-hand column corresponding to $K = 1$. Similarly, the answer to part (b) is 0.9856, which is the value in the right-hand column corresponding to $K = 2$. In both instances, the differences between the results obtained here and before are due to rounding.

The Poisson distribution also has many important applications which have no direct connection with the binomial distribution. In that case, *np* is replaced by the parameter λ (Greek lowercase *lambda*) and we calculate the probability of getting *x* "successes" by means of the formula

Poisson distribution
(parameter λ)

$$f(x) = \frac{\lambda^x \cdot e^{-\lambda}}{x!} \qquad for\ x = 0, 1, 2, 3, \ldots$$

where λ is interpreted as the expected, or average, number of successes (see discussion on page 221).

This formula applies to many situations where we can expect a fixed number of "successes" per unit time (or for some other kind of unit); say, when 1.8 accidents can be expected per day at a busy intersection, when eight small pieces of meat can be expected in a frozen meat pie, when 4.5 imperfections can be expected per roll of cloth, when 0.12 complaint per passenger can be expected by an airline, and so on.

EXAMPLE   If a bank receives on the average $\lambda = 6$ bad checks per day, what is the probability that it will receive four bad checks on any given day?

Solution   Substituting into the formula, we get

$$f(4) = \frac{6^4 \cdot e^{-6}}{4!} = \frac{(1{,}296)(0.0025)}{24} = 0.135$$

where the value of e^{-6} was read from Table XI.

EXERCISES   (Exercises 6.34, 6.36, and 6.40 are practice exercises; their complete solutions are given on page 231.)

6.34 Check in each case whether the conditions for the Poisson approximation to the binomial distribution are satisfied:
(a) $n = 150$ and $p = \frac{1}{20}$;
(b) $n = 950$ and $p = \frac{1}{100}$;
(c) $n = 300$ and $p = \frac{1}{25}$.

6.35 Check in each case whether the conditions for the Poisson approximation to the binomial distribution are satisfied:
(a) $n = 180$ and $p = \frac{1}{10}$;
(b) $n = 200$ and $p = \frac{1}{40}$;
(c) $n = 800$ and $p = \frac{1}{100}$.

6.36 It is known from experience that 2% of the calls received by a switchboard are wrong numbers. Use the Poisson approximation to the binomial distribution to determine the probability that among 250 calls received by the switchboard, four will be wrong numbers.

6.37 It is known that 1.7% of the inhabitants of a border city are illegal immigrants. Use the Poisson approximation to the binomial distribution to determine the probability that in a random sample of 100 inhabitants of the city, two will be illegal immigrants.

6.38 If 1.8% of the fuses delivered to an arsenal are defective, use the Poisson approximation to the binomial distribution to determine the probability that in a random sample of 150 fuses, four or five will be defective.

6.39 Records show that the probability is 0.0012 that a person will get food poisoning spending a day at a certain state fair. Use the Poisson approximation to the binomial distribution to find the probability that among 1,000 persons attending the fair, at most two will get food poisoning.

6.40 The number of minor injuries a football coach can expect during the course of a game is a random variable having the Poisson distribution with $\lambda = 4.4$. Find the probability that during the course of a game there will be at most three minor injuries.

6.41 If the number of complaints which a dry cleaning establishment receives per day is a random variable having the Poisson distribution with $\lambda = 3.3$, what is the probability that it will receive only two complaints on any given day?

6.42 In the inspection of a fabric produced in continuous rolls, the number of imperfections spotted by an inspector during a 5-minute period is a random variable having the Poisson distribution with $\lambda = 2.8$. What are the probabilities that during a 5-minute period an inspector will spot
  (a) one imperfection;
  (b) two imperfections?

6.43 If the number of wild pigs seen on a two-hour jeep trip in the Sonora desert is a random variable having the Poisson distribution with $\lambda = 0.8$, find the probabilities that on such a jeep trip one will see
  (a) no wild pigs;
  (b) one wild pig;
  (c) two wild pigs;
  (d) more than two wild pigs.

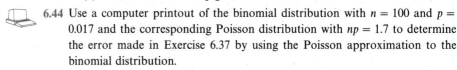

6.44 Use a computer printout of the binomial distribution with $n = 100$ and $p = 0.017$ and the corresponding Poisson distribution with $np = 1.7$ to determine the error made in Exercise 6.37 by using the Poisson approximation to the binomial distribution.

6.45 Use a computer printout of the binomial distribution with $n = 150$ and $p = 0.018$ and the corresponding Poisson distribution with $np = 2.7$ to determine the error made in Exercise 6.38 by using the Poisson approximation to the binomial distribution.

6.46 With reference to Exercise 6.40, use a computer printout of the Poisson distribution with $\lambda = 4.4$ to determine the probabilities that during the course of a game there will be
  (a) at most 6 minor injuries;
  (b) at least 4 minor injuries.

6.47 With reference to Exercise 6.41, use a computer printout of the Poisson distribution with $\lambda = 3.3$ to determine the probabilities that on any given day the dry cleaning establishment will receive
  (a) at most 3 complaints;
  (b) at least 5 complaints;
  (c) anywhere from 2 to 6 complaints.

6.48 With reference to Exercise 6.42, use a computer printout of the Poisson distribution with $\lambda = 2.8$ to determine the probabilities that during a 5-minute interval an inspector will spot
  (a) 3 or 4 imperfections;
  (b) at most 5 imperfections;
  (c) at least 6 imperfections;
  (d) anywhere from 1 to 3 imperfections.

6.5

The Multinomial Distribution

An important generalization of the binomial distribution arises when there are more than two possible outcomes for each trial, the probabilities of the various outcomes remain the same for each trial, and the trials are all independent. This is the case, for example, when we repeatedly roll a die, where each trial has six possible outcomes; when students are asked whether they like a certain new recording, dislike it, or don't care; or when a U.S. Department of Agriculture inspector grades beef as prime, choice, good, commercial, or utility.

If there are k possible outcomes for each trial and their probabilities are $p_1, p_2, \ldots,$ and p_k, it can be shown that the probability of x_1 outcomes of the first kind, x_2 outcomes of the second kind, ..., and x_k outcomes of the kth kind in n trials is given by

Multinomial distribution

$$\frac{n!}{x_1! x_2! \cdot \ldots \cdot x_k!} \, p_1^{x_1} \cdot p_2^{x_2} \cdot \ldots \cdot p_k^{x_k}$$

EXAMPLE   In a certain city, Channel 3 has 50% of the viewing audience on Saturday nights, Channel 12 has 30%, and Channel 10 has 20%. What is the probability that among eight television viewers randomly selected in that city on a Saturday night, five will be watching Channel 3, two will be watching Channel 12, and one will be watching Channel 10?

Solution   Substituting $n = 8$, $x_1 = 5$, $x_2 = 2$, $x_3 = 1$, $p_1 = 0.50$, $p_2 = 0.30$, and $p_3 = 0.20$ into the formula for the multinomial distribution, we get

$$\frac{8!}{5! \cdot 2! \cdot 1!} (0.50)^5 (0.30)^2 (0.20)^1 = 0.0945$$

EXERCISES   (Exercise 6.49 is a practice exercise; its complete solution is given on page 231.)

6.49 At a supermarket, the probabilities that a shopper will choose the poorest grade of ground beef, the third best, the second best, or the best are, respectively, 0.10, 0.20, 0.50, and 0.20. What is the probability that among 12 randomly chosen shoppers buying ground beef at this supermarket, one will buy the poorest grade, three will buy the third best grade, seven will buy the second best grade, and one will buy the best grade?

6.50 The probabilities that a floodlight will last less than 100 hours, from 100 to 150 hours, or more than 150 hours are, respectively, 0.60, 0.30, and 0.10. What is the probability that among six such floodlights three will last less than 100 hours, two will last from 100 to 150 hours, and one will last more than 150 hours?

6.51 It can easily be shown that the probabilities of getting two heads, a head and a tail, and two tails when flipping a pair of balanced coins are, respectively, $\frac{1}{4}$, $\frac{1}{2}$, and $\frac{1}{4}$. What is the probability of getting two heads once, a head and a tail twice, and two tails twice in five flips of a pair of balanced coins?

6.52 The probabilities are 0.60, 0.20, 0.10, and 0.10 that a state income tax form will be filled out correctly, that it will contain only errors favoring the taxpayer, that it will contain only errors favoring the government, and that it will contain both kinds of errors. What is the probability that among ten such tax forms (randomly selected for audit) seven will be filled out correctly, one will contain only errors favoring the taxpayer, one will contain only errors favoring the government, and one will contain both kinds of errors?

6.6

The Mean of a Probability Distribution

On page 127 we said that a child in the age group from 6 to 16 can expect to go to the dentist 1.9 times a year, and we pointed out that this figure is a mathematical expectation, obtained by multiplying 0, 1, 2, . . . , by the respective probabilities that a child in this age group will visit a dentist that many times a year. On the same page, we also showed how similar calculations led to the result that a certain police chief can expect 1.43 car thefts per day.

If we apply the same argument to the first two examples of Section 6.1, we find that the number of points we can expect on one roll of a die is

$$1 \cdot \frac{1}{6} + 2 \cdot \frac{1}{6} + 3 \cdot \frac{1}{6} + 4 \cdot \frac{1}{6} + 5 \cdot \frac{1}{6} + 6 \cdot \frac{1}{6} = 3\frac{1}{2}$$

and the number of heads we can expect in three flips of a balanced coin is

$$0 \cdot \frac{1}{8} + 1 \cdot \frac{3}{8} + 2 \cdot \frac{3}{8} + 3 \cdot \frac{1}{8} = 1\frac{1}{2}$$

Of course, we cannot roll $3\frac{1}{2}$ with a die or get $1\frac{1}{2}$ heads; like all mathematical expectations, these figures must be looked upon as averages.

In general, if a random variable takes on the values $x_1, x_2, x_3, \ldots$, or x_k, with the probabilities $f(x_1), f(x_2), f(x_3), \ldots$, and $f(x_k)$, its expected value (or its mathematical expectation) is given by

$$x_1 \cdot f(x_1) + x_2 \cdot f(x_2) + x_3 \cdot f(x_3) + \cdots + x_k \cdot f(x_k)$$

We refer to this quantity as the **mean of the probability distribution** of the random variable, and, in the $\sum$ notation, we write

Mean of probability distribution

$$\mu = \sum x \cdot f(x)$$

where the summation extends over all values taken on by the random variable. Like the mean of a population, it is denoted by the Greek letter μ (lowercase *mu*). The notation is the same, for as we pointed out in connection with the binomial distribution, when we observe a value of a random variable, we refer to its probability distribution as the population we are sampling. For instance, Figure 6.1 on page 201 pictures the population we are sampling when we observe a value of a random variable having the binomial distribution with $n = 5$ and $p = 0.60$, and Figure 6.3 on page 210 pictures the population we are sampling when we observe a value of a random variable having the hypergeometric distribution with $a = 5$, $b = 11$, and $n = 3$.

EXAMPLE   Use the probabilities obtained in the example on page 200 to determine how many of five persons shopping at the given market can be expected to spend at least $25.

Solution   Substituting $x = 0, 1, 2, 3, 4$, and 5, and the corresponding probabilities, into the formula for μ, we get

$$\mu = 0(0.010) + 1(0.077) + 2(0.230) + 3(0.346) + 4(0.259) + 5(0.078)$$
$$= 3.001$$

or approximately 3. Thus, three out of five shoppers can be expected to spend at least $25 at the given market.

___

When a random variable can take on many different values, the calculation of μ becomes very laborious. For instance, if we want to know how many of 400 persons attending a movie can be expected to buy popcorn (when the probability is, say, 0.30 that any one of them will buy popcorn), we will first have to calculate the 401 probabilities corresponding to 0, 1, 2, ..., or 400 of them buying popcorn. However, if we think for a moment, we might argue that in the long run 30% of all movie goers buy popcorn, 30% of 400 is $400(0.30) = 120$, and, hence, we can expect that 120 of the 400 persons attending the movie will buy popcorn. Similarly, if a balanced coin is flipped 500 times, we can argue that in the long run heads will come up 50% of the time, and, hence, that we can expect to get $500(0.50) = 250$ heads in 500 flips of a balanced coin. These two values are, indeed, correct; both problems deal with random variables having binomial distributions, and it can be shown that in general

Mean of binomial distribution

$$\mu = n \cdot p$$

for the mean of a binomial distribution. In words,

The mean of a binomial distribution is the product of the number of trials and the probability of success on an individual trial.

EXAMPLE   With reference to the preceding example, use this formula to find the mean of the probability distribution of the number of persons, among five, who will spend at least $25 at the given market.

Solution   Since we are dealing with a binomial distribution with $n = 5$ and $p = 0.60$, we find that $\mu = 5(0.60) = 3$, and it should be apparent that the small difference of 0.001 between this exact value and the one obtained before is due to rounding the probabilities to three decimals.

EXAMPLE   Find the mean of the probability distribution of the number of heads obtained in three flips of a balanced coin.

Solution   Since we are dealing with a binomial distribution with $n = 3$ and $p = \frac{1}{2}$, we get $\mu = 3 \cdot \frac{1}{2} = 1\frac{1}{2}$, and this agrees with the result obtained in the beginning of this section.

It is important to remember that the formula $\mu = n \cdot p$ applies only to binomial distributions. There are other formulas for other distributions; for instance, for the hypergeometric distribution the formula for the mean is $\mu = \dfrac{n \cdot a}{a + b}$, and for the Poisson distribution it is $\mu = \lambda$. Proofs of these special formulas may be found in most textbooks on mathematical statistics.

6.7

The Standard Deviation of a Probability Distribution

In Chapter 3 we saw that there are many situations in which we must not only calculate the mean or some other measure of location, but also describe the variability (spread, or dispersion) of a set of data. As we indicated in that chapter, the most widely used measure of variation is the variance and its square root, the standard deviation. For probability distributions, we measure variability in almost the same way, but instead of averaging the squared deviations from the mean, we calculate their expected value. If x is a value of a random variable whose probability distribution has the mean μ, its deviation from the mean is $x - \mu$ and we define the **variance of the probability distribution** as the expected value (mathematical expectation) of the squared deviation from the mean, namely, as

*Variance
of probability
distribution*

$$\sigma^2 = \sum (x - \mu)^2 \cdot f(x)$$

where the summation extends over all values taken on by the random variable. As in the preceding section, and for the same reason, we denote descriptions of probability distributions with the same symbols as the corresponding descriptions of populations.

The square root of the variance defines the **standard deviation of a probability distribution,** and we write

*Standard deviation
of probability
distribution*

$$\sigma = \sqrt{\sum (x - \mu)^2 \cdot f(x)}$$

Note that this formula is like that of the standard deviation of a population as defined on page 62, with the probability $f(x)$ substituted for $\frac{1}{N}$.

EXAMPLE   Use the probabilities obtained in the example on page 201, to determine the standard deviation of the distribution of the number of persons, among five, who will spend at least \$25 at the given market.

Solution   Since we have already shown that $\mu = 3$, we can arrange the necessary calculations as in the following table:

Number of persons x	$x - \mu$	$(x - \mu)^2$	Probability $f(x)$	$(x - \mu)^2 \cdot f(x)$
0	-3	9	0.010	0.090
1	-2	4	0.077	0.308
2	-1	1	0.230	0.230
3	0	0	0.346	0.000
4	1	1	0.259	0.259
5	2	4	0.078	0.312
				$\sigma^2 = 1.199$

The values in the column on the right were obtained by multiplying each squared deviation from the mean by its probability, and their sum is the variance of the distribution. To find the standard deviation, we must look up the square root of 1.199, and the nearest value in Table XII at the end of the book is $\sqrt{1.20} = 1.095$ rounded to three decimals.

As in the case of the mean, the calculation of the variance or the standard deviation of a probability distribution can generally be simplified when we deal with special kinds of probability distributions. For instance, for the binomial distribution we have the formula

Standard deviation of binomial distribution

$$\sigma = \sqrt{np(1-p)}$$

EXAMPLE   Use this formula to verify the result obtained in the preceding example.

Solution   Since we are dealing with a binomial distribution with $n = 5$ and $p = 0.60$, we get

$$\sigma = \sqrt{5(0.60)(0.40)} = \sqrt{1.20} = 1.095$$

This agrees with the results obtained above (except for the rounding error of 0.001 in the value of the variance).

EXAMPLE   Find the variance of the probability distribution of the number of heads obtained in three flips of a balanced coin.

Solution   Since we are dealing with a binomial distribution with $n = 3$ and $p = \frac{1}{2}$, we get

$$\sigma^2 = 3 \cdot \frac{1}{2} \cdot \frac{1}{2} = \frac{3}{4}$$

In Exercise 6.66 on page 226 the reader will be asked to verify this result using the formula on page 223 which defines the variance of a probability distribution.

There also exist formulas for the standard deviation of other special probability distributions, and they may be found in more advanced texts.

(Exercises 6.53, 6.58, and 6.59 are practice exercises; their complete solutions are given on page 231.)

6.53 The probabilities that a building inspector will observe 0, 1, 2, 3, 4, or 5 violations of the building code in a home built in a large development are, respectively, 0.41, 0.22, 0.17, 0.13, 0.05, and 0.02. Find the mean of this probability distribution.

6.54 Suppose that the probabilities are 0.4, 0.3, 0.2, and 0.1 that 0, 1, 2, or 3 hurricanes hit a certain coast area in any given year.
(a) Find the mean of this probability distribution.
(b) Find the variance of this probability distribution.

6.55 The following table gives the probabilities that a probation officer will receive 0, 1, 2, 3, 4, 5, or 6 reports of probation violations on any given day:

Number of violations	0	1	2	3	4	5	6
Probability	0.15	0.22	0.31	0.18	0.09	0.04	0.01

Find the mean and the standard deviation of this probability distribution.

6.56 The probabilities of 0, 1, 2, 3, 4, or 5 armed robberies in a Western city in any given month are 0.1, 0.4, 0.2, 0.1, 0.1, and 0.1. Find the mean and the variance of this probability distribution.

6.57 On page 220 we showed that $\mu = 3\frac{1}{2}$ for the number of points we roll with a balanced die. Use this result and the probabilities on page 196 to determine the standard deviation of this probability distribution.

6.58 Find the mean of the binomial distribution with $n = 4$ and $p = 0.10$
(a) by looking up the probabilities in Table I and then using the formula which defines the mean of a probability distribution;
(b) by using the special formula for the mean of the binomial distribution.

6.59 Find the standard deviation of the binomial distribution with $n = 4$ and $p = 0.10$.
(a) by using the value of μ obtained in part (b) of the preceding exercise, the probabilities looked up in part (a) of that exercise, and the formula which defines the standard deviation of a probability distribution;
(b) by using the special formula for the standard deviation of the binomial distribution.

6.60 As can easily be verified by listing all possibilities or by using the formula for the binomial distribution, the probabilities of getting 0, 1, 2, 3, 4, or 5 heads in five flips of a balanced coin are, respectively, $\frac{1}{32}, \frac{5}{32}, \frac{10}{32}, \frac{10}{32}, \frac{5}{32}$, and $\frac{1}{32}$. Find the mean of this probability distribution using
(a) the formula which defines μ;
(b) the special formula for the mean of a binomial distribution.

6.61 With reference to the preceding exercise, find the variance of the probability distribution using

    (a) the formula which defines σ^2;

    (b) the special formula for the variance of a binomial distribution.

6.62 A study shows that 80% of all patients coming to a certain medical building have to wait in their doctor's waiting room for at least 30 minutes. Find the mean of the distribution of the number of patients, among 10 coming to this medical building, who have to wait in their doctor's waiting room for at least 30 minutes

    (a) by looking up the probabilities in Table I and then using the formula which defines the mean of a probability distribution;

    (b) by using the special formula for the mean of the binomial distribution.

6.63 With reference to the preceding exercise, find the standard deviation of the number of patients, among 10 coming to this medical building, who have to wait in their doctor's waiting room for at least 30 minutes

    (a) by using the value of μ obtained in part (b) of the preceding exercise, the probabilities looked up in part (a) of that exercise, and the formula which defines the standard deviation of a probability distribution;

    (b) by using the special formula for the standard deviation of the binomial distribution.

6.64 If 95% of certain radial tires last at least 30,000 miles, find the mean and the standard deviation of the distribution of the number of these tires, among 20 selected at random, that last at least 30,000 miles, using

    (a) Table V, the formula which defines μ, and the formula which defines σ (with μ rounded to the nearest tenth);

    (b) the special formulas for the mean and the standard deviation of a binomial distribution.

6.65 Find the mean and the standard deviation of the distribution of each of the following binomial random variables:

    (a) The number of heads obtained in 900 flips of a balanced coin.

    (b) The number of 4's obtained in 405 rolls of a balanced die.

    (c) The number of persons among 756 invited, who will attend the opening of a new car dealership, if the probability is 0.30 that any one of them will attend.

6.66 Use Table I and the formula defining the variance of a probability distribution to verify the result given on page 224 for the variance of the binomial distribution with $n = 3$ and $p = \frac{1}{2}$.

6.67 Use the probabilities shown in Figure 6.3 on page 210 to calculate the mean of the given hypergeometric distribution. Verify the result by using the special formula $\mu = \dfrac{n \cdot a}{a + b}$ mentioned on page 222.

6.68 The probabilities that there will be 0, 1, 2, 3, 4, or 5 fires caused by lightning during a summer storm are, respectively, 0.449, 0.360, 0.144, 0.038, 0.008, and 0.001. Calculate the mean of this Poisson distribution with $\lambda = 0.8$, and use the result to verify the special formula $\mu = \lambda$ mentioned on page 222.

6.8
Chebyshev's Theorem

Intuitively speaking, the variance and the standard deviation of a probability distribution measure the expected size of the chance fluctuations of a corresponding random variable. When σ is small, the probability is high that we will get a value close to the mean, and when σ is large, we are more likely to get a value far away from the mean. This important idea is expressed formally by Chebyshev's theorem, which we introduced in Section 3.8, as it pertains to numerical data. For probability distributions, **Chebyshev's theorem** can be stated as follows:

Chebyshev's theorem

> *The probability that a random variable will take on a value within k standard deviations of the mean is at least $1 - 1/k^2$.*

Thus, the probability of getting a value within two standard deviations of the mean (a value between $\mu - 2\sigma$ and $\mu + 2\sigma$) is at least $1 - \dfrac{1}{2^2} = \dfrac{3}{4}$, the probability of getting a value within five standard deviations of the mean (a value between $\mu - 5\sigma$ and $\mu + 5\mu$) is at least $1 - \dfrac{1}{5^2} = \dfrac{24}{25}$, and so forth.

EXAMPLE   The number of telephone calls which an answering service receives between 9 A.M. and 10 A.M. is a random variable whose distribution has the mean $\mu = 27.5$ and the standard deviation $\sigma = 3.2$. What does Chebyshev's theorem with $k = 3$ tell us about the number of telephone calls which the answering service may receive between 9 A.M. and 10 A.M.?

Solution   Since $\mu - 3\sigma = 27.5 - 3(3.2) = 17.9$ and $\mu + 3\sigma = 27.5 + 3(3.2) = 37.1$, we can assert with a probability of at least $1 - \dfrac{1}{3^2} = \dfrac{8}{9}$, or approximately 0.89, that the answering service will receive between 17.9 and 37.1 calls (that is, anywhere from 18 to 37 calls).

227

EXAMPLE   What does Chebyshev's theorem with $k = 6$ tell us about the number of heads we may get in 400 flips of a balanced coin?

Solution   Since $n = 400$ and $p = \frac{1}{2}$, the mean and the standard deviation of the number of heads are $\mu = n \cdot p = 400 \cdot \frac{1}{2} = 200$ and $\sigma = \sqrt{np(1-p)} = \sqrt{400 \cdot \frac{1}{2} \cdot \frac{1}{2}} = 10$, so that $\mu - 6\sigma = 200 - 6(10) = 140$ and $\mu + 6\sigma = 200 + 6(10) = 260$. Thus, we can assert with a probability of at least

$$1 - \frac{1}{6^2} = \frac{35}{36},$$ or approximately 0.97, that we will get between 140 and 260 heads.

If we convert the numbers of heads into proportions in the preceding example, we can assert with a probability of at least $\frac{35}{36}$ that the proportion of heads we get in 400 flips of a balanced coin will lie between $\frac{140}{400} = 0.35$ and $\frac{260}{400} = 0.65$. To continue this argument, the reader will be asked to show in Exercise 6.73 on page 229 that the probability is at least $\frac{35}{36}$ that for 10,000 flips of a balanced coin the proportion of heads will lie between 0.47 and 0.53, and that for 1,000,000 flips of a balanced coin it will lie between 0.497 and 0.503. All this provides support for the Law of Large Numbers, which we mentioned in Section 4.5 in connection with the frequency interpretation of probability.

In actual practice, Chebyshev's theorem is very rarely used, since the probability "at least $1 - 1/k^2$" is usually unnecessarily small. For instance, in the preceding example we showed that the probability of getting a value within six standard deviations of the mean is at least 0.97, whereas the actual probability that this will happen for a random variable having the binomial distribution with $n = 400$ and $p = \frac{1}{2}$ is about 0.999999998. "At least 0.97" is not wrong, but for most practical purposes it does not tell us quite enough.

EXERCISES   (Exercise 6.69 is a practice exercise; its complete solution is given on page 231.)

6.69 The number of customers to whom a restaurant serves breakfast on a weekday morning is a random variable with $\mu = 142$ and $\sigma = 12$. According to Chebyshev's theorem, with what probability can we assert that between 94 and 190 customers will have breakfast there on a weekday morning?

6.70 A student answers the 144 questions on a true-false test by flipping a balanced coin.

(a) Use the special formulas for the mean and the standard deviation of the binomial distribution to find the values of μ and σ for the distribution of the number of correct answers he or she will get.

(b) According to Chebyshev's theorem, what can we assert with a probability of at least 0.96 about the number of correct answers he or she will get?

6.71 The number of marriage licenses issued in a certain city during the month of June is a random variable with $\mu = 146$ and $\sigma = 7.5$.

(a) What does Chebyshev's theorem with $k = 8$ tell us about the number of marriage licenses issued there during a month of June?

(b) According to Chebyshev's theorem, with what probability can we assert that between 71 and 221 marriage licenses will be issued there during a month of June?

6.72 The annual number of rainy days in a certain city is a random variable with $\mu = 126$ and $\sigma = 9$.

(a) What does Chebyshev's theorem with $k = 12$ tell us about the number of days it will rain in the given city in any one year?

(b) According to Chebyshev's theorem, with what probability can we assert that it will rain in the given city between 96 and 156 days in any one year?

6.73 Use Chebyshev's theorem to show that the probability is at least $\frac{35}{36}$ that

(a) in 10,000 flips of a balanced coin there will be between 4,700 and 5,300 heads, and hence the proportion of heads will be between 0.47 and 0.53;

(b) in 1,000,000 flips of a balanced coin there will be between 497,000 and 503,000 heads, and hence the proportion of heads will be between 0.497 and 0.503.

SOLUTIONS OF PRACTICE EXERCISES

6.1 (a) The values are all on the interval from 0 to 1, but since $0.37 + 0.35 + 0.30 = 1.02$, and not 1, the values cannot serve as the values of a probability distribution.

(b) The values are all on the interval from 0 to 1, and since their sum is 1, they can serve as the values of a probability distribution.

(c) Since one of the values is negative, they cannot serve as the values of a probability distribution.

6.5 Substituting $n = 6$, $p = 0.80$, and $x = 4$ into the formula for the binomial distribution, we get

$$f(4) = \binom{6}{4}(0.80)^4(1 - 0.80)^{6-4}$$

$$= 15(0.80)^4(0.20)^2$$

$$= 0.246$$

6.6 (a) Substituting $n = 4$, $p = 0.50$, and $x = 2$ into the formula for the binomial distribution, we get

$$f(2) = \binom{4}{2}(0.50)^2(1 - 0.50)^{4-2}$$

$$= 6(0.50)^4$$

$$= 0.375$$

(b) The value in Table I is 0.375.

6.7 (a) $0.257 + 0.356 + 0.229 = 0.842$;

(b) $0.001 + 0.008 + 0.035 = 0.044$.

6.18 (a) The probability of not getting 10 items in perfect condition is $1 - 0.599 = 0.401$.

(b) The probability of getting 10 items in perfect condition is 0.349.

(c) The probability of getting 10 items in perfect condition is 0.107.

6.23 Substituting $a = 12$, $b = 6$, $n = 5$, and $x = 2$ into the formula for the hypergeometric distribution, we get

$$f(2) = \frac{\binom{12}{2}\binom{6}{3}}{\binom{18}{5}} = \frac{66 \cdot 20}{8,568} = 0.154$$

rounded to three decimals.

6.29 (a) Since $n = 15$ exceeds $0.05(40 + 160) = 10$, the condition is not satisfied.

(b) Since $n = 22$ does not exceed $0.05(380 + 120) = 25$, the condition is satisfied.

(c) Since $n = 30$ does not exceed $0.05(400 + 240) = 32$, the condition is satisfied.

6.30 (a) Substituting $a = 6$, $b = 194$, $n = 3$, and $x = 1$ into the formula for the hypergeometric distribution, we get

$$f(1) = \frac{\binom{6}{1}\binom{194}{2}}{\binom{200}{3}} = \frac{6 \cdot 18,721}{1,313,400} = 0.086$$

(b) Substituting $n = 3$, $p = \frac{6}{200} = 0.03$, and $x = 1$ into the formula for the binomial distribution, we get

$$f(1) = \binom{3}{1}(0.03)^1(1 - 0.03)^{3-1}$$

$$= 3(0.03)(0.97)^2$$

$$= 0.085$$

6.34 (a) Since $n = 150$ is not less than 100 and $np = 150 \cdot \frac{1}{20} = 7.5$ is less than 10, the conditions are satisfied.

(b) Since $n = 950$ is not less than 100 and $np = 950 \cdot \frac{1}{100} = 9.5$ is less than 10, the conditions are satisfied.

(c) $n = 300$ is less than 100, but since $np = 300 \cdot \frac{1}{25} = 12$ is not less than 10, the conditions are not satisfied.

6.36 Substituting $np = 250(0.02) = 5$ and $x = 4$ into the formula for the Poisson distribution, we get

$$f(4) = \frac{5^4 \cdot e^{-5}}{4!} = \frac{625(0.0067)}{24} = 0.174$$

rounded to three decimals.

6.40 Substituting $\lambda = 4.4$ and, respectively, $x = 0, 1, 2,$ and 3 into the formula for the Poisson distribution, we get

$$f(0) = \frac{4.4^0 \cdot e^{-4.4}}{0!} = 0.012 \qquad f(1) = \frac{4.4^1(0.012)}{1!} = 0.053$$

$$f(2) = \frac{4.4^2(0.012)}{2!} = 0.116 \qquad f(3) = \frac{4.4^3(0.012)}{3!} = 0.170$$

and the answer is $0.012 + 0.053 + 0.116 + 0.170 = 0.351$.

6.49 Substituting $n = 12$, $x_1 = 1$, $x_2 = 3$, $x_3 = 7$, $x_4 = 1$, $p_1 = 0.10$, $p_2 = 0.20$, $p_3 = 0.50$, and $p_4 = 0.20$ into the formula for the multinomial distribution, we get

$$\frac{12!}{1!3!7!1!}(0.10)^1(0.20)^3(0.50)^7(0.20)^1 = 0.020$$

rounded to three decimals.

6.53 $\mu = 0(0.41) + 1(0.22) + 2(0.17) + 3(0.13) + 4(0.05) + 5(0.02) = 1.25$.

6.58 (a) $\mu = 0(0.656) + 1(0.292) + 2(0.049) + 3(0.004) = 0.402$;

(b) $\mu = 4(0.10) = 0.40$.

6.59 (a) $\sigma^2 = (0 - 0.4)^2(0.656) + (1 - 0.4)^2(0.292) + (2 - 0.4)^2(0.049) +$ $(3 - 0.4)^2(0.004) = 0.363$; $\sigma = \sqrt{0.363} = 0.602$;

(b) $\sigma = \sqrt{4(0.10)(0.90)} = 0.60$.

6.69 Since $142 - 12k = 94$ and $142 + 12k = 190$, it follows that $12k = 48$ and $k = 4$; thus, the probability is at least $1 - \frac{1}{4^2} = \frac{15}{16}$, or approximately 0.94.

The Normal Distribution

Continuous sample spaces and continuous random variables arise whenever we deal with quantities that are measured on a continuous scale—for instance, when we measure the amount of alcohol in a person's blood, the net weight of a package of frozen food, the amount of tar in a cigarette, the speed of a car, and so forth.

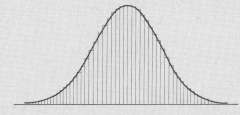

7.1

Continuous distribution curve.

In the continuous case, the place of histograms is taken by continuous curves. This may be pictured as in Figure 7.1, where we can think of histograms with narrower and narrower classes approaching the continuous curve. Among the many continuous distribution curves used in statistics, by far the most important is the

> **NORMAL CURVE** ('nȯr-məl 'kərv) A bell-shaped curve showing a distribution of probability associated with different values of a random variable.

This dictionary definition leaves a good deal to be desired; actually, a normal curve is a special kind of bell-shaped curve given by a special mathematical equation. It dates back to the work of the French–English mathematician Abraham de Moivre (1667–1745), who studied it as a curve which closely approximates the binomial distribution when the number of trials is very large.

In this chapter we shall study continuous random variables. The concept of a continuous distribution will be introduced in Section 7.1, followed by that of a normal distribution in Section 7.2. Various applications of the normal distribution will be discussed in Sections 7.3 and 7.4.

233

7.1

Continuous Distributions

In all the histograms shown in preceding chapters, the frequencies, percentages, or probabilities are represented by the heights of the rectangles, but so long as the class intervals are all equal, we can also say that the frequencies, percentages, or probabilities are represented by the areas of the rectangles. The latter is preferable for the work of this chapter, for in the continuous case we also represent probabilities by areas—not by areas of rectangles, but as is illustrated by Figure 7.2, by areas under continuous curves. The diagram on the left shows a histogram of the probability distribution of a random variable which takes on only the values 0, 1, 2, ..., and 10, and the probability of getting a 3, for example, is given by the area of the white rectangle. The diagram on the right refers to a continuous random variable which can take on any value on the interval from 0 to 10, and the probability of getting a value on the interval from 2.5 to 3.5 is given by the area of the white region under the curve. Similarly, the area of the dark region under the curve gives the probability of getting a value greater than 8.

Continuous curves such as the one shown in the right-hand diagram of Figure 7.2 are the graphs of functions which we refer to as **probability densities,** or more informally as **continuous distributions.** The first of these terms is borrowed from the language of physics, where the terms "weight" and "density"

7.2

Histogram of a probability distribution and graph of a continuous distribution.

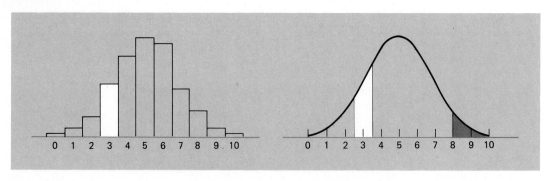

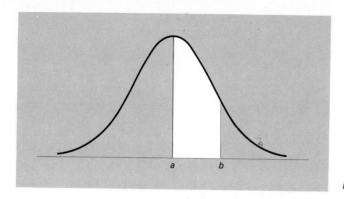

7.3
Probability density.

are used in very much the same way in which we use the terms "probability" and "probability density" in statistics. What characterizes a probability density is the fact that

The area under the curve between any two values a and b (see Figure 7.3) gives the probability that a random variable having this continuous distribution will take on a value on the interval from a to b.

It follows that the values of a probability density must not be negative, and that the total area under the curve, representing the certainty that a random variable must take on one of its values, is always equal to 1. This corresponds to the two rules about probability distributions given near the end of Section 6.1.

EXAMPLE   If a continuous random variable has the probability density shown in Figure 7.4, find the probabilities that it will take on a value
(a) between -2 and 3;
(b) greater than 1;
(c) greater than or equal to 1.

Solution   (a) Since this is the total area under the curve, it must be equal to 1; also, multiplying the base of the rectangle by its height we get $[3 - (-2)] \cdot \frac{1}{5} = 5 \cdot \frac{1}{5} = 1$.
(b) Multiplying the base of the corresponding rectangle by its height, we get $(3 - 1) \cdot \frac{1}{5} = 2 \cdot \frac{1}{5} = \frac{2}{5}$.
(c) The answer is the same as that to part (b), namely, $\frac{2}{5}$.

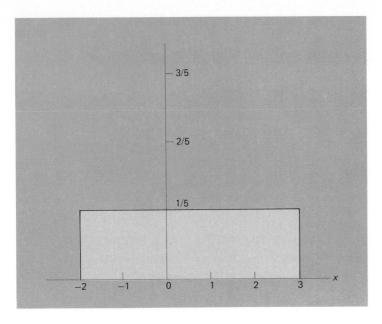

3/5

2/5

1/5

x

−2   −1   0   1   2   3

7.4

A special probability density.

Parts (b) and (c) of the preceding example illustrate the fact that the probability that a random variable having a continuous distribution will take on a particular value is always zero. In that case, the base of the rectangle has shrunk to zero, and so has the area under the curve.

Note that a zero probability does not imply that an event cannot occur. In actual practice, we also assign zero probabilities to events which are so unlikely that we are practically certain that they will not occur. For instance, we would assign a probability of zero to the event that a monkey set loose on a typewriter will by chance type Plato's *Republic* word for word without a single mistake.

Statistical descriptions of continuous distributions, such as the mean and the standard deviation, are as important as descriptions of probability distributions or distributions of observed data. Informally, we can always picture continuous distributions as being approximated by histograms of probability distributions (see Figure 7.1) for which the mean and the standard deviation can be determined by means of the formulas of Sections 6.6 and 6.7. Then, if we choose histograms with narrower and narrower classes, the means and the standard deviations of the corresponding probability distributions will approach the mean and the standard deviation of the continuous distribution.

Since formal definitions of the mean and the standard deviation of a continuous distribution cannot be given without the use of calculus, they will be omitted in this text. Again informally, the mean μ of a continuous distribution is a measure of its "center" or "middle," and the standard deviation σ of a continuous distribution is a measure of its "dispersion" or "spread."

7.2
The Normal Distribution

The normal distribution is often referred to as the cornerstone of modern statistics. This is due partly to its role in the development of statistical theory, and partly to the fact that distributions of observed data often have the same general pattern as normal distributions.

This pattern is shown in Figure 7.5, but it cannot be seen from such a small drawing that the bell-shaped curve extends indefinitely in both directions, coming closer and closer to the horizontal axis without ever reaching it. Fortunately, it is seldom necessary to extend the "tails" of a normal distribution very far because the area under the curve becomes negligible once we go more than four or five standard deviations away from the mean.

The normal distribution has a mathematical equation, but since there is nothing to be gained by giving it here, let us merely point out that this equation is completely determined if we know the values of μ and σ; in other words, there is one and only one normal distribution with a given mean μ and a given standard deviation σ. For instance, part (a) of Figure 7.6 shows two normal distributions with the same mean but different standard deviations, part (b) shows two normal distributions with the same standard deviation but different means, and part (c) shows two normal distributions with different means and different standard deviations.

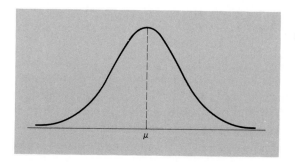

7.5
Normal distribution.

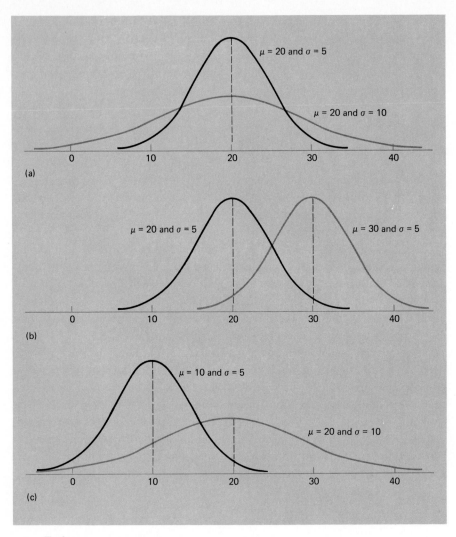

$\mu = 20$ and $\sigma = 5$

$\mu = 20$ and $\sigma = 10$

(a)

$\mu = 20$ and $\sigma = 5$

$\mu = 30$ and $\sigma = 5$

(b)

$\mu = 10$ and $\sigma = 5$

$\mu = 20$ and $\sigma = 10$

(c)

7.6

Three pairs of normal distributions.

In practice, we find areas under normal curves in special tables, such as Table II at the end of the book. As it is physically impossible, and also unnecessary, to construct separate tables of normal-curve areas for all conceivable pairs of values of μ and σ, we tabulate these areas only for the normal distribution with $\mu = 0$ and $\sigma = 1$, called the **standard normal distribution.** Then, we obtain areas under any normal curve by performing the change of scale (see

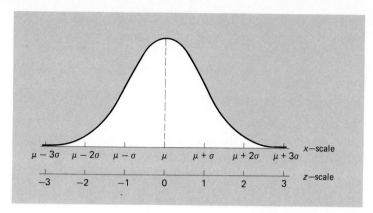

Change of scale to standard units.

Figure 7.7) which converts the units of measurement into standard units by means of the formula

Standard units

$$z = \frac{x - \mu}{\sigma}$$

As we already pointed out on page 69, a value of z simply tells us how many standard deviations the corresponding x-value lies above or below the mean.

The entries in Table II are the areas under the standard normal curve (that is, the graph of the standard normal distribution) between the mean, $z = 0$, and $z = 0.00, 0.01, 0.02, \ldots, 3.08,$ and 3.09, and also $z = 4.00, z = 5.00,$ and $z = 6.00$. In other words, the entries in Table II are areas under the standard normal curve like that of the white region in Figure 7.8.

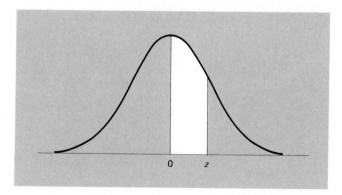

7.8

Tabulated areas under the graph of the standard normal distribution.

Table II has no entries corresponding to negative values of z, for these are not needed by virtue of the symmetry of any normal curve about its mean.

EXAMPLE   Find the area under the standard normal curve between $z = -1.20$ and $z = 0$.

Solution   As can be seen from Figure 7.9, the area under the curve between $z = -1.20$ and $z = 0$ equals the area under the curve between $z = 0$ and $z = 1.20$. So, we look up the entry for $z = 1.20$ and get 0.3849.

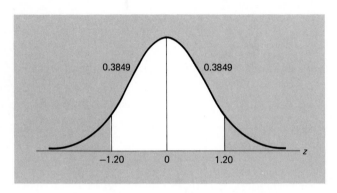

7.9

Area under normal curve.

Questions concerning areas under normal curves arise in various ways, and the ability to find any desired area quickly can be a big help. Although the table gives only areas between $z = 0$ and selected positive values of z, we often have to find areas to the left or to the right of given positive or negative values of z. This should not cause any difficulties, provided that we remember exactly what areas are represented by the entries in Table II, and also that the standard normal curve is symmetrical about $z = 0$, so that the area under the curve to the left of $z = 0$ and the area under the curve to the right of $z = 0$ are both equal to 0.5000.

EXAMPLE   Find the area under the standard normal curve which lies
(a) to the left of 0.94;
(b) to the right of $z = -0.65$;
(c) to the right of $z = 1.76$;
(d) to the left of $z = -0.85$;
(e) between $z = 0.87$ and $z = 1.28$;
(f) between $z = -0.34$ and $z = 0.62$.

Solution (a) The area to the left of $z = 0.94$ is 0.5000 plus the entry in Table II corresponding to $z = 0.94$, namely, $0.5000 + 0.3264 = 0.8264$ (see Figure 7.10).

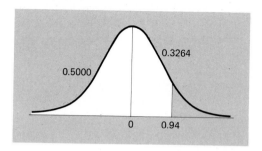

7.10

Area to the left of z = 0.94.

(b) The area to the right of $z = -0.65$ is 0.5000 plus the entry in Table II corresponding to $z = 0.65$, namely, $0.5000 + 0.2422 = 0.7422$ (see Figure 7.11).

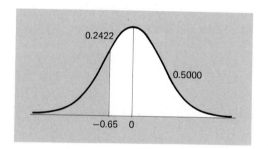

7.11

Area to the right of z = −0.65.

(c) The area to the right of $z = 1.76$ is 0.5000 minus the entry corresponding to $z = 1.76$, namely, $0.5000 - 0.4608 = 0.0392$ (see Figure 7.12).

7.12

Area to the right of z = 1.76.

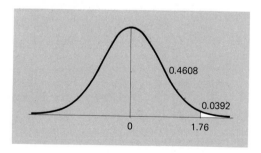

(d) The area to the left of $z = -0.85$ is 0.5000 minus the entry corresponding to $z = 0.85$, namely, $0.5000 - 0.3023 = 0.1977$ (see Figure 7.13).

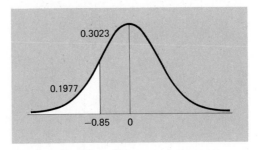

7.13

Area to the left of z = −0.85.

(e) The area between $z = 0.87$ and $z = 1.28$ is the difference between the entries corresponding to $z = 1.28$ and $z = 0.87$, namely, $0.3997 - 0.3078 = 0.0919$ (see Figure 7.14).

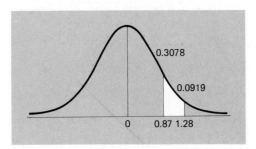

7.14

Area between z = 0.87 and z = 1.28.

(f) The area between $z = -0.34$ and $z = 0.62$ is the sum of the entries corresponding to $z = 0.34$ and $z = 0.62$, namely, $0.1331 + 0.2324 = 0.3655$ (see Figure 7.15).

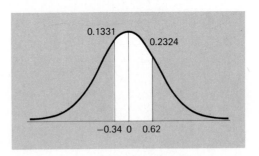

7.15

Area between z = −0.34 and z = 0.62.

In both of the preceding examples we dealt directly with the standard normal distribution. Now let us consider an example where μ and σ are not 0 and

1, so that we must first use the formula $z = \dfrac{x - \mu}{\sigma}$ to convert to standard units.

EXAMPLE   Find the probabilities that a random variable will take on a value between 12 and 15 given that it has a normal distribution with
(a) $\mu = 10$ and $\sigma = 5$;
(b) $\mu = 20$ and $\sigma = 10$.

Solution   (a) The probability is given by the area of the white region of the upper diagram of Figure 7.16. The values of z corresponding to 12 and 15 are

$$z = \frac{12 - 10}{5} = 0.40 \quad \text{and} \quad z = \frac{15 - 10}{5} = 1.00$$

7.16
Areas under normal curves.

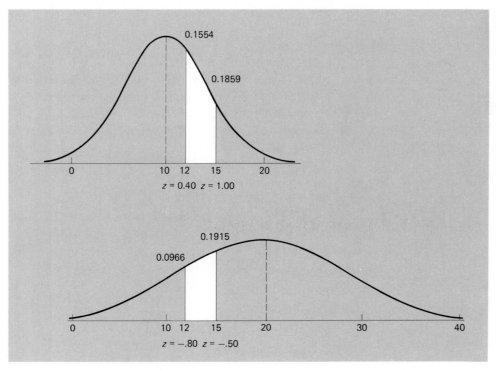

the corresponding entries in Table II are 0.1554 and 0.3413, and the probability that the random variable will take on a value between 12 and 15 is $0.3413 - 0.1554 = 0.1859$.

(b) The probability is given by the area of the white region of the lower diagram of Figure 7.16. The values of z corresponding to 12 and 15 are

$$z = \frac{12 - 20}{10} = -0.80 \quad \text{and} \quad z = \frac{15 - 20}{10} = -0.50$$

the entries in Table II corresponding to $z = 0.80$ and $z = 0.50$ are 0.2881 and 0.1915, and the probability that the random variable will take on a value between 12 and 15 is $0.2881 - 0.1915 = 0.0966$.

There are also problems in which we are given areas under normal curves and are asked to find the corresponding values of z. The results of the example which follows will be used extensively in subsequent chapters.

EXAMPLE   If z_α denotes the value of z for which the area under the standard normal curve to its right is equal to α (Greek lowercase *alpha*), find

(a) $z_{0.01}$;
(b) $z_{0.05}$.

Solution   (a) As can be seen from Figure 7.17, $z_{0.01}$ corresponds to an entry of $0.5000 - 0.0100 = 0.4900$ in Table II; since the nearest entry is 0.4910 corresponding to $z = 2.33$, we let $z_{0.01} = 2.33$.

(b) As can be seen from Figure 7.17, $z_{0.05}$ corresponds to an entry of $0.5000 - 0.0500 = 0.4500$ in Table II; since the two nearest entries

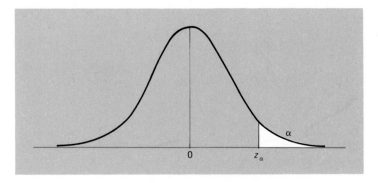

7.17

Diagram for determination of z_α.

are 0.4495 and 0.4505 corresponding to $z = 1.64$ and $z = 1.65$, we let $z_{0.05} = 1.645$.

Table II also enables us to verify the remark on page 69 that for frequency distributions having the general shape of the cross section of a bell, about 68% of the values will lie within one standard deviation of the mean, about 95% will lie within two standard deviations of the mean, and about 99.7% will lie within three standard deviations of the mean. These figures apply to frequency distributions having the general shape of normal distributions, and in Exercise 7.11 on page 247 the reader will be asked to verify these percentages with the use of Table II.

EXERCISES   (Exercises 7.1, 7.6, and 7.13 are practice exercises; their complete solutions are given on page 260.)

7.1 Suppose that a continuous random variable takes on values on the interval from 1 to 5 and that the graph of its probability density is given by the horizontal line of Figure 7.18.

(a) What probability is represented by the white region of the diagram and what is its value?

(b) What is the probability that the random variable will take on a value greater than 4.5? Would the answer be the same if we asked for the probability that the random variable will take on a value greater than or equal to 4.5?

(c) What is the probability that the random variable will take on a value between 1.8 and 4.2?

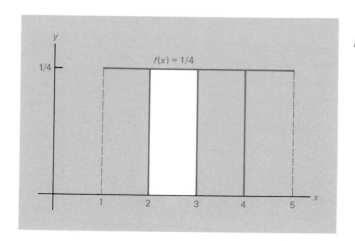

7.18
Diagram for Exercise 7.1.

7.2 Suppose that a continuous random variable takes on values on the interval from 0 to 4 and that the graph of its probability density is given by the white line of Figure 7.19.

    (a) Verify that the total area under the curve is equal to 1.

    (b) What is the probability that this random variable will take on a value less than 3?

    (c) What is the probability that this random variable wil take on a value between 1 and 2?

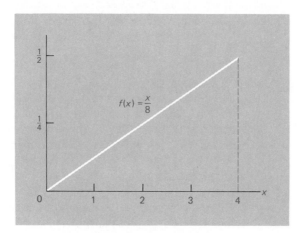

7.19

Diagram for Exercise 7.2.

7.3 Find the area under the standard normal curve which lies

    (a) between $z = -0.78$ and $z = 0$;   (c) to the left of $z = -1.55$;

    (b) to the left of $z = 2.50$;   (d) between $z = 0.33$ and $z = 0.66$.

7.4 Find the area under the standard normal curve which lies

    (a) between $z = 0$ and $z = 0.95$;

    (b) to the right of $z = -0.75$;

    (c) to the right of $z = 1.66$;

    (d) between $z = -1.12$ and $z = -1.08$.

7.5 Find the area under the standard normal curve which lies

    (a) between $z = -0.25$ and $z = 0.25$;

    (b) between $z = -1.88$ and $z = 1.09$;

    (c) between $z = 2.16$ and $z = 2.54$;

    (d) between $z = -2.05$ and $z = -1.24$.

7.6 Find z if

    (a) the normal-curve area between 0 and z is 0.3340;

    (b) the normal-curve area to the left of z is 0.6517;

    (c) the normal-curve area to the left of z is 0.3085;

    (d) the normal-curve area between $-z$ and z is 0.9700.

7.7 Find z if

    (a) the normal-curve area between 0 and z is 0.2019;

    (b) the normal-curve area to the right of z is 0.8810;

    (c) the normal-curve area to the right of z is 0.0336;

    (d) the normal-curve area between $-z$ and z is 0.2662.

7.8 Find z if

    (a) the normal-curve area between 0 and $-z$ is 0.4573;

    (b) the normal-curve area to the left of z is 0.0838;

    (c) the normal-curve area to the left of z is 0.9713;

    (d) the normal-curve area between $-z$ and z is 0.5878.

7.9 A random variable has a normal distribution with $\mu = 75.0$ and $\sigma = 4.8$. What are the probabilities that this random variable will take on a value

    (a) less than 82.2;   (c) between 75.0 and 76.2;

    (b) greater than 71.4;   (d) between 66.6 and 83.4?

7.10 A random variable has a normal distribution with $\mu = 54.2$ and $\sigma = 4.4$. What are the probabilities that this random variable will take on a value

    (a) less than 63.0;   (c) between 43.2 and 65.2;

    (b) less than 46.5;   (d) between 48.7 and 64.1?

7.11 Find the probabilities that a random variable having a normal distribution will take on a value within

    (a) one standard deviation of the mean;

    (b) two standard deviations of the mean;

    (c) three standard deviations of the mean;

    (d) four standard deviations of the mean.

7.12 According to the definition given on page 244, $z_{\alpha/2}$ denotes the value of z for which the area under the standard normal curve to its right is equal to $\alpha/2$ and, hence, the area between $-z_{\alpha/2}$ and $z_{\alpha/2}$ is equal to $1 - \alpha$ (see Figure 7.20). Verify that

    (a) $z_{0.025} = 1.96$ corresponding to $\alpha = 0.05$;

    (b) $z_{0.005} = 2.575$ corresponding to $\alpha = 0.01$.

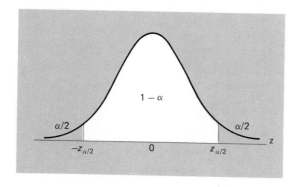

7.20
Diagram for Exercise 7.12.

7.13 A normal distribution has the mean $\mu = 61.6$. Find its standard deviation if 20% of the total area under the curve lies to the right of 70.0.

7.14 A normal distribution has the mean $\mu = 74.4$. Find its standard deviation if 10% of the area under the curve lies to the right of 100.0.

7.15 A random variable has a normal distribution with the standard deviation $\sigma = 10$. Find its mean if the probability is 0.8264 that it will take on a value less than 77.5.

7.3

Some Applications

Let us now consider some applied problems in which it will be assumed that the random variables under consideration have normal distributions.

EXAMPLE The amount of cosmic radiation to which a person is exposed while flying by jet across the United States is a random variable having a normal distribution with $\mu = 4.35$ mrem and $\sigma = 0.59$ mrem.[†] Find the probabilities that a person on such a flight will be exposed to
  (a) more than 5.00 mrem of cosmic radiation;
  (b) anywhere from 3.00 to 4.00 mrem of cosmic radiation.

Solution (a) This probability is given by the area of the white region of the upper diagram of Figure 7.21, namely, the area under the curve to the right of

$$z = \frac{5.00 - 4.35}{0.59} = 1.10$$

Since the entry in Table II corresponding to $z = 1.10$ is 0.3643, we find that the probability is $0.5000 - 0.3643 = 0.1357$, or approximately 0.14, that a person will be exposed to more than 5.00 mrem of cosmic radiation on such a flight.

  (b) This probability is given by the area of the white region of the lower diagram of Figure 7.21, namely, the area under the curve between

$$z = \frac{3.00 - 4.35}{0.59} = -2.29 \quad \text{and} \quad z = \frac{4.00 - 4.35}{0.59} = -0.59$$

[†] The unit of radiation mrem stands for "milliroentgen equivalent man."

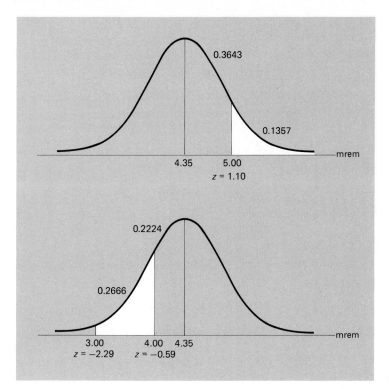

7.21

Diagrams for cosmic radiation example.

Since the entries in Table II corresponding to $z = 2.29$ and $z = 0.59$ are, respectively, 0.4890 and 0.2224, we find that the probability is $0.4890 - 0.2224 = 0.2666$, or approximately 0.27, that a person will be exposed to anywhere from 3.00 to 4.00 mrem of cosmic radiation on such a flight.

EXAMPLE   The actual amount of instant coffee which a filling machine puts into "4-ounce" jars varies from jar to jar, and it may be looked upon as a random variable having a normal distribution with $\sigma = 0.04$ ounce. If only 2% of the jars are to contain less than 4 ounces of coffee, what must be the mean fill of these jars?

Solution   We are given $\sigma = 0.04$, a normal-curve area (that of the white region of Figure 7.22), and we are asked to find μ. Since the value of z for which the entry in Table II is closest to $0.5000 - 0.0200 = 0.4800$ is 2.05, we

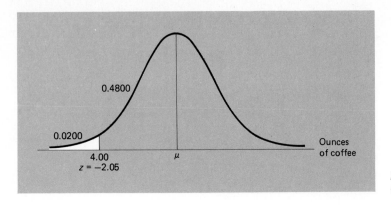

7.22

Diagram for instant-coffee-filling example.

have

$$-2.05 = \frac{4.00 - \mu}{0.04}$$

and, solving for μ, we get

$$4.00 - \mu = -2.05(0.04) = -0.082$$

and then

$$\mu = 4.00 + 0.082 = 4.082 \text{ ounces}$$

or 4.08 ounces to the nearest hundredth of an ounce.

Although the normal distribution is a continuous distribution which applies to continuous random variables, it is often used to approximate distributions of random variables which can take on only a finite number of values or as many values as there are positive integers. There are many situations where this will yield satisfactory results, provided that we make the **continuity correction** illustrated in the following example.

EXAMPLE   In a study of aggressive behavior, male white mice, returned to the group in which they live after four weeks of isolation, averaged 18.6 fights in the first five minutes with a standard deviation of 3.3 fights. If it can be assumed that the distribution of this random variable (the number of fights into which such a mouse gets under the stated conditions) can be approximated closely with a normal distribution, what is the probability that such a mouse will get into at least 15 fights in the first five minutes?

Solution The answer is given by the area of the white region of Figure 7.23; the area to the right of 14.5, not 15. The reason for this is that the number of fights in which such a mouse gets involved is a whole number. Hence, if we want to approximate the distribution of this random variable with a normal curve, we must "spread" its values over a continuous scale, and we do this by representing each whole number k by the interval from $k - \frac{1}{2}$ to $k + \frac{1}{2}$. For instance, 5 is represented by the interval from 4.5 to 5.5, 10 is represented by the interval from 9.5 to 10.5, 20 is represented by the interval from 19.5 to 20.5, and the probability of 15 or more is given by the area under the curve to the right of 14.5. Accordingly, we get

$$z = \frac{14.5 - 18.6}{3.3} = -1.24$$

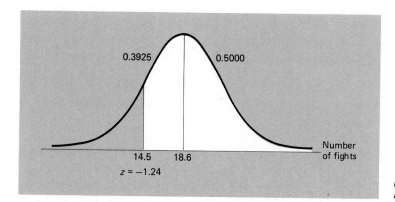

7.23

Diagram for example dealing with aggressive behavior of mice.

and it follows from Table II that the area of the white region of Figure 7.23—the probability that such a mouse will get into at least 15 fights in the first five minutes—is $0.5000 + 0.3925 = 0.8925$, or approximately 0.89.

All the examples of this section dealt with random variables having normal distributions, or distributions which can be approximated closely with normal

251

curves. Whenever we observe a value of a random variable having a normal distribution, we say that we are **sampling a normal population;** this is consistent with the terminology introduced at the end of Section 6.2.

EXERCISES

(Exercises 7.16 and 7.23 are practice exercises; their complete solutions are given on page 260.)

7.16 In an experiment to determine the amount of time required to assemble an "easy to assemble" toy, the assembly time was found to be a random variable having approximately a normal distribution with $\mu = 27.8$ minutes and $\sigma = 4.0$ minutes. What are the probabilities that this kind of toy can be assembled in
(a) less than 25.0 minutes;
(b) anywhere from 26.0 to 29.6 minutes?

7.17 A salesman who frequently drives from Boston to New York finds that his driving time is a random variable having roughly a normal distribution with $\mu = 4.3$ hours and $\sigma = 0.2$ hour. Find the probabilities that such a trip will take
(a) more than 4.5 hours;
(b) less than 4.0 hours.

7.18 With reference to the preceding exercise, below what value (number of hours) are the fastest 25% of his trips?

7.19 Suppose that during periods of transcendental meditation the reduction of a person's oxygen consumption may be looked upon as a random variable having a normal distribution with $\mu = 38.6$ cubic centimeters (cc) per minute and $\sigma = 4.3$ cc per minute. Find the probabilities that during a period of transcendental meditation a person's oxygen consumption will be reduced by
(a) at least 40.0 cc per minute;
(b) anywhere from 35.0 to 45.0 cc per minute.

7.20 The lengths of the sardines received by a cannery have a mean of 4.64 inches and a standard deviation of 0.25 inch. If the distribution of these lengths can be approximated closely with a normal distribution, what percentage of all these sardines are
(a) shorter than 4.00 inches;
(b) from 4.40 to 4.80 inches long?

7.21 With reference to the preceding exercise, above which length lies the longest 10% of the sardines?

7.22 With reference to the filling-machine example on page 249, verify that about 97.7% of the jars will contain at least 4 ounces of coffee if the machine is modified so that $\mu = 4.02$ ounces and $\sigma = 0.01$ ounce.

7.23 In a very large class in world history, the final examination grades have a mean of 66.5 and a standard deviation of 12.6. Assuming that it is reasonable to approximate the distribution of these grades with a normal distribution, what percentage of the grades should exceed 74?

7.24 An airline knows from experience that the number of suitcases that get lost each week on a certain route is a random variable having approximately a normal distribution with the mean $\mu = 21.4$ and the standard deviation $\sigma = 4.5$. What are the probabilities that in any given week they will lose
(a) exactly 20 suitcases;   (b) at most 20 suitcases?

7.25 With reference to the example on page 250, what is the probability that such a mouse will get into exactly 15 fights in the first five minutes?

7.26 The annual number of tornadoes in a certain state is a random variable with $\mu = 28.2$ and $\sigma = 5.1$. Approximating the distribution of this random variable with a normal distribution, find the probability that there will be at least 24 tornadoes in the given state in any given year.

7.27 A cab driver knows from experience that the number of fares he will pick up in an evening is a random variable with $\mu = 21.3$ and $\sigma = 3.4$. Assuming that the distribution of this random variable can be approximated closely with a normal curve, find the probabilities that in an evening the driver will pick up
(a) at least 30 fares;
(b) anywhere from 20 to 25 fares.

7.4

The Normal Approximation to the Binomial Distribution

The normal distribution is a continuous distribution which provides a close approximation to the binomial distribution when n, the number of trials, is large and p, the probability of a success on an individual trial, is close to $\frac{1}{2}$. Figure 7.24 shows the histograms of binomial distributions with $p = \frac{1}{2}$ and $n = 2, 5, 10$, and 25, and it can be seen that with increasing n these distributions approach the symmetrical bell-shaped pattern of a normal curve. In fact, a normal distribution with $\mu = np$ and $\sigma = \sqrt{np(1 - p)}$ can often be used to approximate a binomial distribution even when n is fairly small and p differs from $\frac{1}{2}$. A good rule of thumb is to use this approximation only when np and $n(1 - p)$ are both greater than five;[†] symbolically, when

Conditions for normal approximation to binomial distribution

$$np > 5 \quad and \quad n(1 - p) > 5$$

[†] Note that there are values of n and p (for instance, $n = 100$ and $p = 0.08$) for which either the normal approximation to the binomial distribution or the Poisson approximation (see page 213) can be used. This matter is discussed further on page 258.

253

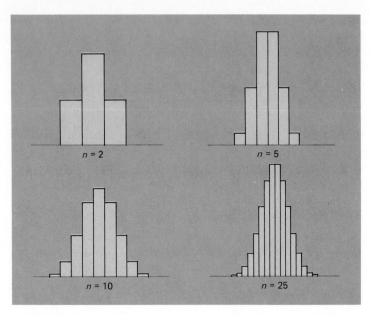

EXAMPLE   Check in each case whether the conditions for the normal approximation to the binomial distribution are satisfied.

  (a) $n = 40$ and $p = \frac{1}{4}$;
  (b) $n = 100$ and $p = \frac{1}{25}$;
  (c) $n = 150$ and $p = 0.98$.

Solution   (a) Since $np = 40 \cdot \frac{1}{4} = 10$ and $n(1 - p) = 40 \cdot \frac{3}{4} = 30$ both exceed 5,
the conditions are satisfied.

  (b) Since $np = 100 \cdot \frac{1}{25} = 4$ does not exceed 5, the conditions are not
satisfied.

  (c) Since $n(1 - p) = 150(0.02) = 3$ does not exceed 5, the conditions are
not satisfied.

EXAMPLE   Find the exact probability of getting 6 heads and 10 tails in 16 flips
of a balanced coin, and also use a normal curve to approximate this
binomial probability.

Solution   Substituting $n = 16$, $p = \frac{1}{2}$, and $x = 6$ into the formula for the binomial distribution, we find that the probability of getting six heads and

ten tails in 16 flips of a balanced coin is

$$f(6) = \binom{16}{6}\left(\frac{1}{2}\right)^6\left(1 - \frac{1}{2}\right)^{16-6}$$

$$= 8{,}008\left(\frac{1}{2}\right)^{16}$$

$$= \frac{8{,}008}{65{,}536}$$

or 0.1222 rounded to four decimals. To find the normal-curve approximation to this probability, we use the continuity correction and represent 6 heads by the interval from 5.5 to 6.5 (see Figure 7.25). Since $\mu = 16 \cdot \frac{1}{2} = 8$ and $\sigma = \sqrt{16 \cdot \frac{1}{2} \cdot \frac{1}{2}} = 2$, converting 5.5 and 6.5 into standard units yields

$$z = \frac{5.5 - 8}{2} = -1.25 \quad \text{and} \quad z = \frac{6.5 - 8}{2} = -0.75$$

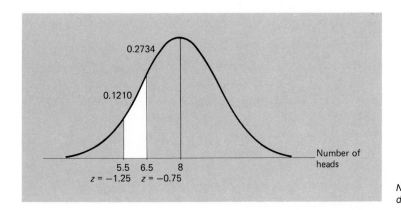

7.25

Normal-curve approximation to binomial distribution.

The entries in Table II corresponding to $z = 1.25$ and $z = 0.75$ are 0.3944 and 0.2734, and we get $0.3944 - 0.2734 = 0.1210$ for the normal-curve approximation to the probability of getting 6 heads and 10 tails in 16 flips of a balanced coin. This differs by only 0.0012 from the value obtained with the formula for the binomial distribution.

The normal-curve approximation to the binomial distribution is particularly useful in problems where we would otherwise have to use the formula for the binomial distribution repeatedly to obtain the values of many different terms.

EXAMPLE   What is the probability that at least 26 of 50 mosquitos will be killed by a new insect spray when the probability is 0.60 that any one of them will be killed by the spray?

Solution   If we tried to find the answer by using the formula for the binomial distribution, we would have to find the sum of the probabilities corresponding to 26, 27, 28, . . . , 49, and 50 successes in 100 trials. Without a computer this would obviously involve a great deal of work, but by using the normal-curve approximation we need only find the area of the white region of Figure 7.26, namely, that to the right of 25.5. Here again, we are using the continuity correction since we are dealing with whole numbers. Accordingly, 26 is represented by the interval from 25.5 to 26.5, 27 is represented by the interval from 26.5 to 27.5, and so on.

Since $\mu = 50(0.60) = 30$ and $\sigma = \sqrt{50(0.60)(0.40)} = 3.464$, we find that in standard units 25.5 becomes

$$z = \frac{25.5 - 30}{3.464} = -1.30$$

The entry in Table II corresponding to $z = 1.30$ is 0.4032, and the probability of at least 26 successes in 50 trials when $p = 0.60$ is $0.5000 + 0.4032 = 0.9032$.

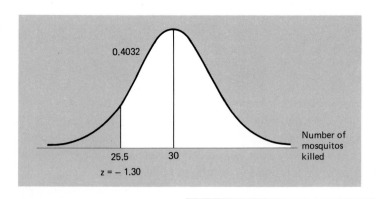

7.26

Normal-curve approximation to binomial distribution.

To determine the error of the approximation made in this example, let us refer to the computer printout of Figure 7.27, which shows the binomial probabilities for $n = 50$ and $p = 0.60$ as well as the cumulative "or less" probabilities. Since "at least 26 successes" is 1 minus the probability of "25 or less," we find that the desired probability is $1 - 0.0978 = 0.9022$, where 0.0978 is the

7.27

Computer printout of the binomial distribution with $n = 50$ and $p = 0.60$.

```
MTB > BINOMIAL N=50 P=.6

   BINOMIAL PROBABILITIES FOR N =   50   AND P =   .600000

        K            P( X = K)         P(X LESS OR = K)
        0             .0000               .0000
        1             .0000               .0000
        2             .0000               .0000
        3             .0000               .0000
        4             .0000               .0000
        5             .0000               .0000
        6             .0000               .0000
        7             .0000               .0000
        8             .0000               .0000
        9             .0000               .0000
       10             .0000               .0000
       11             .0000               .0000
       12             .0000               .0000
       13             .0000               .0000
       14             .0000               .0000
       15             .0000               .0000
       16             .0000               .0001
       17             .0001               .0002
       18             .0003               .0005
       19             .0009               .0014
       20             .0020               .0034
       21             .0043               .0076
       22             .0084               .0160
       23             .0154               .0314
       24             .0259               .0573
       25             .0405               .0978
       26             .0584               .1562
       27             .0778               .2340
       28             .0959               .3299
       29             .1091               .4390
       30             .1146               .5535
       31             .1109               .6644
       32             .0987               .7631
       33             .0808               .8439
       34             .0606               .9045
       35             .0415               .9460
       36             .0260               .9720
       37             .0147               .9867
       38             .0076               .9943
       39             .0035               .9978
       40             .0014               .9992
       41             .0005               .9998
       42             .0002               .9999
```

SEC. 7.4: The Normal Approximation to the Binomial Distribution

value in the right-hand column of the printout corresponding to $K = 25$. Thus, the error of the normal approximation is only $0.9032 - 0.9022 = 0.0010$.

With the method of this section, we now have four ways of determining binomial probabilities:

1. **We refer to a table of binomial probabilities or a computer printout of binomial probabilities.**
2. **We use the formula for the binomial distribution.**
3. **We use the Poisson approximation to the binomial distribution.**
4. **We use the normal approximation to the binomial distribution.**

In actual practice, the first alternative is by far most preferable, even if it requires that we refer to a table other than the one given at the end of this book. In contrast, the formula for the binomial distribution is used only when absolutely necessary (say, when $p = \frac{1}{3}$ and we do not want to round this probability to 0.33).

So far as the approximations are concerned, they are not only subject to the rules of thumb stated on pages 213 and 233, but their use requires a good deal of professional judgment. For instance, if either approximation can be used, we may be inclined to use the Poisson approximation to determine a probability associated with a single value of a binomial random variable, and we may be inclined to use the normal approximation to determine a probability associated with the "tail" (and, hence, many values) of a binomial distribution.

In any case, we have given the Poisson approximation to the binomial distribution mainly to *introduce* the Poisson distribution, and the normal approximation because it will be needed in Chapter 10 for large-sample inferences concerning proportions.

EXERCISES

(Exercise 7.28 is a practice exercise; its complete solution is given on page 261.)

7.28 If 20% of the loan applications received by a bank are refused, what is the probability that among 225 loan applications at least 50 will be refused?

7.29 Check in each case whether the conditions for the normal approximation to the binomial distribution are satisfied.
   (a) $n = 12$ and $p = \frac{1}{3}$;
   (b) $n = 55$ and $p = 0.10$;
   (c) $n = 120$ and $p = 0.97$.

7.30 Use the normal-curve approximation to find the probability of getting 5 heads and 7 tails in 12 flips of a balanced coin, and compare the result with the corresponding value given in Table I.

7.31 If 62% of all clouds seeded with silver iodide show spectacular growth, what is the probability that among 24 clouds thus seeded exactly 15 will show spectacular growth? If the value given for this probability in a table of binomial probabilities is 0.1661, what is the error of the approximation?

7.32 A dating service finds that 15% of the couples that it matches eventually get married. In the next 50 matches that the service makes, find the probabilities that
    (a) at least 6 couples marry;
    (b) at most 10 couples marry.

7.33 If 75% of all persons flying across the Atlantic Ocean feel the effect of the time difference for at least 24 hours, what is the probability that among 50 persons flying across the Atlantic Ocean at least 35 will feel the effect of the time difference for at least 24 hours?

7.34 A student answers each of the 48 questions on a multiple-choice test, each with four possible answers, by randomly drawing a card from an ordinary deck of 52 playing cards and checking the first, second, third, or fourth answer depending on whether the card drawn is a spade, heart, diamond, or club. Use the normal-curve approximation to find the probabilities that the student will get
    (a) exactly 15 correct answers;
    (b) at least 15 correct answers.

7.35 What is the probability of getting fewer than 70 responses to 1,000 invitations sent out to promote a new restaurant if the probability that any one person will respond is 0.08?

7.36 To avoid accusations of sexism or worse, the author of a mathematics text decides by the flip of a balanced coin whether to use "he" or "she" whenever the occasion arises in exercises and examples. If he runs into this problem 80 times while revising one of his books, what is the probability that he will use "she" at least 48 times?

7.37 To illustrate the Law of Large Numbers which we mentioned in connection with the frequency interpretation of probability and also on page 228, find the probabilities that the proportion of heads will be anywhere from 0.49 to 0.51 when a balanced coin is flipped
    (a) 100 times;   (b) 1,000 times;   (c) 10,000 times.

 7.38 Use a computer printout of the binomial distribution with $n = 50$ and $p = 0.15$ to determine the errors of the approximations of both parts of Exercise 7.32.

7.39 Use a computer printout of the binomial distribution with $n = 48$ and $p = 0.25$ to determine the errors of the approximations of both parts of Exercise 7.34.

7.40 Use a computer printout of the binomial distribution with $n = 40$ and $p = 0.30$ to determine the errors we would make by using the normal-curve approximation to find the probabilities that a random variable having this binomial distribution will take on
  (a) a value greater than or equal to 15;
  (b) a value less than or equal to 13;
  (c) the value 9.

SOLUTIONS OF PRACTICE EXERCISES

7.1 (a) The white region represents the probability of getting a value on the interval from 2 to 3, and its area is $\frac{1}{4} \cdot 1 = 0.25$.
  (b) The probability is $\frac{1}{4} \cdot (5 - 4.5) = 0.125$; the answer would be the same, since the probability is zero that the value of the random variable will actually equal 4.5.
  (c) The probability is $\frac{1}{4} \cdot (4.2 - 1.8) = 0.60$.

7.6 (a) Since 0.3340 is the entry in Table II corresponding to 0.97, the answer is $z = 0.97$.
  (b) Since $0.6517 - 0.5000 = 0.1517$ is the entry in Table II corresponding to 0.39, the answer is $z = 0.39$.
  (c) Since $0.5000 - 0.3085 = 0.1915$ is the entry in Table II corresponding to 0.50, the answer is $z = -0.50$.
  (d) Since $\dfrac{0.9700}{2} = 0.4850$ is the entry in Table II corresponding to 2.17, the answer is $z = 2.17$ or $z = -2.17$.

7.13 Since the entry in Table II nearest to $0.5000 - 0.2000 = 0.3000$ is 0.2995 corresponding to 0.84, we get

$$z = \frac{70.0 - 61.6}{\sigma} = 0.84$$

and, hence, $0.84\sigma = 70.0 - 61.6 = 8.4$ and $\sigma = \dfrac{8.4}{0.84} = 10$.

7.16 (a) $z = \dfrac{25.0 - 27.8}{4.0} = -0.70$, and since the entry in Table II corresponding to 0.70 is 0.2580, the probability is $0.5000 - 0.2580 = 0.2420$.
  (b) $z = \dfrac{29.6 - 27.8}{4.0} = 0.45$ and $z = \dfrac{26.0 - 27.8}{4.0} = -0.45$, and since the entry in Table II corresponding to 0.45 is 0.1736, the probability is $0.1736 + 0.1736 = 0.3472$.

7.23 Using the continuity correction, we must find the area under the curve to the right of 74.5; since $z = \dfrac{74.5 - 66.5}{12.6} = 0.63$ and the entry in Table II corresponding to 0.63 is 0.2357, it follows that the answer is $0.5000 - 0.2357 = 0.2643$, or 26.43%.

7.28 Using the continuity correction, we must find the area under the curve to the right of 49.5; since $\mu = 225(0.20) = 45$ and $\sigma = \sqrt{225(0.20)(0.80)} = 6$, so that $z = \dfrac{49.5 - 45}{6} = 0.75$, it follows that the entry in Table II corresponding to 0.75 is 0.2734, and that the required probability is $0.5000 - 0.2734 = 0.2266$.

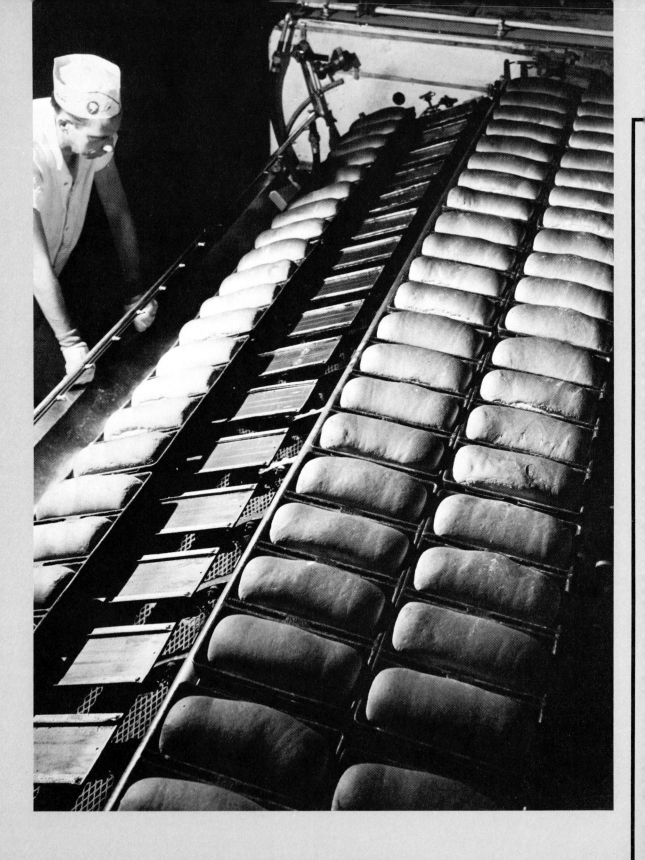

Sampling and Sampling Distributions

The main objective of most statistical studies is to make sound generalizations based on samples about the parameters of the populations from which they came. Note the word "sound," because the question of when and under what conditions samples permit such generalizations is not easily answered. For instance, if we want to estimate the average amount of money that persons spend on their vacations, would we take as our sample the amounts spent by deluxe-class passengers on an around-the-world cruise? Or would we attempt to predict changes in the retail prices of farm products on the basis of changes in the price of peaches alone? Obviously not, but just which vacationers and which farm products we should include in our samples, and how many of them, is neither intuitively clear nor self-evident.

In most of the methods we shall study in this book, it will be assumed that we are

> **RANDOM SAMPLING** (ran-dəm sam-pliŋ) A method of selecting a sample in such a way that every possible sample has the same probability of being selected.

We pay this attention to random sampling because it permits valid, or logical generalizations, and hence is widely used in practice.

In this chapter we begin with a formal definition of random sampling in Section 8.1. Then, in Section 8.2 we introduce the related concept of a sampling distribution, which tells us how quantities determined from samples may vary from sample to sample, and in Sections 8.3 and 8.4 we learn how such variations can be measured.

263

8.1
Random Sampling

In Section 3.1 we distinguished between populations and samples, saying that a population consists of all conceivably or hypothetically possible observations (instances, or occurrences) of a given phenomenon, while a sample is simply part of a population. For the work which follows, let us also distinguish between **finite populations** and **infinite populations.**

A population is finite if it consists of a finite, or fixed number of elements (items, objects, measurements, or observations). Examples of finite populations are the one which consists of the net weights of 5,000 cans of baked beans in a production lot, the one which consists of the SAT scores of all the freshman admitted to a certain university in 1984, and the one which consists of the daily low temperatures recorded at a weather station during the years 1980–1985.

In contrast, a population is infinite if there is no limit to the number of items, or values, that can be observed. This is the case, for example, when we observe repeated flips of a coin, when we sample with replacement from a finite population, or when we observe values of a continuous random variable, say, when we sample a normal population.

It should be apparent from this distinction between finite and infinite populations that the definition of "random sampling" on page 263 applies only to finite populations—if there are infinitely many possibilities, we cannot very well speak of "every possible sample." Indeed, let us state formally that

> A sample of size n from a finite population of size N is random if it is chosen in such a way that each of the $\binom{N}{n}$ possible samples has the same probability, $\dfrac{1}{\binom{N}{n}}$, of being selected.

For instance, if a finite population consists of the $N = 5$ elements a, b, c, d, and e (which might be the incomes of five persons, the weights of five guinea pigs, or the prices of five commodities), there are $\binom{5}{3} = 10$ possible samples of size $n = 3$ consisting, respectively, of the elements abc, abd, abe, acd, ace, ade, bcd, bce, bde, and cde. If we choose one of these samples in such a way that each sample has the probability $\frac{1}{10}$ of being selected, we call this sample a random sample.

This leaves the question of how random samples can be drawn in actual practice. In a simple case like the one described in the preceding paragraph, we could write each of the ten possible samples on a slip of paper, put the slips of paper into a hat, shuffle them thoroughly, and then draw one without looking. Obviously, though, this would be impractical in a more realistically complex situation where N, and hence $\binom{N}{n}$, is quite large. In fact, we have mentioned it here only to stress the point that the selection of a random sample must depend entirely on chance.

Fortunately, we can take random samples from finite populations without having to go through the tedious process of listing all possible samples. One possibility is to write each of the N elements on a slip of paper, and then take a random sample by choosing the elements to be included in the sample one at a time, making sure that in each of the successive drawings the remaining elements all have the same chance of being selected. It is not very difficult to show mathematically that this also leads to the same probability of $\dfrac{1}{\binom{N}{n}}$ for each possible sample. For instance, to take a random sample of size $n = 12$ from the population which consists of the sales tax collected by a city's 247 drugstores in December, 1984, we could write each of the 247 figures on a slip of paper, mix them up thoroughly in a bag, a box, or a hat, and then draw (without looking) twelve of the slips of paper one after the other without replacement.

Even this relatively easy procedure can be simplified further. Usually, the simplest way of taking a random sample from a finite population is to refer to a table of **random numbers,** which consists of pages on which the digits 0, 1, 2, 3, 4, 5, 6, 7, 8, and 9 are set down in much the same fashion as they might appear if they had been generated by a gambling device giving each digit the same probability of $\frac{1}{10}$ of appearing at any given place in the table. Table XIII at the end of the book is excerpted from such a table.

EXAMPLE   With reference to the sales tax example, number the 247 drugstores 001, 002, 003, . . . , 246, and 247. Then refer to the table on page 266 and read off the digits in the 26th, 27th, and 28th columns, starting with the sixth row and going down the page, to obtain a random sample of twelve of the drugstores and, hence, a random sample of size twelve of the sales tax figures for December, 1984.

Solution   Ignoring numbers greater than 247, it can be seen from Figure 8.1 that we get

046   230   079   022   119   150   056   064   193   232   040   146

48611	62866	33963	14045	79451	04934	45576
78812	03509	78673	73181	29973	18664	04555
19472	63971	37271	31445	49019	49405	46925
51266	11569	08697	91120	64156	40365	74297
55806	96275	26130	47949	14877	69594	83041
77527	81360	18180	97421	55541	90275	18213
77680	58788	33016	61173	93049	04694	43534
15404	96554	88265	34537	38526	67924	40474
14045	22917	60718	66487	46346	30949	03173
68376	43918	77653	04127	69930	43283	35766
93385	13421	67957	20384	58731	53396	59723
09858	52104	32014	53115	03727	98624	84616
93307	34116	49516	42148	57740	31198	70336
04794	01534	92058	03157	91758	80611	45357
86265	49096	97021	92582	61422	75890	86442
65943	79232	45702	67055	39024	57383	44424
90038	94209	04055	27393	61517	23002	96560
97283	95943	78363	36498	40662	94188	18202
21913	72958	75637	99936	58715	07943	23748
41161	37341	81838	19389	80336	46346	91895
23777	98392	31417	98547	92058	02277	50315
59973	08144	61070	73094	27059	69181	55623
82690	74099	77885	23813	10054	11900	44653
83854	24715	48866	65745	31131	47636	45137
61980	34997	41825	11623	07320	15003	56774
99915	45821	97702	87125	44488	77613	56823
48293	86847	43186	42951	37804	85129	28993
33225	31280	41232	34750	91097	60752	69783
06846	32828	24425	30249	78801	26977	92074
32671	45587	79620	84831	38156	74211	82752
82096	21913	75544	55228	89796	05694	91552
51666	10433	10945	55306	78562	89630	41230
54044	67942	24145	42294	27427	84875	37022
66738	60184	75679	38120	17640	36242	99357
55064	17427	89180	74018	44865	53197	74810
69599	60264	84549	78007	88450	06488	72274
64756	87759	92354	78694	63638	80939	98644
80817	74533	68407	55862	32476	19326	95558
39847	96884	84657	33697	39578	90197	80532
90401	41700	95510	61166	33757	23279	85523
78227	90110	81378	96659	37008	04050	04228
87240	52716	87697	79433	16336	52862	69149
08486	10951	26832	39763	02485	71688	90936
39338	32169	03713	93510	61244	73774	01245
21188	01850	69689	49426	49128	14660	14143
13287	82531	04388	64693	11934	35051	68576
53609	04001	19648	14053	49623	10840	31915
87900	36194	31567	53506	34304	39910	79630
81641	00496	36058	75899	46620	70024	88753
19512	50277	71508	20116	79520	06269	74173

8.1

Part of table of random numbers.

These are the numbers of the drugstores, and the corresponding sales tax figures constitute the desired random sample. Note that if any number had recurred when reading the values off the table, it would also have been ignored.

Some early tables of random numbers were copied from pages of census data and from tables of 20-place logarithms, but nowadays they are prepared with the use of computers. Indeed, it is fairly easy to program a computer so that a person can generate his or her own random numbers.

When lists are available and items are readily numbered, it is easy to draw samples from finite populations with the aid of published or computer-generated random numbers. Unfortunately, though, it is often impossible to proceed as in the example immediately above. For instance, if we want to use a sample to estimate the mean diameter of thousands of ball bearings packed in a crate, or if we want to estimate the mean height of the trees in a forest, it would be impossible to number the ball bearings or the trees, choose random numbers, and then locate and measure the corresponding ball bearings or trees. In these and in many similar situations, we may have no choice but to proceed according to the alternative dictionary definition of the word "random," which says that the selection should be "haphazard, without aim or purpose." With some reservations, such samples can often be treated as if they were, in fact, "real" random samples.

So far we have discussed random sampling only in connection with finite populations.

> **For infinite populations, we say that a sample is random if it consists of values of independent random variables having the same distribution.**

By "independent" we mean that the distribution of any one of the random variables is in no way affected by the values taken on by the others. For example, this definition applies when we observe "honest" flips of a balanced coin. If we get, say, HTHTTTHHTHHHHT, where H and T denote heads and tails, this result is a random sample so long as the H's and T's are values of independent random variables, each having the binomial distribution with $n = 1$ and $p = \frac{1}{2}$. As we already pointed out on page 264, the population we are sampling here is infinite because there is no limit to the number of times we can flip the coin.

For another example of random sampling from an infinite population, suppose that the weight loss of persons on a certain two-week diet is a random variable having the normal distribution with $\mu = 7.4$ pounds and $\sigma = 1.3$ pounds. If we get, say, 8.3, 5.9, 7.0, 10.5, and 6.8 pounds for the weight loss of five persons who have been on the diet, these figures constitute a random sample so long as they are values of independent random variables, each having the normal distribution with $\mu = 7.4$ pounds and $\sigma = 1.3$ pounds. To judge whether this is actually the case, we would have to see how the persons were chosen; namely, how the data were collected.

EXERCISES (Exercises 8.1 and 8.2 are practice exercises; their complete solutions are given on page 284.)

8.1 With reference to the example on page 264, where a random sample of size $n = 3$ is drawn from the finite population which consists of the elements a, b, c, d, and e, what is the probability that any specific element, say, the element b, will be contained in the sample?

8.2 A newspaper reporter wants to interview a random sample of 15 of the 6,285 persons who signed a petition for the recall of a local government official. If these persons are numbered 0001, 0002, 0003, ..., 6284, and 6285, which ones (by number) will the reporter select for an interview if she selects the sample using the first four columns of the table on page 529, going down the page beginning with the tenth row?

8.3 How many different samples of size $n = 2$ can be drawn from a finite population of size

(a) $N = 8$; (b) $N = 12$; (c) $N = 20$?

8.4 How many different samples of size $n = 3$ can be drawn from a finite population of size

(a) $N = 6$; (b) $N = 10$; (c) $N = 25$?

8.5 What is the probability of each possible sample if a random sample of size 4 is to be drawn from a finite population of size 18?

8.6 What is the probability of each possible sample if a random sample of size 5 is to be drawn from a finite population of size 25?

8.7 List the $\binom{6}{2} = 15$ possible samples of size $n = 2$ that can be drawn from the finite population whose elements are denoted a, b, c, d, e, and f.

8.8 With reference to the preceding exercise, what is the probability that a random sample of size $n = 2$ from the given finite population will include the element denoted by the letter f?

8.9 List all possible choices of four of the following six airlines: TWA, American, Western, Continental, PSA, and Delta. If a person randomly selects four of these airlines to study their safety records, find
   (a) the probability of each possible sample;
   (b) the probability that Western will be included in the sample.

8.10 A sociologist wants to include ten of the 83 counties in Michigan in a survey. If he numbers these counties 01, 02, ..., 82, and 83, which ones (by number) will he include in the survey if he selects them by using the 21st and 22nd columns of the table on page 531, going down the page starting with the sixth row?

8.11 A bacteriologist wants to doublecheck a sample of $n = 8$ of the 754 blood specimens analyzed by a medical laboratory in a given month. If he numbers the specimens 001, 002, . . . , 753, and 754, which ones (by number) will he select if he chooses them by using the sixth, seventh, and eighth columns of the table on page 532, going down the page starting with the sixteenth row?

8.12 A county assessor wants to reassess a random sample of 20 of 8,312 one-family homes. If he numbers them 0001, 0002, . . . , 8311, and 8312, which ones (by number) will he select if he chooses them by using the 11th, 12th, 13th, and 14th columns of the table on page 531, going down the page starting with the fourth row?

8.13 The employees of a company have badges numbered serially from 1 through 544. Use the 26th, 27th, and 28th columns of the table on page 532, going down the page starting at the top, to select a random sample of twelve of the employees to serve on a grievance committee.

8.2
Sampling Distributions

The sample mean, the sample median, and the sample standard deviation are examples of random variables, whose values will vary from sample to sample. Their distributions, which reflect such chance variations, play an important role in statistics and they are referred to as **sampling distributions.**

To illustrate the concept of a sampling distribution, let us construct the one for the mean of a random sample of size $n = 2$ from the finite population of size $N = 5$, whose elements are the numbers 1, 3, 5, 7, and 9. The mean and the standard deviation of this population, which we shall need later, are

$$\mu = \frac{1 + 3 + 5 + 7 + 9}{5} = 5$$

and

$$\sigma = \sqrt{\frac{(1 - 5)^2 + (3 - 5)^2 + (5 - 5)^2 + (7 - 5)^2 + (9 - 5)^2}{5}}$$
$$= \sqrt{8}$$

in accordance with the formulas on pages 43 and 62.

Now, if we take a random sample of size $n = 2$ from this population, there are $\binom{5}{2} = 10$ possibilities, and they are

1 and 3, 1 and 5, 1 and 7, 1 and 9, 3 and 5,

3 and 7, 3 and 9, 5 and 7, 5 and 9, 7 and 9

The means of these samples are $\dfrac{1 + 3}{2} = 2, \dfrac{1 + 5}{2} = 3, 4, 5, 4, 5, 6, 6, 7,$ and $8,$ and if sampling is random so that each sample has the probability $\frac{1}{10}$, we arrive at the following sampling distribution of the mean:

$\bar{x}$	Probability
2	$\frac{1}{10}$
3	$\frac{1}{10}$
4	$\frac{2}{10}$
5	$\frac{2}{10}$
6	$\frac{2}{10}$
7	$\frac{1}{10}$
8	$\frac{1}{10}$

A histogram of this probability distribution is shown in Figure 8.2.

An examination of this sampling distribution reveals some pertinent information about the chance variations of the mean of a random sample of size $n = 2$ from the given population. For instance, we find that the probability is

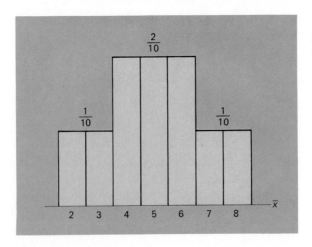

8.2

Sampling distribution of the mean.

$\frac{6}{10}$ that a sample mean will differ from the population mean $\mu = 5$ by 1 or less, and that the probability is $\frac{8}{10}$ that a sample mean will differ from the population mean $\mu = 5$ by 2 or less. (The first case corresponds to $\bar{x} = 4, 5,$ or 6, and the second case corresponds to $\bar{x} = 3, 4, 5, 6,$ or 7.) Thus, if we did not know the mean of the given population and wanted to estimate it with the mean of a random sample of two observations, this would give us some idea about the potential size of our error.

Further useful information about this sampling distribution of the mean can be obtained by calculating its mean $\mu_{\bar{x}}$ and its standard deviation $\sigma_{\bar{x}}$, where the subscripts serve to distinguish between these parameters and those of the original population. Following the definitions of the mean and the variance of a probability distribution on pages 221 and 223, we get

$$\mu_{\bar{x}} = 2 \cdot \frac{1}{10} + 3 \cdot \frac{1}{10} + 4 \cdot \frac{2}{10} + 5 \cdot \frac{2}{10} + 6 \cdot \frac{2}{10} + 7 \cdot \frac{1}{10} + 8 \cdot \frac{1}{10}$$

$$= 5$$

and

$$\sigma_{\bar{x}}^2 = (2-5)^2 \cdot \frac{1}{10} + (3-5)^2 \cdot \frac{1}{10} + (4-5)^2 \cdot \frac{2}{10} + (5-5)^2 \cdot \frac{2}{10}$$

$$+ (6-5)^2 \cdot \frac{2}{10} + (7-5)^2 \cdot \frac{1}{10} + (8-5)^2 \cdot \frac{1}{10}$$

$$= 3$$

so that $\sigma_{\bar{x}} = \sqrt{3}$.

Observe that, at least for this example,

1. $\mu_{\bar{x}}$, the mean of the sampling distribution of $\bar{x}$, equals μ, the mean of the population.
2. $\sigma_{\bar{x}}$, the standard deviation of the sampling distribution of $\bar{x}$, is smaller than σ, the standard deviation of the population.

These relationships are of fundamental importance in statistics, and we shall return to them in Section 8.3.

In the preceding example we took a very small sample from a very small population, but it would be difficult to use the same method to construct the sampling distribution of the mean of a large sample from a large population— we would have to enumerate too many possibilities. To construct the sampling

271

distribution of the mean even for random samples of size $n = 5$ from a finite population of size $N = 100$, we would already have to list $\binom{100}{5} = 75{,}287{,}520$ possibilities.

So, to get an idea about the sampling distribution of the mean of a larger random sample from a larger finite population, we shall use a **computer simulation.** This means that we shall use a computer to take a number of random samples from the given population, calculate their means, and display the results in the form of a histogram. Hopefully, this histogram will give us some idea about the overall shape and some other key features of the actual sampling distribution of the mean for random samples of the given size from the given population.

EXAMPLE   Use a computer to generate 100 random samples of size $n = 15$ from the finite population which consists of the integers from 1 to 1,000. Also, calculate the mean of each of the 100 samples, find their mean and their standard deviation, and group them into a distribution with the class marks $25.5, 75.5, \ldots$, and 975.5.[†]

Solution   Without a computer, the reader may picture this kind of simulation as follows: First, the numbers from 1 to 1,000 are written on 1,000 slips of paper (poker chips, small balls, or whatever may lend itself to drawing random samples). Then, a random sample of size $n = 15$ is drawn from this population and the values are recorded; it is replaced before the next sample is drawn; and this process is repeated until 100 random samples have been obtained.

Actually using an appropriate computer package, we obtained the printout shown in Figure 8.3. Here "GENE 1 1000, C1" is the instruction to the computer to generate our population, the integers from 1 to 1,000, and put them in column 1. In the next step, this population is described as having $N = 1{,}000$, $\mu = 500.5$, and $\sigma = 289$ (to the nearest integer). Then, the 100 samples of size $n = 15$, are generated and their means are put into column C4. The printout following the instructions "MEAN C4" and "STDEV C4" tell us that the mean of the 100 means is 506.3 and that their standard deviation is 63.9.

Finally, the 100 sample means are grouped into a distribution with the required classes, and this distribution is presented graphically in a form which might be described as a histogram lying on its side. Asked

---

[†] These class marks were chosen so that the corresponding class boundaries are "impossible" values; namely, values which cannot be taken on by a mean of 15 positive integers.

```
MTB > GENE 1 1000, C1
MTB > MEAN C1
   MEAN      =       500.5
MTB > STDEV C1
   STDEV     =       289.
MTB > SET C3
DATA> 4
DATA> END
MTB > STORE
STOR> SAMPLE 15 C1 C2
STOR> LET K1=AVER(C2)
STOR> JOIN K1 TO C3 PUT IN C3
STOR> ERASE C2
STOR> END
MTB > EXECUTE 100 TIMES
MTB > OMIT ROW WITH 4 IN C3 AND PUT REST IN C4
MTB > MEAN C4
   MEAN      =       506.3
MTB > STDEV C4
   STDEV     =       63.9
MTB > HIST C4 25.5 50

   C4

   MIDDLE OF      NUMBER OF
   INTERVAL       OBSERVATIONS
      25.5        0
      75.5        0
     125.5        0
     175.5        0
     225.5        0
     275.5        0
     325.5        1     *
     375.5        4     ****
     425.5       16     ***************
     475.5       23     ***********************
     525.5       33     *********************************
     575.5       16     ****************
     625.5        5     *****
     675.5        2     **
```

8.3

Computer simulation of a sampling distribution of the mean.

to describe the distribution, we might say that it is fairly symmetrical and bell-shaped; in fact, the overall pattern seems to follow quite closely that of a normal curve. All this applies to the distribution which we obtained by means of the computer simulation involving only 100 random samples of size $n = 15$ from the population which consists of the integers from 1 to 1,000. Hopefully, it applies also to the actual sampling distribution of the mean which pertains to all possible random samples of size $n = 15$ from this population.

Note also that the results of this "experiment" support the two points which we made on page 271. Although the mean of the 100 $\bar{x}$'s does not equal $\mu = 500.5$, its value, 506.3, is very close. Also, the standard deviation of the 100 $\bar{x}$'s, 63.9, is smaller than the population standard deviation, $\sigma = 289$.

8.3

The Standard Error of the Mean

In most practical situations we can determine how close a sample mean might be to the mean of the population from which the sample came, by referring to two theorems which express essential facts about sampling distributions. The first of these theorems expresses formally what we discovered in connection with the first sampling distribution of the preceding section, namely, that the mean of the sampling distribution of $\bar{x}$ equals the mean of the population, and that its standard deviation is smaller than the standard deviation of the population. It may be phrased as follows: For random samples of size n taken from a population having the mean μ and the standard deviation σ, the sampling distribution of $\bar{x}$ has the mean

Mean of sampling distribution of $\bar{x}$

$$\mu_{\bar{x}} = \mu$$

and the standard deviation

Standard error of the mean

$$\sigma_{\bar{x}} = \frac{\sigma}{\sqrt{n}} \quad or \quad \sigma_{\bar{x}} = \frac{\sigma}{\sqrt{n}} \cdot \sqrt{\frac{N-n}{N-1}}$$

depending on whether the population is infinite or finite of size N.

It is customary to refer to $\sigma_{\bar{x}}$, the standard deviation of the sampling distribution of the mean, as the **standard error of the mean.** Its role in statistics is fundamental, since it measures the extent to which sample means can be expected to fluctuate, or vary, due to chance. What determines the size of $\sigma_{\bar{x}}$, and hence the goodness of an estimate, can be seen from the formulas above. Both formulas show that the standard error of the mean increases as the variability of the population increases (in fact, it is directly proportional to σ), and that it decreases as the sample size increases.

EXAMPLE   When we sample from an infinite population, what happens to the standard error of the mean (and, hence, to the size of the error we are exposed to when we use $\bar{x}$ as an estimate of μ) if the sample size is increased from $n = 50$ to $n = 200$?

Solution   The ratio of the two standard errors is

$$\frac{\dfrac{\sigma}{\sqrt{200}}}{\dfrac{\sigma}{\sqrt{50}}} = \frac{\sqrt{50}}{\sqrt{200}} = \sqrt{\frac{50}{200}} = \sqrt{\frac{1}{4}} = \frac{1}{2}$$

so that the standard error of the mean is divided by 2.

The factor $\sqrt{\dfrac{N-n}{N-1}}$ in the second formula for $\sigma_{\bar{x}}$ is called the **finite population correction factor,** for without it the two formulas for $\sigma_{\bar{x}}$ (for infinite and finite populations) are the same. In practice, it is omitted unless the sample constitutes at least 5% of the population, for otherwise it is so close to 1 that it has little effect on the value of $\sigma_{\bar{x}}$.

EXAMPLE   Find the value of the finite population correction factor for $n = 100$ and
(a) $N = 10,000$;
(b) $N = 200$.

Solution   (a) Substituting $n = 100$ and $N = 10,000$, we get

$$\sqrt{\frac{N-n}{N-1}} = \sqrt{\frac{10,000 - 100}{10,000 - 1}} = 0.995$$

and this is so close to 1 that the correction factor can be omitted for most practical purposes.
(b) Substituting $n = 100$ and $N = 200$, we get

$$\sqrt{\frac{N-n}{N-1}} = \sqrt{\frac{200 - 100}{200 - 1}} = 0.71$$

rounded to two decimals, so that the finite population correction factor will substantially reduce the standard error of the mean.

This reflects the fact that in this case we have a great deal of information about the population; in fact, the sample constitutes half the population.

Since we did not actually prove the two formulas for the standard error of the mean, let us verify the second one with reference to the examples on pages 271 and 272.

EXAMPLE   With reference to the example on page 271, verify that the formula for $\sigma_{\bar{x}}$ for a random sample from a finite population will also yield $\sigma_{\bar{x}} = \sqrt{3}$.

Solution   Substituting $n = 2$, $N = 5$, and $\sigma = \sqrt{8}$ into the second of the two formulas for $\sigma_{\bar{x}}$, we get

$$\sigma_{\bar{x}} = \frac{\sqrt{8}}{\sqrt{2}} \cdot \sqrt{\frac{5-2}{5-1}} = \frac{\sqrt{8}}{\sqrt{2}} \cdot \sqrt{\frac{3}{4}} = \sqrt{\frac{8}{2} \cdot \frac{3}{4}} = \sqrt{3}$$

EXAMPLE   With reference to the computer simulation, where we had $n = 15$, $N = 1,000$, and $\sigma = 289$, what value might we have expected for the standard deviation of the 100 sample means?

Solution   Substituting $n = 15$, $N = 1,000$, and $\sigma = 289$ into the second of the two formulas for $\sigma_{\bar{x}}$, we get

$$\sigma_{\bar{x}} = \frac{289}{\sqrt{15}} \cdot \sqrt{\frac{1,000 - 15}{1,000 - 1}} = 74.1$$

and this is fairly close to 63.9, the value which we actually obtained in the printout of Figure 8.3.

EXERCISES   (Exercises 8.14 and 8.15 are practice exercises; their complete solutions are given on pages 284 and 285.)

8.14 Supposing that in the example on page 269 sampling is with replacement, list the 25 possible ordered samples 1 and 1, 1 and 3, 1 and 5, 1 and 7, 1 and 9, 3 and 1, 3 and 3, 3 and 5, ... ; calculate their means; and assuming that each ordered pair has the probability $\frac{1}{25}$, construct the sampling distribution of the mean for random samples of size $n = 2$, taken with replacement from the given population.

CHAP. 8: Sampling and Sampling Distributions

8.15 Calculate the standard deviation of the sampling distribution obtained in the preceding exercise, and verify the result by substituting $n = 2$ and $\sigma = \sqrt{8}$ into the first of the two standard error formulas on page 274.

8.16 Random samples of size $n = 2$ are drawn from the finite population which consists of the numbers 2, 4, 6, and 8.

(a) Calculate the mean and the standard deviation of this population.

(b) List the six possible random samples of size $n = 2$ that can be drawn from this population and calculate their means.

(c) Use the results of part (b) to construct the sampling distribution of the mean for random samples of size $n = 2$ from the given population.

(d) Calculate the standard deviation of the sampling distribution obtained in part (c) and verify the result by substituting $n = 2$, $N = 4$, and the value of σ obtained in part (a) into the second of the two standard error formulas on page 274.

8.17 Rework the preceding exercise for sampling with replacement from the given population.

8.18 When we sample from an infinite population, what happens to the standard error of the mean if the sample size is

(a) increased from 25 to 225;   (b) decreased from 480 to 30?

8.19 When we sample from an infinite population, what happens to the standard error of the mean if the sample size is

(a) increased from 20 to 45;   (b) decreased from 250 to 40?

8.20 What is the value of the finite population correction factor when

(a) $n = 5$ and $N = 150$;

(b) $n = 10$ and $N = 150$;

(c) $n = 10$ and $N = 400$?

State in each case whether the finite population factor may reasonably be omitted in the formula for $\sigma_{\bar{x}}$.

8.21 Use a computer package to

(a) generate the population of integers from 1 to 800 and determine its mean and its standard deviation;

(b) take 100 random samples of size $n = 5$ from this population;

(c) determine the means of these 100 samples and calculate their mean and their standard deviation;

(d) construct a histogram of the means obtained in part (c), using the class marks 40.5, 80.5, 120.5, 160.5, ...

Also,

(e) compare the mean of the 100 sample means with that of the population;

(f) compare the standard deviation of the 100 sample means with the value we would expect in accordance with the theory on page 274;

(g) discuss the overall shape of the histogram obtained in part (d).

8.22 Repeat Exercise 8.21 using sampling with replacement, so that the whole procedure may be looked upon as sampling from an infinite population.

8.23 Use a computer package to

    (a) generate the population of integers from 1 to 1,200 and determine its mean and its standard deviation;

    (b) take 150 random samples of size $n = 25$ from this population;

    (c) determine the means of these 150 samples and calculate their mean and their standard deviation;

    (d) construct a histogram of the means obtained in part (c), using the class marks 30.5, 90.5, 150.5, 210.5, ...

Also,

    (e) compare the mean of the 150 sample means with that of the population;

    (f) compare the standard deviation of the 150 sample means with the value we would expect in accordance with the theory on page 274;

    (g) discuss the overall shape of the histogram obtained in part (d).

 8.24 Repeat Exercise 8.23 using sampling with replacement, so that the whole procedure may be looked upon as sampling from an infinite population.

8.4
The Central Limit Theorem

When we use a sample mean to estimate the mean of a population, we usually express our confidence in the estimate by attaching a probability statement about the size of our error. For instance, if we apply Chebyshev's theorem (as formulated on page 227) to the sampling distribution of the mean, we can assert with a probability of at least $1 - \dfrac{1}{k^2}$ that the mean of a random sample will differ from the mean of the population from which it came by less than $k \cdot \sigma_{\bar{x}}$. In other words, if we use the mean of a random sample to estimate the mean of the population from which it came, we can assert with a probability of at least $1 - \dfrac{1}{k^2}$ that our error will be less than $k \cdot \sigma_{\bar{x}}$.

EXAMPLE   Based on Chebyshev's theorem with $k = 2$, what can we assert about the maximum size of our error if we use the mean of a random sample of size $n = 64$ to estimate the mean of an infinite population with $\sigma = 20$?

Solution   Substituting $\sigma = 20$ and $n = 64$ into the first of the two formulas for the standard error of the mean, we get

$$\sigma_{\bar{x}} = \frac{20}{\sqrt{64}} = 2.5$$

and it follows that we can assert with a probability of at least $1 - \dfrac{1}{2^2} =$ 0.75 that the error will be less than $k \cdot \sigma_{\bar{x}} = 2 \cdot 2.5 = 5$.

Chebyshev's theorem can be used in problems like this, but if we want to be more specific about the probabilities, we have to refer to another theorem, which is called the **Central Limit theorem.** It states that for large samples the sampling distribution of the mean can be approximated closely with a normal distribution.

To apply this result, namely, approximate the sampling distribution of the mean with a normal distribution, we must be able to convert $\bar{x}$'s to standard units. Since a value of a random variable is converted to standard units by subtracting from it the mean of the distribution of the random variable and then dividing by its standard deviation, we write

$$z = \frac{\bar{x} - \mu_{\bar{x}}}{\sigma_{\bar{x}}}$$

where we used the same notation as in Section 8.3. Recalling from page 274 that $\mu_{\bar{x}} = \mu$ and $\sigma_{\bar{x}} = \dfrac{\sigma}{\sqrt{n}}$ for random samples from infinite populations, we can now say formally that

Central Limit theorem

> *If $\bar{x}$ is the mean of a random sample of size n from an infinite population with the mean μ and the standard deviation σ and n is large, then*
>
> $$z = \frac{\bar{x} - \mu}{\sigma/\sqrt{n}}$$
>
> *has approximately the standard normal distribution.*

This theorem is of fundamental importance in statistics, as it justifies the use of normal-curve methods in a wide range of problems; it applies to infinite populations, and also to finite populations when n, though large, constitutes but a small portion of the population. We cannot say precisely how large n must be so that the Central Limit theorem can be applied, but unless the population distribution has a very unusual shape, $n = 30$ is usually regarded as sufficiently large. When the population we are sampling has, itself, roughly the shape of a normal curve, the sampling distribution of the mean can be approximated closely with a normal distribution regardless of the size of n.

279

EXAMPLE Returning to the example on page 278 but basing the argument on the Central Limit theorem, with what probability can we assert that our error will be less than 5 when we use the mean of a random sample of size $n = 64$ to estimate the mean of an infinite population with $\sigma = 20$?

Solution The desired probability is given by the area of the white region under the curve in Figure 8.4, namely, the normal-curve area between $z = \dfrac{-5}{20/\sqrt{64}} = -2$ and $z = \dfrac{5}{20/\sqrt{64}} = 2$. Since the entry in Table II corresponding to $z = 2.00$ is 0.4772, we find that the answer is $0.4772 + 0.4772 = 0.9544$. Compared to "at least 0.75," we can thus make the much stronger statement that the probability is 0.9544 that the mean of a random sample of size $n = 64$ from the given population will differ from the population mean by less than 5.

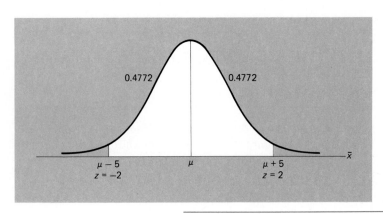

8.4

Sampling distribution of the mean.

EXAMPLE Suppose that $\sigma = 5.5$ for the daily sulfur oxides emission of a certain industrial plant. What is the probability that the mean of a random sample of size $n = 40$ will differ from the mean of the population by less than 1.0 ton?

Solution The desired probability is given by the area of the white region under the curve in Figure 8.5, namely, the normal-curve area between $z = \dfrac{-1.0}{5.5/\sqrt{40}} = -1.15$ and $z = 1.15$. Since the entry in Table II corresponding to $z = 1.15$ is 0.3749, we find that the answer is $0.3749 + 0.3749 = 0.7498$, or approximately 0.75.

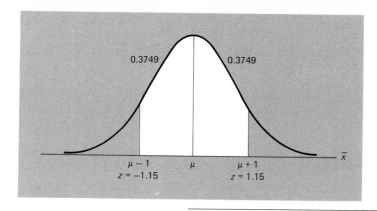

8.5

Sampling distribution of the mean.

In both of the preceding examples we were concerned with the size of the error; in the example which follows, μ will be given and we will be asked for the probability that the mean of a sample will fall within a certain range.

EXAMPLE   The time it takes students in a cooking school to learn how to prepare a particular meal is a random variable with the mean $\mu = 3.2$ hours and the standard deviation $\sigma = 1.8$ hours. Find the probability that the average time it will take 36 students to learn how to prepare the meal is less than 3.4 hours.

Solution   Again, we are concerned with the sampling distribution of the mean, and since n is greater than 30 we can approximate it with a normal curve. Thus, the desired probability is given by the area of the white region under the curve in Figure 8.6, namely, the normal-curve area

8.6

Sampling distribution of the mean.

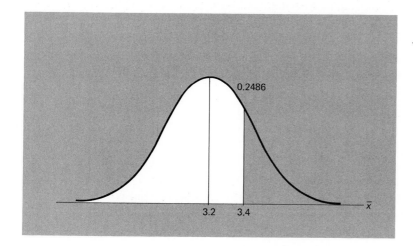

to the left of

$$z = \frac{3.4 - 3.2}{1.8/\sqrt{36}} = \frac{0.2}{0.3} = 0.67$$

Since the entry in Table II corresponding to $z = 0.67$ is 0.2486, the answer is $0.5000 + 0.2486 = 0.7486$.

The primary goal of this chapter has been to introduce the concept of a sampling distribution, and the one which we chose for this purpose was the sampling distribution of the mean. Observe, however, that instead of the mean we could have studied the median or some other statistic and investigated its chance fluctuations. So far as the corresponding theory is concerned, this would, of course, have required different formulas for the standard errors, namely, different formulas for the standard deviations of the respective sampling distributions. For instance, for infinite populations, the **standard error of the median** is approximately $\sigma_{\tilde{x}} = 1.25 \cdot \dfrac{\sigma}{\sqrt{n}}$, where n is the size of the sample and σ is the population standard deviation. Note that comparison of the two formulas $\sigma_{\bar{x}} = \dfrac{\sigma}{\sqrt{n}}$ and $\sigma_{\tilde{x}} = 1.25 \cdot \dfrac{\sigma}{\sqrt{n}}$ reflects the fact that the mean is generally more reliable than the median, namely, that it is subject to smaller chance fluctuations (see also Exercise 8.33).

Let us also point out that there have been three occasions so far to work with standard units. In Section 3.8 we used them to illustrate an important application of the standard deviation; in Section 7.2 we used them in connection with the standard normal distribution; and in this section we used them in the formulation of the Central Limit theorem and in its applications. In general, the conversion of a value of a random variable to standard units depends on the distribution with which we are concerned, and we use the formula

$$z = \frac{\text{(value of random variable)} - \text{(mean of random variable)}}{\text{(standard deviation of random variable)}}$$

The mean which we subtract in the numerator may be μ or $\mu_{\bar{x}}$, and the standard deviation by which we divide may be σ or $\sigma_{\bar{x}}$, depending on whether we are dealing with one observation or with a mean. They may also be $\mu_{\tilde{x}}$ and $\sigma_{\tilde{x}}$, for example, when we are dealing with the sampling distribution of the median.

8.25 A biologist wants to estimate the mean weight of a certain kind of animal. He knows that the standard deviation of their weights is 4.8 ounces and decides to use a random sample of size 100. Based on the Central Limit theorem, with what probability can he assert that his error will be

    (a) less than 0.60 ounce;

    (b) less than 1.20 ounces?

8.26 The dean of a college would like to estimate how many points applicants can be expected to get on an entrance examination. If she uses a random sample of 100 applicants and assumes that the standard deviation is 20 points, what can she assert about the probability that her error will be less than three points if she uses

    (a) Chebyshev's theorem;

    (b) the Central Limit theorem?

8.27 An automobile club's emergency repair service wants to use a sample of 81 calls to estimate the average time it takes for the club to respond to a motorist's request for help. If the standard deviation of the response times is assumed to be 7.5 minutes, what can the club assert about the probability that its error will be less than 2.5 minutes if it uses

    (a) Chebyshev's theorem;

    (b) the Central Limit theorem?

8.28 A laboratory technician working for a soft contact lens manufacturer wants to estimate the mean lifetime of one of the manufacturer's new lenses. Suppose that he uses a sample of 50 of the lenses and knows from experience that the standard deviation is 6 months for such data. Based on the Central Limit theorem, with what probability can he assert that the error will be

    (a) less than one month;

    (b) less than two months?

8.29 A company that makes tabletop water purifiers wants to estimate the mean number of grams per gallon of an impurity that its machine removes. Suppose that it uses a sample of 150 one-gallon jugs and that the standard deviation is 2 grams for such data. If it uses the Central Limit theorem, with what probability can the company assert that its error will be

    (a) less than 0.3 gram;

    (b) less than 0.5 gram?

8.30 During the last week of the semester, students at a certain college spend on the average 4.2 hours using the school's computer terminals with a standard deviation of 1.8 hours. For a random sample of 36 students at that college, find the probabilities that the average time spent using the computer terminals

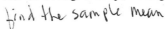

during the last week of the semester is

    (a) at least 4.8 hours;

    (b) between 4.1 and 4.5 hours.

8.31 A particular make of car is known to show rust when it is 2.4 years old on the average with a standard deviation of 0.8 year. If a car rental agency purchases 64 new cars of this kind, what are the probabilities that the average time it will take for these cars to show rust is

    (a) at most 2.6 years;

    (b) between 1.9 and 2.3 years?

8.32 The time it takes salespersons in a certain company to meet their yearly quotas is 10.6 months on the average with a standard deviation of 1.8 months. Find the probabilities that in a sample of 36 of the company's salespersons the average time it will take to meet their yearly quotas is

    (a) at most 11 months;

    (b) between 10.0 and 10.5 months.

8.33 Show that the mean of a random sample of size $n = 64$ is as reliable an estimate of the mean of a symmetrical infinite population as the median of a random sample of size $n = 100$. (For symmetrical populations, the means of the sampling distributions of $\bar{x}$ and $\tilde{x}$ are both equal to the population mean μ.)

8.34 How large a random sample do we have to take so that its mean is as reliable an estimate of the mean of a symmetrical infinite population as the median of a random sample of size $n = 225$?

8.35 How large a random sample do we have to take so that its median is as reliable an estimate of the mean of a symmetrical infinite population as the mean of a random sample of size $n = 400$?

8.36 Show that if the mean of a random sample of size n is used to estimate the mean of an infinite population with the standard deviation σ, there is a fifty-fifty chance that the error will be less than the quantity $0.6745 \cdot \dfrac{\sigma}{\sqrt{n}}$, which is called the **probable error of the mean.** (*Hint:* Interpolate between the entries in Table II corresponding to $z = 0.67$ and $z = 0.68$.)

SOLUTIONS OF PRACTICE EXERCISES

8.1 Since 6 of the 10 samples *abc*, *abd*, *abe*, *acd*, *ace*, *ade*, *bcd*, *bce*, *bde*, and *cde* include *b*, the probability is $\frac{6}{10} = \frac{3}{5}$. The same result holds for any other letter.

8.2 Reading the values off the table, we get 3057, 0241, 1896, 2852, 4428, 5697, 5550, 4701, 3645, 2676, 4261, 1745, 0370, 2153, and 5717.

8.14 The ordered samples are 1 and 1, 1 and 3, 1 and 5, 1 and 7, 1 and 9, 3 and 1, 3 and 3, 3 and 5, 3 and 7, 3 and 9, 5 and 1, 5 and 3, 5 and 5, 5 and 7, 5 and 9, 7 and 1, 7 and 3, 7 and 5, 7 and 7, 7 and 9, 9 and 1, 9 and 3, 9 and 5, 9 and 7, and 9 and 9; the means are 1, 2, 3, 4, 5, 2, 3, 4, 5, 6, 3, 4, 5, 6, 7, 4, 5, 6, 7, 8,

5, 6, 7, 8, and 9; and the sampling distribution of the mean is

$\bar{x}$	Probability
1	$\frac{1}{25}$
2	$\frac{2}{25}$
3	$\frac{3}{25}$
4	$\frac{4}{25}$
5	$\frac{5}{25}$
6	$\frac{4}{25}$
7	$\frac{3}{25}$
8	$\frac{2}{25}$
9	$\frac{1}{25}$

8.15 Since $\mu_{\bar{x}} = \mu = 5$, we get $\sigma_{\bar{x}}^2 = (1-5)^2 \cdot \frac{1}{25} + (2-5)^2 \cdot \frac{2}{25} + (3-5)^2 \cdot \frac{3}{25} + (4-5)^2 \cdot \frac{4}{25} + (5-5)^2 \cdot \frac{5}{25} + (6-5)^2 \cdot \frac{4}{25} + (7-5)^2 \cdot \frac{3}{25} + (8-5)^2 \cdot \frac{2}{25} + (9-5)^2 \cdot \frac{1}{25} = 4$, so that $\sigma_{\bar{x}} = 2$; substitution into the formula yields

$$\sigma_{\bar{x}} = \frac{\sqrt{8}}{\sqrt{2}} = \sqrt{\frac{8}{2}} = \sqrt{4} = 2$$

and the two answers agree.

8.25 (a) $\sigma_{\bar{x}} = \dfrac{4.8}{\sqrt{100}} = 0.48$, $z = \dfrac{0.60}{0.48} = 1.25$, the corresponding entry in Table II is 0.3944, and the answer is $0.3944 + 0.3944 = 0.7888$.

(b) $z = \dfrac{1.20}{0.48} = 2.50$, the corresponding entry in Table II is 0.4938, and the answer is $0.4938 + 0.4938 = 0.9876$.

8.33 $\sigma_{\bar{x}} = \dfrac{\sigma}{\sqrt{64}} = \dfrac{\sigma}{8}$ and $\sigma_{\bar{x}} = 1.25 \cdot \dfrac{\sigma}{\sqrt{100}} = 1.25 \cdot \dfrac{\sigma}{10} = \dfrac{\sigma}{8}$; since the two standard errors are equal, the mean of a random sample of size $n = 64$ is as reliable as the median of a random sample of size $n = 100$.

Achievements

Having read and studied these chapters, and having worked a good portion of the exercises, you should be able to

1. Explain what is meant by "random variable" and "probability distribution."

2. State the two rules which must be satisfied by the values of a probability distribution.

3. List the assumptions which underlie the binomial distribution.

4. Use the formula for the binomial distribution.

5. Use the table of binomial probabilities.

6. Use the formula for the hypergeometric distribution.

7. Approximate hypergeometric probabilities with binomial probabilities and know when this approximation may be used.

8. Approximate binomial probabilities with Poisson probabilities and know when this approximation may be used.

9. Use the formula for the Poisson distribution with the parameter λ.

10. Calculate multinomial probabilities.

11. Determine the mean and the standard deviation of a probability distribution.

12. Find the mean and the standard deviation of a binomial distribution.

13. Find the mean of a hypergeometric distribution.

14. Find the mean of a Poisson distribution.

15. Use Chebyshev's theorem as it applies to the distribution of a random variable.

16. Explain what is meant by "probability density" or "continuous distribution."

17. Discuss the normal distribution and explain what is meant by "standard normal distribution."

18. Use the table of normal-curve areas.

19. Apply the continuity correction where needed.

20. Use the normal distribution to approximate binomial probabilities and know when this approximation may be used.

21. Explain what is meant by "random sample from a finite population."

22. Use random numbers to select random samples from finite populations.

23. Explain what is meant by "sampling distribution" and "sampling distribution of the mean."

24. State and apply the theorem about the mean and the standard deviation of the sampling distribution of the mean.

25. Explain what is meant by "standard error."

26. Determine the finite population correction factor and know when it should be used.

27. State the Central Limit theorem and use it to calculate probabilities relating to the sampling distribution of the mean.

Checklist of Key Terms (with page references to their definitions)

Binomial distribution, 200
Binomial population, 204
Central Limit theorem, 279
Chebyshev's theorem, 227, 278
Continuity correction, 250

Continuous distribution, 234
Finite population, 264
Finite population correction factor, 275
Hypergeometric distribution, 208

REVIEW EXERCISES

R.82 In a photographic process, the developing time of prints may be looked upon as a random variable having a normal distribution with $\mu = 12.26$ seconds and $\sigma = 0.24$ second. Find the probabilities that it will take
   (a) at least 11.86 seconds to develop one of the prints;
   (b) from 12.32 to 12.80 seconds to develop one of the prints.

R.83 The management of a bookstore knows from experience that the probabilities are 0.50, 0.40, and 0.10 that a person browsing at the store will not make a purchase, buy one book, or buy two or more. What is the probability that among nine persons browsing at the store, five will not make a purchase, three will buy one book, and one will buy two or more?

R.84 What is the value of the finite population correction factor when
   (a) $n = 30$ and $N = 120$;
   (b) $n = 20$ and $N = 500$?

R.85 The weekly number of car thefts in a certain precinct is a random variable with $\mu = 15.3$ and $\sigma = 2.5$. Assuming that the distribution of this random variable can be appoximated closely with a normal curve, find the probabilities that in any given week there will be
   (a) exactly 17 car thefts;
   (b) at least 17 car thefts.

R.86 How many different samples of size $n = 4$ can be drawn from a finite population of size
   (a) $N = 10$;
   (b) $N = 20$?

R.87 Suppose that a continuous random variable takes on values on the interval from 0 to 2 and that the graph of its probability density is as shown in Figure R.3. What are the probabilities that the random variable will take on a value
   (a) less than 0.5;
   (b) between 1.2 and 1.6?

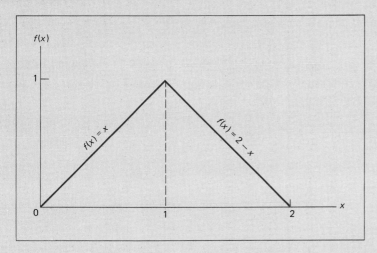

R.3

Diagram for Exercise R.87.

R.88 An agricultural cooperative claims that 95% of the watermelons that are shipped out are ripe and ready to eat. Use Table I to find the probabilities that among ten watermelons that are shipped out
(a) at least eight are ripe and ready to eat;
(b) anywhere from seven through nine are ripe and ready to eat;
(c) at most nine are ripe and ready to eat.

R.89 Use the normal-curve approximation to find the probability of getting 6 heads in 15 flips of a balanced coin, and compare the result with the value given in Table I.

R.90 The Dean of Students of a college wants to interview 15 entering freshmen of a class of 632, who are listed in the registrar's office in alphabetic order. Which ones (by numbers) will she select, if she numbers the freshmen from 001 through 632 and chooses a random sample by means of random numbers, using the last three columns of the table on page 530, going down the table starting with the 16th row?

R.91 Suppose that the number of minutes which a tourist spends in a museum is a random variable with $\mu = 73.4$ minutes and $\sigma = 6.8$ minutes. Assuming that the distribution of this random variable can be approximated closely with a normal curve, find the probabilities that a tourist will spend
(a) at most 66.0 minutes in the museum;
(b) from 70.0 to 80.0 minutes in the museum.

R.92 With reference to the preceding exercise, find the value of the random variable for which 25% of the tourists spend less time in the museum and 75% spend more time in the museum.

289

R.93 If the number of blossoms on a rare cactus is a random variable having the Poisson distribution with $\lambda = 2.4$, what are the probabilities that such a cactus will have

    (a) no blossoms;

    (b) two blossoms;

    (c) five blossoms?

R.94 Find the area under the standard normal curve which lies

    (a) between $z = 0$ and $z = 1.83$;

    (b) to the left of $z = 2.15$;

    (c) to the right of $z = -0.44$;

    (d) to the right of $z = 1.24$;

    (e) to the left of $z = -0.71$.

R.95 Among ten faculty members considered for promotions there are six men and four women.

    (a) If two of them are chosen at random, find the probabilities that 0, 1, or 2 women will be included.

    (b) Use the probabilities obtained in part (a) to calculate the mean and the standard deviation of this probability distribution.

    (c) Use the special formula for the mean of a hypergeometric distribution to verify the value obtained for μ in part (b).

R.96 Find the mean and the variance of the binomial distribution with $n = 5$ and $p = 0.30$

    (a) by looking up the probabilities in Table I and then using the formulas for the mean and the variance of a probability distribution;

    (b) by using the special formulas for the mean and the variance of the binomial distribution.

R.97 The mean of a random sample of size $n = 36$ is used to estimate the mean time it takes clerks to locate a file from a standard file cabinet. If $\sigma = 0.12$ minute, what can we assert about the probability that the error will be less than 0.03 minute, if we use

    (a) Chebyshev's theorem;

    (b) the Central Limit theorem?

R.98 Check in each case whether the conditions on page 213 for the Poisson approximation to the binomial distribution are satisfied:

    (a) $n = 80$ and $p = \frac{1}{20}$;

    (b) $n = 250$ and $p = \frac{1}{50}$;

    (c) $n = 400$ and $p = \frac{1}{20}$.

R.99 An automobile battery which is guaranteed for five years lasts on the average 62 months with a standard deviation of 18 months. Find the probability that a random sample of 100 of these batteries will last on the average at least 61 months.

R.100 Suppose that a certain surgical procedure is successful 90% of the time. Use the formula for the binomial distribution to determine the probability that four of five operations of the given kind will be a success.

R.101 Among 25 workers on a picket line, 14 are men and 11 are women. If a television news reporter randomly picks four of them to be shown on camera, what are the probabilities that this will include
(a) two men and two women;
(b) one man and three women?

R.102 What is the probability of each sample if a random sample of size 3 is to be drawn from a finite population of size 16?

R.103 In each case determine whether the given values can be looked upon as the values of a probability distribution of a random variable which can take on the values 1, 2, 3, 4, and 5, and explain your answers:
(a) $f(1) = 0.18, f(2) = 0.20, f(3) = 0.22, f(4) = 0.20, f(5) = 0.18$;
(b) $f(1) = 0.05, f(2) = 0.05, f(3) = 0.05, f(4) = -0.05, f(5) = 0.90$;
(c) $f(1) = 0.07, f(2) = 0.23, f(3) = 0.50, f(4) = 0.19, f(5) = 0.01$.

R.104 A random variable has a normal distribution with $\sigma = 10$. If the probability that the random variable will take on a value less than 71.3 is 0.8264, what is the probability that it will take on a value greater than 46.9?

R.105 If 65% of all scorpion stings cause extensive discomfort, what is the probability that among 200 scorpion stings at most 125 will cause extensive discomfort?

R.106 Determine whether the following can be probability distributions (defined in each case only for the given values of x) and explain your answers:

(a) $f(x) = \dfrac{1}{5}$ for $x = 0, 1, 2, 3, 4, 5$;

(b) $f(x) = \dfrac{x + 1}{14}$ for $x = 1, 2, 3, 4$;

(c) $f(x) = \dfrac{x(x - 2)}{2}$ for $x = 1, 2, 3$.

R.107 Check in each case whether the conditions for the normal approximation to the binomial distribution are satisfied:
(a) $n = 50$ and $p = \frac{1}{5}$;
(b) $n = 80$ and $p = \frac{1}{40}$;
(c) $n = 100$ and $p = 0.96$;
(d) $n = 200$ and $p = \frac{24}{25}$.

R.108 The probabilities of 0, 1, 2, 3, or 4 armed robberies in a Western city on any given day are 0.33, 0.37, 0.20, 0.08, and 0.02. Find the mean and the standard deviation of this probability distribution.

R.109 Find z if
(a) the normal-curve area between 0 and z is 0.4909;

(b) the normal-curve area to the right of z is 0.6985;

(c) the normal-curve area to the right of z is 0.0089;

(d) the normal-curve area between $-z$ and z is 0.1742.

R.110 The mean of a random sample of size $n = 81$ is used to estimate the mean annual growth of certain plants. If the standard deviation of their annual growth is $\sigma = 45$ mm, what are the probabilities that our estimate will be "off" either way by

(a) less than 8.5 mm;

(b) less than 2.0 mm?

R.111 Check in each case whether the condition for the binomial approximation to the hypergeometric distribution is satisfied:

(a) $a = 150$, $b = 150$, and $n = 12$;

(b) $a = 100$, $b = 300$, and $n = 25$;

(c) $a = 400$, $b = 200$, and $n = 25$.

R.112 Random samples of size $n = 2$ are drawn from the finite population which consists of the numbers 1, 2, 3, 4, 5, and 6.

(a) Verify that the mean of this finite population is $\mu = 3\frac{1}{2}$ and that its variance is $\sigma^2 = \frac{35}{12}$.

(b) List the 15 possible samples of size $n = 2$ that can be drawn without replacement from the given population, calculate their means, and, assigning each of these values the probability $\frac{1}{15}$, construct the sampling distribution of the mean for random samples of size $n = 2$ from the given population.

(c) Calculate the mean and the standard deviation of the sampling distribution obtained in part (b), and verify the result obtained for the standard deviation by substituting $n = 2$, $N = 6$, and $\sigma = \sqrt{\frac{35}{12}}$ into the second of the two standard error formulas on page 274.

R.113 Among 500 plants exposed to excessive radiation, 90 show abnormal growth. If a scientist collects the seed of three of the plants chosen at random, find the probability that she will get the seed from one plant with abnormal growth and two plants with normal growth by using

(a) the formula for the hypergeometric distribution;

(b) the binomial distribution as an approximation.

R.114 During the month of August, the daily number of persons visiting a certain tourist attraction is a random variable with $\mu = 1,200$ and $\sigma = 80$.

(a) What does Chebyshev's theorem with $k = 7$ tell us about the number of persons who will visit the tourist attraction on an August day?

(b) According to Chebyshev's theorem, with what probability can we assert that between 1,000 and 1,400 persons will visit the tourist attraction on an August day?

R.115 Suppose that we draw cards, one at a time and with replacement, from an ordinary deck of 52 playing cards. To determine the probability of getting 5 red kings among 156 cards thus drawn, can we use

    (a) the Poisson approximation;

    (b) the normal approximation?

R.116 When we sample from an infinite population, what happens to the standard error of the mean if the sample size is

    (a) increased from 20 to 500;

    (b) decreased from 490 to 40?

R.117 If 4% of all persons who drive cars are not properly licensed, use the Poisson approximation to the binomial distribution to determine the probability that among 100 drivers stopped at a road block, three will not be properly licensed.

R.118 The average time required to perform job A is 78.5 minutes with a standard deviation of 16.2 minutes, and the average time required to perform job B is 103.2 minutes with a standard deviation of 11.3 minutes. Assuming normal distributions, what proportion of the time will job A take longer than the average job B, and what proportion of the time will job B take less time than the average job A?

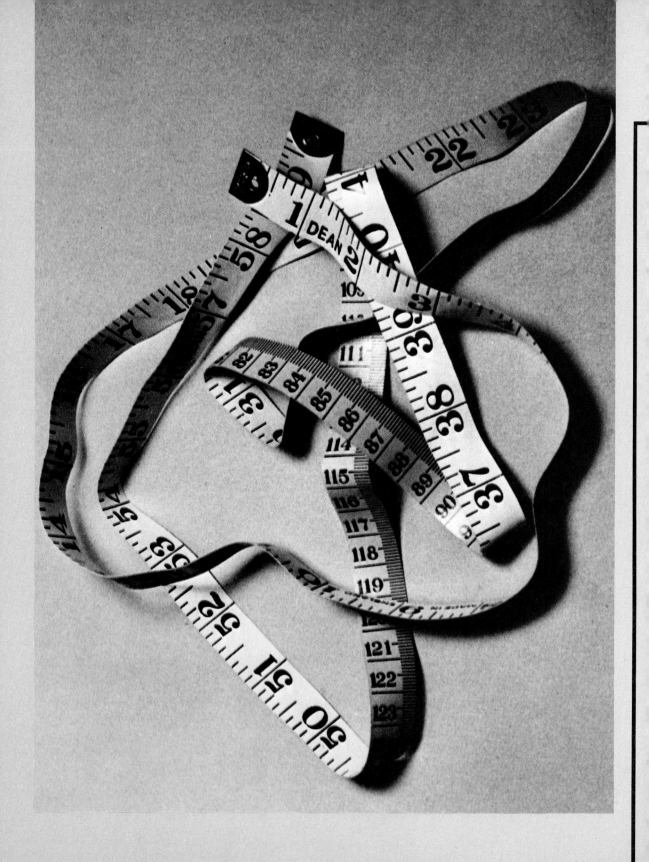

The Analysis of Measurements

In statistical inference we make generalizations based on samples, and, traditionally, such inferences have been divided into problems of

ESTIMATION (ˌes-tə-ˈmā-shən) Assigning a numerical value to a population parameter on the basis of sample data.

and problems of

HYPOTHESIS TESTING (hi-ˈpäth-ə-səs ˈtest-iŋ) Accepting or rejecting assumptions concerning the parameters or the form of a population or populations.

Problems of estimation arise everywhere—in science, in business, and in everyday life. In science, a biologist may want to determine the average wingspan of an insect; in business, a retailer may want to determine the average income of all families living within a mile of a proposed new shopping center; and in everyday life, we may want to know how long it takes on the average to iron a shirt or mend a pair of socks. These are all problems of estimation, but they would have been tests of hypotheses if the biologist had wanted to check another scientist's claim that the average wingspan of the insects is 13.4 mm, if the retailer had wanted to find out whether the average income of all families living within a mile of the proposed shopping center is at least $14,000, or if we wanted to check whether it really takes only 2.6 minutes to iron a shirt.

Sections 9.1 through 9.3 are devoted to the estimation of means, and Section 9.4 to the estimation of standard deviations. After a general introduction to tests of hypotheses in Sections 9.5 and 9.6, Sections 9.7 and 9.8 are devoted to tests of hypotheses concerning the mean of one population, Sections 9.9 through 9.11 are devoted to tests of hypotheses concerning the means of two populations, and Sections 9.12 and 9.13 are devoted to tests of hypotheses concerning the means of more than two populations.

9.1

The Estimation of Means

To illustrate some of the problems we face in the estimation of means, let us refer to a study in which a doctor wants to determine the true average increase in the pulse rate of a person performing a certain strenuous task. The following are the data (increases in pulse rate in beats per minute) which the doctor obtained for 32 persons who performed the given task:

27	25	19	28	35	23	24	22
14	30	32	34	23	26	29	27
27	24	31	22	23	38	25	16
32	29	26	25	28	26	21	28

The mean of this sample is $\bar{x} = 26.2$ beats per minute, and in the absence of any other information this figure may well have to serve as an estimate of the population mean μ, the true average increase in the pulse rate of persons performing the given task.

An estimate like this is called a **point estimate,** since it consists of a single number, or a single point on the real number scale. Although this is the most common way in which estimates are expressed, it leaves room for many questions. For instance, it does not tell us on how much information the estimate is based, and it does not tell us anything about the possible size of the error. And of course, we must expect an error. This should be clear from our discussion of the sampling distribution of the mean in Chapter 8, where we saw that the chance fluctuations of the mean (and, hence, its reliability as an estimate of μ) depend on two things—the size of the sample and the size of the population standard deviation σ. Thus, we might supplement $\bar{x} = 26.2$, the foregoing estimate of the true average increase in the pulse rate of persons performing the given task, with the information that the sample size is $n = 32$ and that the sample standard deviation is $s = 5.15$, as can easily be verified. Although this does not tell us the actual value of σ, the sample standard deviation can serve as an estimate of this quantity.

Scientific reports often present sample means in this way, together with the values of n and s, but this does not supply the reader of the report with a coherent picture unless he or she has had some formal training in statistics. To make the supplementary information meaningful also to the layman, let us refer

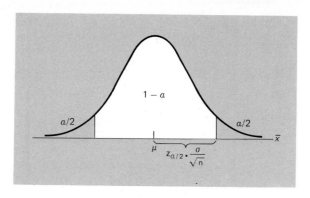

9.1

Sampling distribution of the mean.

to the theory of Sections 8.3 and 8.4 and the definition of Exercise 7.12 (see also Figure 7.20), according to which $z_{\alpha/2}$ is such that the area under the standard normal curve between $-z_{\alpha/2}$ and $z_{\alpha/2}$ is equal to $1 - \alpha$. Thus, making use of the fact that the sampling distribution of the mean of a large random sample from an infinite population is approximately a normal distribution with $\mu_x = \mu$ and $\sigma_{\bar{x}} = \dfrac{\sigma}{\sqrt{n}}$, we find from Figure 9.1 that the probability is $1 - \alpha$ that a sample mean will differ from the population mean μ by at most $z_{\alpha/2} \cdot \dfrac{\sigma}{\sqrt{n}}$. In other words,

Maximum error of estimate ($n \geq 30$)

> *When we use $\bar{x}$ as an estimate of μ, the probability is $1 - \alpha$ that this estimate will be "off" either way by at most*
>
> $$E = z_{\alpha/2} \cdot \frac{\sigma}{\sqrt{n}}$$

This result applies when n is large ($n \geq 30$) and the population is infinite, or large enough so that the finite population correction factor need not be used. The two values which are most commonly used for $1 - \alpha$ are 0.95 and 0.99, and as the reader was asked to verify in Exercise 7.12 on page 247 $z_{0.025} = 1.96$ corresponding to $1 - \alpha = 0.95$ and $z_{0.005} = 2.575$ corresponding to $1 - \alpha = 0.99$.

EXAMPLE   An efficiency expert intends to use the mean of a random sample of size $n = 40$ to estimate the average time it takes automobile mechanics to perform a certain task. If, based on experience, the efficiency expert can assume that $\sigma = 2.9$ minutes for such data, what can he assert with probability 0.99 about the maximum size of his error?

Solution   Substituting $n = 40$, $\sigma = 2.9$, and $z_{0.005} = 2.575$ into the formula for E, we get

$$E = 2.575 \cdot \frac{2.9}{\sqrt{40}} = 1.18$$

Thus, the efficiency expert can assert with probability 0.99 that his error will be at most 1.18 minutes.

Suppose now that the efficiency expert of this example collects his data and gets $\bar{x} = 27.36$ minutes. Can he still assert with probability 0.99 that the error of his estimate, $\bar{x} = 27.36$ minutes, is at most 1.18? After all, $\bar{x} = 27.36$ differs from the true mean by at most 1.18 minutes or it does not, and he does not know which. Actually, he can, but it must be understood that the 0.99 probability applies to the method which he used to determine the maximum error (getting the sample data and using the formula for E) and not directly to the parameter he is trying to estimate.

To make this distinction, it has become the custom to use the word "confidence" here instead of "probability."

> In general, we make probability statements about future values of random variables (say, the potential error of an estimate) and confidence statements once the data have been obtained.

Accordingly, we would say in our example that the efficiency expert can be 99% confident that the error of his estimate, $\bar{x} = 27.36$ minutes, is at most 1.18 minutes.

The result which we have obtained involves one complication. To be able to judge the size of the error we might make when we use $\bar{x}$ as an estimate of μ, we must know the value of the population standard deviation σ. Since this is not the case in most practical situations, we have no choice but to replace σ with an estimate, usually the sample standard deviation s. In general, this is considered to be reasonable provided the sample is sufficiently large, and by sufficiently large we mean again $n \geq 30$.

EXAMPLE   With reference to the pulse rates on page 296, where we had $n = 32$, $\bar{x} = 26.2$, and $s = 5.15$, what can we assert with 95% confidence about the maximum error if we use $\bar{x} = 26.2$ as an estimate of the true average increase in the pulse rate of a person performing the given task?

Solution   Substituting $n = 32$, $s = 5.15$ for σ, and $z_{0.025} = 1.96$ into the formula for E, we find that we can assert with 95% confidence that the error is at most

$$E = 1.96 \cdot \frac{5.15}{\sqrt{32}} = 1.78$$

beats per minute.

The formula for E can also be used to determine the sample size that is needed to attain a desired degree of precision. Suppose that we want to use the mean of a large random sample to estimate the mean of a population, and we want to be able to assert with probability $1 - \alpha$ that the error of this estimate will be at most some prescribed quantity E. As before, we write $E = z_{\alpha/2} \cdot \dfrac{\sigma}{\sqrt{n}}$, and upon solving this equation for n we get

Sample size for estimating μ ($n \geq 30$)

$$n = \left[\frac{z_{\alpha/2} \cdot \sigma}{E} \right]^2$$

EXAMPLE   The dean of a college wants to use the mean of a random sample to estimate the average amount of time students take to get from one class to the next, and she wants this estimate to be in error by at most 0.25 minute with probability 0.95. If she knows from previous studies of a similar kind that it is reasonable to let $\sigma = 1.50$ minutes, how large a sample will she have to take?

Solution   Substituting $z_{0.025} = 1.96$, $\sigma = 1.50$, and $E = 0.25$ into the formula for n, we get

$$n = \left[\frac{1.96 \cdot 1.50}{0.25} \right]^2 = 139$$

rounded up to the nearest whole number. Thus, a random sample of size $n = 139$ is required for the estimate.

As can be seen from the formula and also from the example, this method has the shortcoming that it cannot be used unless we know (at least approximately) the value of the standard deviation of the population whose mean we want to estimate.

299

(Exercises 9.1, 9.2, and 9.10 are practice exercises; their complete solutions are given on pages 352 and 353.)

9.1 A district official intends to use the mean of a random sample of 100 fifth graders to estimate the mean score which all the fifth graders in the district would get if they took a certain arithmetic achievement test. If, based on experience, the official knows that $\sigma = 9.2$ for such data, what can he assert with probability 0.95 about the maximum error?

9.2 A random sample of 50 cans of peach halves has a mean weight of 16.1 ounces and a standard deviation of 0.4 ounce. If $\bar{x} = 16.1$ ounces is used as an estimate of the mean weight of all the cans of peach halves in the large lot from which the sample came, with what confidence can we assert that the error of this estimate is at most 0.1 ounce?

9.3 A distributor of soft-drink vending machines plans to use the mean number of drinks dispensed by 40 of her machines during one week to estimate the average number of drinks she can expect any one of her machines to dispense per week. If she knows from experience that $\sigma = 43.6$ for such data and she can look upon the data as if they constitute a random sample, what can she assert with probability 0.99 about the maximum error?

9.4 An actuary wants to estimate his insurance company's average annual losses for a certain kind of liability coverage using the mean of a random sample of 200 policies with that kind of coverage from the company's very extensive files. If he knows from experience that $\sigma = \$43.52$ for such data, what can he say with probability 0.95 about the maximum error?

9.5 To estimate the average weight loss of persons on a certain diet, a research worker recorded the numbers of pounds lost by a random sample of 60 persons on the diet, obtaining a mean of 22.9 pounds and a standard deviation of 4.2 pounds. What can the research worker say with 98% confidence about the maximum error, if she uses the sample mean as an estimate of the true average weight loss of persons on the given diet?

9.6 A study of the annual growth of certain cacti showed that 64 of them, selected at random in a certain desert region, grew on the average 22.9 mm with a standard deviation of 2.5 mm during twelve months' time. What can we say with 99% confidence about the maximum error, if 22.9 mm is used as an estimate of the true average annual growth of the given kind of cactus?

9.7 A sample survey conducted in a large city in 1985 showed that 150 families spent on the average \$93.57 per week on food with a standard deviation of \$8.35. What can we say with 95% confidence about the maximum error if \$93.57 is used as an estimate of the actual average weekly food expenditures of families in the population sampled?

9.8 A study conducted by an airline showed that 120 of its passengers disembarking at Kennedy airport, a random sample, had to wait on the average 9.45 minutes with a standard deviation of 1.84 minutes to get their luggage. What can it say with 99% confidence about the maximum error, if the airline uses $\bar{x} = 9.45$ minutes as an estimate of the true average time it takes its passengers to get their luggage when disembarking at Kennedy airport?

9.9 A power company takes a random sample from its very extensive files and finds that the amounts owed on 200 delinquent accounts have a mean of $25.38 and a standard deviation of $5.36. If it uses $\bar{x} = \$25.38$ as an estimate of the average size of all its delinquent accounts, with what confidence can it assert that this estimate is off by at most $0.50?

9.10 In a study of television viewing habits, it is desired to estimate the average number of hours that teenagers spend watching per week. If it is reasonable to assume that $\sigma = 3.2$ hours for data of this kind, how large a sample is needed so that one will be able to assert with 99% confidence that the sample mean is off by at most half an hour?

9.11 It is desired to estimate the mean number of days of continuous use until a new refrigerator will first require repairs. If it can be assumed that $\sigma = 242$ days, how large a sample is needed so that it can be asserted with probability 0.95 that the sample mean will be off by less than 30 days?

9.12 Suppose that we want to estimate the mean score that adults living in a rural area get in a current events test, and we want to be able to assert with probability 0.99 that the error of our estimate, the mean of a random sample, will be less than 2.5. How large a sample will we need, if it can be assumed that $\sigma = 14.5$ for such scores?

9.13 Before bidding on a contract, a contractor wants to be 95% confident that she is in error by less than 20 minutes in estimating the average time it takes a certain kind of cement to dry. If the standard deviation of the time it takes the cement to dry can be assumed to equal 72 minutes, on how large a sample should she base her estimate?

9.2

Confidence Intervals for Means

Let us now introduce a different way of presenting the information provided by a sample mean and an assessment of the error we might make when we use it to estimate the mean of the population from which the sample came. In what follows, we shall make use of the fact that for large random samples from infinite populations, the sampling distribution of the mean is approximately normal with the mean μ and the standard deviation $\dfrac{\sigma}{\sqrt{n}}$, namely, that

$$z = \frac{\bar{x} - \mu}{\sigma/\sqrt{n}}$$

is a random variable having approximately the standard normal distribution. Since the probability is $1 - \alpha$ that a random variable having the standard normal distribution will take on a value between $-z_{\alpha/2}$ and $z_{\alpha/2}$ (see Figure 7.20), namely, that

$$-z_{\alpha/2} < z < z_{\alpha/2}$$

we can substitute into this inequality the foregoing expression for z and get

$$-z_{\alpha/2} < \frac{\bar{x} - \mu}{\sigma/\sqrt{n}} < z_{\alpha/2}$$

If we now apply some relatively simple algebra, we can rewrite the inequality as

Large-sample confidence interval for μ

$$\bar{x} - z_{\alpha/2} \cdot \frac{\sigma}{\sqrt{n}} < \mu < \bar{x} + z_{\alpha/2} \cdot \frac{\sigma}{\sqrt{n}}$$

and we can assert with probability $1 - \alpha$ that it will be satisfied for any given sample. In other words, we can assert with $(1 - \alpha)100\%$ confidence that the interval from $\bar{x} - z_{\alpha/2} \cdot \frac{\sigma}{\sqrt{n}}$ to $\bar{x} + z_{\alpha/2} \cdot \frac{\sigma}{\sqrt{n}}$, determined on the basis of a large random sample, contains the population mean we are trying to estimate. When σ is unknown and n is at least 30, we replace σ by the sample standard deviation s.

An interval like this is called a **confidence interval,** its endpoints are called **confidence limits,** and the probability $1 - \alpha$ is called the **degree of confidence.** As in Section 9.1, the values most commonly used for $1 - \alpha$ are 0.95 and 0.99, the corresponding values of $z_{\alpha/2}$ are 1.96 and 2.575, and the resulting confidence intervals are referred to as 95% and 99% confidence intervals for μ.

EXAMPLE With reference to the pulse rates on page 296, where we had $n = 32$, $\bar{x} = 26.2$, and $s = 5.15$, construct a 95% confidence interval for the true average increase in the pulse rate of persons performing the given task.

Solution Substituting $n = 32$, $\bar{x} = 26.2$, $s = 5.15$ for σ, and $z_{0.025} = 1.96$ into the confidence interval formula, we get

$$26.2 - 1.96 \cdot \frac{5.15}{\sqrt{32}} < \mu < 26.2 + 1.96 \cdot \frac{5.15}{\sqrt{32}}$$

$$24.4 < \mu < 28.0$$

for the true average increase in the pulse rate of persons performing the given task. Of course, the interval from 24.4 to 28.0 contains μ or it does not, and we really don't know which, but the 95% confidence implies that the interval was obtained by a method which "works 95% of the time." To put it differently, the interval may contain μ or it may not, but if we had to bet, 95 to 5 (or 19 to 1) would be fair odds that it does.

Had we wanted to calculate a 99% confidence interval in this example, we would have substituted 2.575 instead of 1.96 for $z_{\alpha/2}$, and we would have obtained $23.9 < \mu < 28.5$. Comparison of the two confidence intervals shows that

> **When we increase the degree of certainty, namely, the degree of confidence, the confidence interval becomes wider and thus tells us less about the quantity we are trying to estimate.**

In other words, "the surer we want to be, the less we have to be sure of."

When we use a confidence interval to estimate the mean of a population, we call this kind of estimate an **interval estimate.** In contrast to point estimates, interval estimates require no further elaboration about their reliability—this is taken care of indirectly by their width and the degree of confidence.

9.3

Confidence Intervals for Means (Small Samples)

In the preceding section we assumed that the sample is large enough, $n \geq 30$, to treat the sampling distribution of the mean as if it were a normal distribution and, when necessary, to replace σ with s. To develop corresponding theory which applies also to small samples, it will be necessary to assume that the population we are sampling has roughly the shape of a normal distribution. We can then base our method on the statistic

$$t = \frac{\bar{x} - \mu}{s/\sqrt{n}}$$

whose sampling distribution is a continuous distribution called the **t distribution.** As is shown in Figure 9.2, this distribution is symmetrical and bell-shaped with zero mean; in fact, its shape is very similar to that of the standard normal distribution. The exact shape of the t distribution depends on a parameter called the **number of degrees of freedom,** or simply the **degrees of freedom,** which, as the distribution will be used here, is given by $n - 1$, the sample size

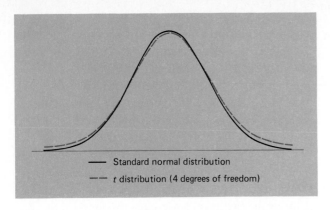

— Standard normal distribution
-- *t* distribution (4 degrees of freedom)

9.2

*Standard normal distribution
and t distribution.*

less one. For the t distribution we define $t_{\alpha/2}$ in the same way in which we defined $z_{\alpha/2}$, so that the area under the curve to the right of $t_{\alpha/2}$ is equal to $\alpha/2$. However, $t_{\alpha/2}$ depends on $n - 1$, the number of degrees of freedom, and its value will have to be looked up in each case in Table III at the end of the book.

Now we proceed as on page 302. Making use of the fact that the t distribution is symmetrical about $t = 0$, we find that the probability is $1 - \alpha$ that a random variable having the t distribution will take on a value between $-t_{\alpha/2}$ and $t_{\alpha/2}$ (see Figure 9.3), namely, that

$$-t_{\alpha/2} < t < t_{\alpha/2}$$

Then, substituting into this inequality the foregoing expression for t, we get

$$-t_{\alpha/2} < \frac{\bar{x} - \mu}{s/\sqrt{n}} < t_{\alpha/2}$$

9.3

t distribution.

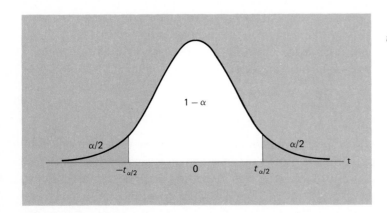

CHAP. 9: The Analysis of Measurements

and using some relatively simple algebra, we arrive at the following **small-sample confidence interval for μ:**

Small-sample confidence interval for μ

$$\bar{x} - t_{\alpha/2} \cdot \frac{s}{\sqrt{n}} < \mu < \bar{x} + t_{\alpha/2} \cdot \frac{s}{\sqrt{n}}$$

The degree of confidence is $1 - \alpha$ and the only difference between this confidence interval formula and the large-sample formula (with s substituted for σ) is that $t_{\alpha/2}$ takes the place of $z_{\alpha/2}$.

EXAMPLE

To test the durability of a new paint for white center lines, a highway department painted test strips across heavily traveled roads in eight different locations, and electronic counters showed that they deteriorated after having been crossed by (to the nearest hundred) 142,600, 167,800, 136,500, 108,300, 126,400, 133,700, 162,000, and 149,400 cars. Construct a 95% confidence interval for the average amount of traffic (car crossings) this paint can withstand before it deteriorates.

Solution

The mean and the standard deviation of these values are $\bar{x} = 140,800$ and $s = 19,200$ (to the nearest hundred), and since $t_{0.025}$ for $8 - 1 = 7$ degrees of freedom equals 2.365, substitution into the formula yields

$$140,800 - 2.365 \cdot \frac{19,200}{\sqrt{8}} < \mu < 140,800 + 2.365 \cdot \frac{19,200}{\sqrt{8}}$$

$$124,700 < \mu < 156,900$$

This is the desired 95% confidence interval estimate of the average amount of traffic (car crossings) the paint can withstand before it deteriorates.

A computer printout of the preceding example is shown in Figure 9.4. The difference between the confidence limits given in the printout and those given above are due to rounding; since the original data were given to the nearest hundred, we rounded the mean, the standard deviation, and the confidence limits to the nearest hundred.

The method which we used in Section 9.1 to determine the maximum error we risk with probability $1 - \alpha$ when we use a sample mean to estimate the mean of a population can easily be adapted to small samples, provided that

305

SEC. 9.3: Confidence Intervals for Means (Small Samples)

```
MTB > SET C1
DATA> 142600   167800   136500   108300   126400   133700   162000   149400
MTB > TINTERVAL WITH 95 PERCENT CONFIDENCE, DATA IN C1

          N      MEAN    STDEV   SE MEAN    95.0 PERCENT C.I.
C1        8     140837    19228     6798   ( 124758,   156917)
```

9.4

Computer printout for small-sample confidence interval for μ.

the population we are sampling has roughly the shape of a normal distribution. All we have to do is substitute s for σ and $t_{\alpha/2}$ for $z_{\alpha/2}$ in the formula for E on page 297.

EXAMPLE   In 12 test runs an experimental engine consumed on the average 12.9 gallons of gasoline per minute with a standard deviation of 1.6 gallons. What can we assert with 99% confidence about the maximum error, if we use $\bar{x} = 12.9$ gallons as an estimate of the true average gasoline consumption of the engine?

Solution   Substituting $s = 1.6$, $n = 12$, and $t_{0.005} = 3.106$ (the entry in Table III for $12 - 1 = 11$ degrees of freedom) into the modified formula for E, we get

$$E = t_{\alpha/2} \cdot \frac{s}{\sqrt{n}} = 3.106 \cdot \frac{1.6}{\sqrt{12}} = 1.43$$

Thus, if we use the mean $\bar{x} = 12.9$ gallons per minute as an estimate of the true average gasoline consumption of the engine and it is reasonable to assume that the data constitute a random sample from a normal population, we can assert with 99% confidence that the error of this estimate is less than 1.43 gallons.

EXERCISES   (Exercises 9.14 and 9.20 are practice exercises; their complete solutions are given on page 353.)

9.14 With reference to Exercise 9.1, where we had $n = 100$ and $\sigma = 9.2$, suppose that fifth graders in the sample averaged 62.7 on the test. Construct a 99% confidence interval for the mean score which all the fifth graders in the district would get if they took the arithmetic achievement test.

CHAP. 9: The Analysis of Measurements

9.15 With reference to Exercise 9.5, where we had $n = 40$, $\bar{x} = 22.85$ minutes, and $s = 2.91$ minutes, construct a 95% confidence interval for the true average time it takes an automobile mechanic to perform the given task.

9.16 With reference to Exercise 9.6, where we had $n = 64$, $\bar{x} = 22.9$ mm, and $s = 2.5$ mm, construct a 95% confidence interval for the true average annual growth of the given kind of cactus.

9.17 With reference to Exercise 9.7, where we had $n = 150$, $\bar{x} = \$93.57$, and $s = \$8.35$, construct a 99% confidence interval for the actual average weekly food expenditures in 1985 of families in the population sampled.

9.18 With reference to Exercise 9.8, where we had $n = 120$, $\bar{x} = 9.45$ minutes, and $s = 1.84$ minutes, construct a 95% confidence interval for the true average time it takes passengers to get their luggage when disembarking at Kennedy airport.

9.19 With reference to Exercise 9.9, where we had $n = 200$, $\bar{x} = 25.38$, and $s = \$5.36$, construct a 90% confidence interval for the actual average size of all of the power company's delinquent accounts.

9.20 In an air pollution study, an experiment station obtained a mean of 2.36 micrograms of suspended benzene-soluble organic matter per cubic meter with a standard deviation of 0.48 from a random sample of size $n = 10$.
  (a) Construct a 99% confidence interval for the mean of the population sampled.
  (b) What can be asserted with 95% confidence about the maximum error, if $\bar{x} = 2.36$ micrograms is used as an estimate of the mean of the population sampled?

9.21 A major truck stop has kept extensive records of various transactions with its customers. If a random sample of 15 of these records shows average sales of 63.9 gallons of diesel fuel with a standard deviation of 2.8 gallons, construct a 95% confidence interval for the truck stop's average sales of diesel fuel.

9.22 With reference to the preceding exercise, what can be asserted with 99% confidence about the maximum error if $\bar{x} = 63.9$ gallons is used as an estimate of the truck stop's average sales of diesel fuel?

9.23 Nine bearings made by a certain process have a mean diameter of 1.005 cm and a standard deviation of 0.004 cm. What can be said with 95% confidence about the maximum error if $\bar{x} = 1.005$ cm is used as an estimate of the mean diameter of bearings made by the given process?

9.24 With reference to the preceding exercise, construct a 99% confidence interval for the mean diameter of bearings made by the given process.

9.25 In six attempts it took a locksmith 9, 14, 7, 8, 11, and 5 seconds to open a certain kind of lock. Verify that $\bar{x} = 9$ and $s = 3.16$ for these data, and construct a 95% confidence interval for the average time it takes the locksmith to open this kind of lock.

9.26 In establishing the authenticity of an ancient coin, its weight is often of critical importance. If four experts independently weighed a Phoenician tetradrachm and obtained 14.28, 14.34, 14.26, and 14.32 grams, verify that $\bar{x} = 14.30$ and $s = 0.037$ for the given data. What can one assert with 99% confidence about the maximum error, if $\bar{x} = 14.30$ grams is used as an estimate of the actual weight of the coin?

9.27 If it took a random sample of six mechanics 13, 14, 18, 14, 16, and 15 minutes to assemble a certain device, construct a 95% confidence interval for the mean time it takes a mechanic to assemble the device.

 9.28 Use a computer package to rework Exercise 9.25.

 9.29 With reference to Exercise 9.26, use a computer package to construct a 99% confidence interval for the true weight of the coin.

 9.30 Use a computer package to rework Exercise 9.27.

9.4

Confidence Intervals for Standard Deviations ⋆

Since there are many occasions where we must estimate population standard deviations, for instance, when we substitute s for σ to calculate a large-sample confidence interval for the mean, let us now show how confidence intervals for σ may be determined on the basis of large samples. There also exists a small-sample technique, but it will not be discussed in this book.

For large samples, $n \geq 30$, we base our method on the theory that the sampling distribution of s is approximately normal with the mean σ and the standard deviation $\sigma_s = \dfrac{\sigma}{\sqrt{2n}}$, which is called the **standard error of s.** Thus, the sampling distribution of

$$z = \frac{s - \sigma}{\sigma/\sqrt{2n}}$$

is approximately the standard normal distribution, and if we substitute this expression for z into the inequality $-z_{\alpha/2} < z < z_{\alpha/2}$ (as on page 302) and use some relatively simple algebra, we arrive at the following **large-sample confidence interval for σ:**

Large-sample confidence interval for σ

$$\frac{s}{1 + \dfrac{z_{\alpha/2}}{\sqrt{2n}}} < \sigma < \frac{s}{1 - \dfrac{z_{\alpha/2}}{\sqrt{2n}}}$$

As before, the degree of confidence is $1 - \alpha$.

EXAMPLE   With reference to the example on page 296, where we obtained $\bar{x} = 26.2$ and $s = 5.15$ for the increase in the pulse rate of $n = 32$ persons performing a certain strenuous task, construct a 95% confidence interval for the standard deviation of the population sampled.

Solution   Substituting $n = 32$, $s = 5.15$, and $z_{0.025} = 1.96$ into the large-sample confidence interval formula for σ, we get

$$\frac{5.15}{1 + \dfrac{1.96}{\sqrt{64}}} < \sigma < \frac{5.15}{1 - \dfrac{1.96}{\sqrt{64}}}$$

$$4.14 < \sigma < 6.82$$

This is an estimate of the true variability of the increase in the pulse rate of persons performing the given task.

EXERCISES   (Exercise 9.31 is a practice exercise; its complete solution is given on page 353.)

⋆ 9.31 With reference to Exercise 9.2, where we had $n = 50$ and $s = 0.4$ ounce, construct a 99% confidence interval for the standard deviation of the weights of all the cans of peach halves from which the sample came.

⋆ 9.32 With reference to Exercise 9.5, where we had $n = 60$ and $s = 4.2$ pounds, construct a 95% confidence interval for the standard deviation of the weight losses of persons on the given diet.

⋆ 9.33 With reference to Exercise 9.6, where we had $n = 64$ and $s = 2.5$ mm, construct a 99% confidence interval for the standard deviation of the annual growth of the given kind of cactus.

⋆ 9.34 With reference to Exercise 9.7, where we had $n = 150$ and $s = \$8.35$, construct a 95% confidence interval for the standard deviation of the weekly food expenditures of all families in the given city in 1985.

⋆ 9.35 With reference to Exercise 9.8, where we had $n = 120$ and $s = 1.84$ minutes, construct a 99% confidence interval for the true standard deviation of the amount of time that passengers disembarking at Kennedy airport have to wait for their luggage.

9.5

Tests of Hypotheses

In Sections 9.1 through 9.3 we learned how to estimate a population mean μ by giving a confidence interval or by accompanying the point estimate $\bar{x}$ with an assessment of the possible error. Now we shall learn how to test a **hypothesis**

SEC. 9.5: Tests of Hypotheses

about a population mean μ; that is, we shall present methods for deciding whether to accept or reject an assertion about a particular value, or particular values, of μ. As we already indicated on page 295, the biologist would be testing a hypothesis if he wanted to check (see whether to accept or reject) the claim that the average wingspan of certain insects is $\mu = 13.4$ mm. In general,

> A statistical hypothesis is an assertion or conjecture about the parameter, or parameters, of a population; it may also concern the type, or nature, of a population.

So far as the second part of this definition is concerned, we shall test in Section 10.7 whether it is reasonable to treat a random variable as having a binomial distribution, or perhaps a Poisson distribution, and whether it is reasonable to treat a set of data as coming from a normal population. In this chapter we shall be concerned only with hypotheses concerning the population mean μ.

In order to know exactly what to expect when a hypothesis is true, we often have to hypothesize the opposite of what we hope to prove. For instance, if we want to find out whether one method of teaching computer programming is more effective than another, we hypothesize that the two methods are equally effective. Also, if we want to determine whether one method of irrigating the soil is more expensive than another, we hypothesize that the two methods are equally expensive; and if we want to see whether a new steel alloy is stronger than ordinary steel, we hypothesize that their strength is the same. Since we hypothesize in each case that there is no difference—no difference in the effectiveness of the two teaching methods, no difference in the cost of the two methods of irrigation, and no difference in the strength of the two kinds of steel—we refer to hypotheses like these as **null hypotheses** and denote them by H_0. Actually, the term "null hypothesis" is used nowadays for any hypothesis set up primarily to see whether it can be rejected, and the idea of setting up a null hypothesis is common even in nonstatistical thinking. It is precisely what we do in criminal proceedings, where an accused is assumed to be innocent unless his guilt is established beyond a reasonable doubt. The assumption that the accused is not guilty is a null hypothesis.

In order to know when to reject a null hypothesis, we must also formulate an **alternative hypothesis.** Denoted by H_A, this is the hypothesis which we accept when the null hypothesis can be rejected. For instance, in the wingspan example the alternative hypothesis might be $\mu \neq 13.4$ mm or $\mu > 13.4$ mm, and in the criminal proceedings referred to above, the alternative hypothesis is that the accused is guilty.

EXAMPLE   A research worker wants to compare the effectiveness of advertising a department store sale with newspaper ads or on television. Formulate a null hypothesis and an alternative hypothesis appropriate for this situation.

Solution H_0: The two methods of advertising are equally effective.

H_A: The two methods of advertising are not equally effective.

There is no question here about H_0, but depending on what the research worker hopes to show, the alternative hypothesis might also be

H_A: Newspaper ads are more effective.

or

H_A: Advertising on television is more effective.

It might even be that newspaper ads are twice as effective as advertising on television.

To illustrate the problems we face when testing a statistical hypothesis, let us refer again to the wingspan example, and let us suppose that the biologist wants to test the null hypothesis

$$H_0: \mu = 13.4 \text{ mm}$$

against the alternative hypothesis

$$H_A: \mu \neq 13.4 \text{ mm}$$

where μ is the mean wingspan of the given kind of insect. To perform this test, he decides to take a random sample of size $n = 40$ with the intention of accepting the null hypothesis if the mean of the sample falls anywhere from 13.2 to 13.6 mm; otherwise, the null hypothesis is rejected.

This provides a clear-cut criterion for accepting or rejecting the null hypothesis, but unfortunately it is not infallible. Since the decision is based on a sample, there is the possibility that the sample mean will be greater than 13.6 mm or less than 13.2 mm even though the true mean is 13.4 mm, and there is also the possibility that the sample mean will fall in the interval from 13.2 mm to 13.6 mm even though the true mean is, say, 13.7 mm. Thus, before adopting the criterion (and, for that matter, any decision criterion) it would seem wise to investigate the chances that it will lead to a wrong decision.

Assuming that it is known from similar studies that $\sigma = 0.8$ mm for this kind of data, let us first investigate the possibility that the sample mean will be greater than 13.6 mm or less than 13.2 mm even though the true mean is 13.4 mm. The probability that this will happen purely due to chance is given by the combined area of the white regions of Figure 9.5, and it can easily be determined by approximating the sampling distribution of the mean with a normal distribution. Assuming that the population sampled is large enough to be treated as if it were infinite, which seems very reasonable in this case, we have $\sigma_{\bar{x}} = \dfrac{\sigma}{\sqrt{n}} = \dfrac{0.8}{\sqrt{40}} = 0.126$, and it follows that the dividing lines of the criterion, in standard units, are

$$z = \frac{13.2 - 13.4}{0.126} = -1.59 \quad \text{and} \quad z = \frac{13.6 - 13.4}{0.126} = 1.59$$

According to Table II, the area in each "tail" of the sampling distribution of Figure 9.5 is $0.5000 - 0.4441 = 0.0559$, and hence the probability of getting a value in either tail of the sampling distribution is $0.0559 + 0.0559 = 0.1118$, or approximately 0.11. This is the probability of erroneously rejecting the null hypothesis; namely, the probability of rejecting it when it is true. Whether this

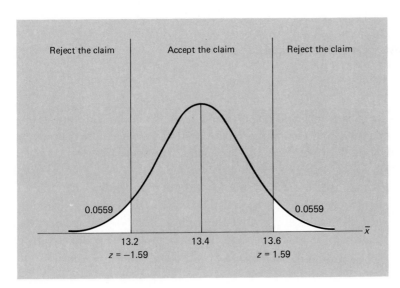

9.5

Test criterion, where "Accept the claim" and "Reject the claim" refer to the corresponding intervals of $\bar{x}$.

is an acceptable risk is for the biologist to decide, and it will have to depend on the consequences of making this kind of error.

Let us now look at the other possibility, where the test fails to detect that the null hypothesis is false; namely, that $\mu \neq 13.4$ mm. Assuming again that $\mu = 13.7$ mm, purely for the sake of argument, we find that the probability that the sample mean will fall anywhere from 13.2 to 13.6 is given by the area of the white region of Figure 9.6, namely, the area under the curve between 13.2 and 13.6. The mean of the sampling distribution is now 13.7, its standard deviation is again $\sigma_{\bar{x}} = \dfrac{0.8}{\sqrt{40}} = 0.126$, so that the dividing lines of the criterion, in standard units, are now

$$z = \frac{13.2 - 13.7}{0.126} = -3.97 \quad \text{and} \quad z = \frac{13.6 - 13.7}{0.126} = -0.79$$

Since the area under the curve to the left of $z = -3.97$ is negligible, it follows from Table II that the area of the white region of Figure 9.6 is $0.5000 - 0.2852 = 0.2148$, or approximately 0.21. This is the probability of erroneously accepting the null hypothesis; namely, the probability of accepting it when it is false. Again, it is up to the biologist to decide whether this is an acceptable risk.

9.6

Test criterion.

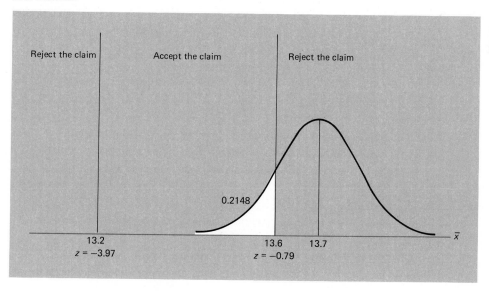

The situation described in this example is typical of testing a statistical hypothesis, and it may be summarized as in the following table:

	Accept H_0	Reject H_0
H_0 is true	Correct decision	Type I error
H_0 is false	Type II error	Correct decision

If the null hypothesis H_0 is true and accepted, or false and rejected, the decision is in either case correct; if it is true and rejected, or false and accepted, the decision is in either case in error.

> The error of rejecting H_0 when it is true is called a Type I error and the probability of committing such an error is denoted by α (Greek lowercase *alpha*).
>
> The error of accepting H_0 when it is false is called a Type II error and the probability of committing such an error is denoted by β (Greek lowercase *beta*).

Thus, in our example we showed that for the given test criterion $\alpha = 0.11$ and $\beta = 0.21$ when the mean wingspan of the given kind of insect is actually 13.7 mm.

EXAMPLE   Suppose that in our example $\mu = 13.0$ mm, which the biologist does not know, and that the mean of his sample is $\bar{x} = 13.3$ mm. What type of error will he commit, if he uses the criterion of Figure 9.5?

Solution   Since $\bar{x} = 13.3$ mm falls in the interval from 13.2 to 13.6 mm, he will accept the null hypothesis $\mu = 13.4$ mm even though it is false; this is a Type II error.

In calculating the probability of a Type II error in our example, we arbitrarily chose the alternative value $\mu = 13.7$. However, in this problem, as in most others, there are infinitely many other alternatives, and for each of them there is a positive probability β of erroneously accepting H_0. So, in practice we choose some key alternative values and calculate the probabilities β of committing a Type II error, as we calculated it for $\mu = 13.7$, or we sidestep the issue by proceeding as will be explained in Section 9.6.

If we do calculate β for various alternative values and plot these probabilities as in Figure 9.7, the resulting curve is called an **operating characteristic**

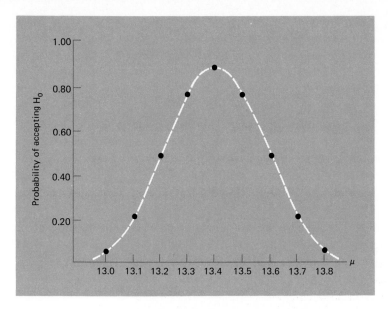

9.7

Operating characteristic curve.

curve, or simply an **OC-curve.** Such a curve provides a good overall picture of the risks to which one is exposed by a test criterion.

The operating characteristic curve shown in Figure 9.7 pertains to the wingspan example, and in Exercise 9.43 the reader will be asked to verify some of the other values of β. Since the probability of a Type II error is the probability of accepting H_0 when it is false, we "completed the picture" in Figure 9.7 by labeling the vertical scale "Probability of accepting H_0" and plotting at $\mu = 13.4$ the probability $1 - \alpha = 1 - 0.11 = 0.89$.

The problems we met in this section are not limited to the particular example dealing with the wingspan of the given kind of insect; the same questions would also have arisen if we had wanted to test the hypothesis that the mean age of divorced women at the time of their divorce is 35.6 years, the hypothesis that an antibiotic is 82% effective, the hypothesis that a computer-assisted method of instruction will on the average raise a student's score on a standard achievement test by 8.2 points, and so forth.

9.6

Significance Tests

In the example of the preceding section, we had less trouble with Type I errors than with Type II errors, because we formulated H_0 as a **simple hypothesis** about μ; that is, we formulated H_0 so that the parameter μ took on a single

SEC. 9.6: Significance Tests

value and the probability of a Type I error could be calculated. Had we formulated instead the **composite hypothesis** $\mu \neq 13.4$ mm, the composite hypothesis $\mu < 13.4$ mm, or the composite hypothesis $\mu > 13.4$ mm, where in each case μ can take on more than one possible value, we could not have calculated the probability of a Type I error without specifying by how much μ differs from, is less than, or is greater than 13.4 mm.

As we saw in Section 9.5, the probability of a Type I error can easily be calculated when we are given a simple hypothesis, an unambiguous criterion, and enough information to apply the sampling theory of Chapter 8. We also saw how the probabilities of Type II errors may have to be calculated for specific alternative values of μ. The following illustrates how we can sidestep Type II errors altogether.

Suppose that a retailer wants to find out whether the average income of families living within a mile of a proposed shopping center is greater than \$14,000. So, he has a number of these families interviewed and bases his decision on the following criterion:

> **Reject the null hypothesis $\mu = \$14,000$ and accept the alternative $\mu > \$14,000$ if the sample mean exceeds, say, \$14,400; if the sample mean does not exceed \$14,400, reserve judgment (perhaps, pending further checks).**

If we reserve judgment, there is no possibility of committing a Type II error—no matter what happens, the null hypothesis is never really accepted. This is all right in our example because the retailer wants to know primarily whether the null hypothesis can be rejected.

The procedure we have outlined here is called a **significance test.** If the difference between what we expect under the null hypothesis and what we observe in a sample is too large to be reasonably attributed to chance, we reject the null hypothesis. If the difference between what we expect and what we observe is so small that it may well be attributed to chance, we say that the results are **not statistically significant.**

With respect to the wingspan example, the biologist could convert the criterion on page 311 into that of a significance test by writing

> **Reject the null hypothesis $\mu = 13.4$ mm (and accept the alternative $\mu \neq 13.4$ mm) if the mean of a random sample of size $n = 40$ is greater than 13.6 mm or less than 13.2 mm; reserve judgment if the mean of the sample falls on the interval from 13.2 to 13.6 mm.**

So far as the rejection of the null hypothesis is concerned, the criterion has remained unchanged and the probability of a Type I error is still 0.11. However, so far as its acceptance is concerned, the biologist is now playing it safe by reserving judgment.

Reserving judgment in a significance test is similar to what happens in court proceedings where the prosecution does not have sufficient evidence to get a conviction, but where it would be going too far to say that the defendent definitely did not commit the crime. In general, whether one can afford the luxury of reserving judgment in any given situation depends entirely on the nature of the problem. If a decision must be reached one way or the other, there is no way of avoiding the risk of committing a Type II error.

Since the general problem of testing hypotheses and constructing statistical decision criteria may seem confusing, at least to the beginner, it will help to proceed systematically as outlined in the following five steps:

1. **We formulate a simple null hypothesis and an appropriate alternative hypothesis which is to be accepted when the null hypothesis must be rejected.**

In the wingspan example the null hypothesis was $\mu = 13.4$ mm, and the alternative hypothesis was $\mu \neq 13.4$ mm (since the biologist wanted to reject 13.4 mm if this value is too high or too low). We refer to this kind of alternative as a **two-sided alternative.** In the family income example the null hypothesis is $\mu = \$14{,}000$, and the alternative hypothesis is $\mu > \$14{,}000$ (since the retailer wants to know whether average family income in the given area exceeds \$14,000). This is called a **one-sided alternative.** We can also write a one-sided alternative with the inequality going the other way. For instance, if we hope to be able to show that the mean time required to do a certain job is less than 20 minutes, we would test the null hypothesis $\mu = 20$ minutes against the alternative hypothesis $\mu < 20$ minutes.

As in the examples of the preceding paragraph, alternative hypotheses usually specify that the population mean (or whatever other parameter may be of concern) is not equal to, greater than, or less than the value assumed under the null hypothesis. For any given problem, the choice of an appropriate alternative hypothesis depends mostly on what we hope to be able to show, or better, perhaps, where we want to put the burden of proof.

EXAMPLE   A dress manufacturer whose sewing machines average 128 dresses per work shift, is considering the purchase of new sewing machines. Against what alternative hypothesis would he test the null hypothesis $\mu = 128$ if

(a) he does not want to buy the new machines unless they will actually increase the output;

(b) he wants to buy the new machines (which have some other nice features) unless they will actually decrease the output.

Solution (a) He would use the alternative hypothesis $\mu > 128$ and buy the new machines only if the null hypothesis can be rejected.
(b) He would use the alternative hypothesis $\mu < 128$ and buy the new machines unless the null hypothesis is rejected.

EXAMPLE   A company uses a production process designed to make machine components that have an average thickness of 5 inches. The company suspects that the process is not maintaining its intended average.
(a) If the company wants to modify its process if the average thickness is smaller than 5 inches, what null hypothesis and alternative hypothesis should it use?
(b) If the company wants to modify its process if the average thickness is different from 5 inches, what null hypothesis and alternative hypothesis should it use?

Solution (a) The words "smaller than" suggest that the hypothesis $\mu < 5$ inches is needed together with the hypothesis $\mu = 5$ inches. Only the second of these hypotheses is a simple hypothesis where μ takes on a single value, so it must be the null hypothesis and we write

$$H_0: \ \mu = 5 \text{ inches}$$

$$H_A: \ \mu < 5 \text{ inches}$$

(b) The words "different from" suggest that the hypothesis $\mu \neq 5$ inches is needed together with the hypothesis $\mu = 5$ inches. Again, only the second of the two hypotheses is a simple hypothesis where μ takes on a single value. Using it as the null hypothesis, we write

$$H_0: \ \mu = 5 \text{ inches}$$

$$H_A: \ \mu \neq 5 \text{ inches}$$

2. We specify the probability of a Type I error; if possible, desired, or necessary, we may also make some specifications about the probabilities of Type II errors for specific alternatives.

The probability of a Type I error is also called the **level of significance,** and it is usually set at $\alpha = 0.05$ or $\alpha = 0.01$. Testing a null hypothesis at the level of significance $\alpha = 0.05$ simply means that we are fixing the probability of rejecting the null hypothesis even though it is true at 0.05. The decision to use $\alpha = 0.05$, $\alpha = 0.01$, or some other value, depends mostly on the consequences of committing a Type I error in the given situation. Observe, however, that we cannot make the probability of a Type I error too small, for this will have the tendency to make the probability of serious Type II errors too large. In actual practice, experimenters choose α depending on the risks; in the exercises in this text, the level of significance will always be specified.

3. **Based on the sampling distribution of an appropriate statistic and the choice of the level of significance, we construct a test criterion for testing the null hypothesis against the given alternative.**
4. **We calculate from the data the value of the statistic on which the decision is to be based.**
5. **We decide whether to reject the null hypothesis, whether to accept it, or whether to reserve judgment.**

So far as step 3 is concerned, in the wingspan example we based the criterion on the normal-curve approximation to the sampling distribution of the mean. In general, this step depends on the statistic upon which we want to base the decision and on its sampling distribution. Looking back at our two illustrations—the wingspan example and the family income example—we find that the construction of a test criterion depends also on the alternative hypothesis we happen to choose. In the wingspan example, we used a **two-sided criterion (two-sided test,** or **two-tailed test)** with the two-sided alternative hypothesis $\mu \neq 13.4$ mm, rejecting the null hypothesis for large or small values of the sample mean; in the family income example, we used a **one-sided criterion (one-sided test,** or **one-tailed test)** with the one-sided alternative hypothesis $\mu > \$14,000$, rejecting the null hypothesis only for large values of the sample mean. In general, a test is called two-sided or two-tailed if the null hypothesis is rejected for values of the **test statistic** falling into either tail of its sampling distribution, and it is called one-sided or one-tailed if the null hypothesis is rejected only for values of the test statistic falling into one specified tail of its sampling distribution. By "test statistic" we mean the statistic (for instance, the sample mean) on which the test is to be based.

In connection with step 5, let us point out that we often accept null hypotheses with the tacit hope that we are not exposed to overly high risks of committing serious Type II errors. Of course, if it is necessary we can calculate enough probabilities of Type II errors to get an overall picture from the operating characteristic curve of the test criterion.

As at the end of Section 9.5, let us also point out that the concepts we have introduced here are not limited to tests concerning population means; they apply equally to tests concerning other parameters, or tests concerning the nature, or form, of populations.

EXERCISES

(Exercises 9.36, 9.40, and 9.44 are practice exercises; their complete solutions are given on pages 353 and 354.)

9.36 A doctor has to examine a school's athletes to check whether they are fit for intercollegiate competition.
   (a) What type of error would he commit if he erroneously rejects the hypothesis that a given athlete is fit for intercollegiate competition?
   (b) What type of error would he commit if he erroneously accepts the hypothesis that a given athlete is fit for intercollegiate competition?

9.37 Suppose we want to test the hypothesis that the average noise level of a vacuum cleaner meets specifications. Explain under what conditions we would be committing a Type I error and under what conditions we would be committing a Type II error.

9.38 A poll wants to test the hypothesis that 60% of the public is against smoking in supermarkets.
   (a) What type of error would the poll commit if it erroneously accepts the hypothesis?
   (b) What type of error would the poll commit if it erroneously rejects the hypothesis?

9.39 A professor of education is concerned with the effectiveness of a method of computer-assisted instruction.
   (a) What hypothesis is she testing if she would commit a Type I error by erroneously concluding that the method is effective?
   (b) What hypothesis is she testing if she would commit a Type II error by erroneously concluding that the method is effective?

9.40 With reference to the wingspan example on page 311, suppose that the biologist increases the sample size to $n = 60$ while everything else remains unchanged.
   (a) Show that this decreases the probability of a Type I error from 0.11 to 0.05.
   (b) Show that this decreases the probability of a Type II error when $\mu = 13.7$ mm from 0.21 to 0.17.

9.41 With reference to the wingspan example on page 311, suppose that the biologist changes the criterion so that the hypothesis $\mu = 13.4$ mm is accepted if the sam-

ple mean falls anywhere from 13.1 to 13.7 mm, while otherwise the hypothesis will be rejected. If everything else remains the same, show that
  (a) this will decrease the probability of a Type I error from 0.11 to 0.02;
  (b) this will increase the probability of a Type II error when $\mu = 13.7$ from 0.21 to 0.50.

9.42 With reference to the wingspan example on page 311, suppose that the biologist increases the sample size to $n = 100$ while everything else remains unchanged.
  (a) What is the probability that the biologist will erroneously reject the null hypothesis $\mu = 13.4$ mm?
  (b) What is the probability that the biologist will erroneously accept the null hypothesis $\mu = 13.4$ mm when actually $\mu = 13.7$ mm?

9.43 With reference to the operating characteristic curve of Figure 9.7, verify that the probabilities of Type II errors are
  (a) 0.78 when $\mu = 13.5$ mm or $\mu = 13.3$ mm;
  (b) 0.50 when $\mu = 13.6$ mm or $\mu = 13.2$ mm;
  (c) 0.06 when $\mu = 13.8$ mm or $\mu = 13.0$ mm.

9.44 A department store has a salesperson whom it suspects of making more mistakes than the average of all its salespersons.
  (a) If the department store decides to let the salesperson go if this suspicion is confirmed, what null hypothesis and what alternative hypothesis should it use?
  (b) If the department store decides to let the salesperson go unless he actually makes fewer mistakes than the average of all its salespersons, what null hypothesis and what alternative hypothesis should it use?

9.45 The average drying time of a manufacturer's paint is 20 minutes. Investigating the effectiveness of a modification in the chemical composition of the paint, the manufacturer wants to test the null hypothesis $\mu = 20$ against a suitable alternative, where μ is the mean drying time of the modified paint.
  (a) What alternative hypothesis should the manufacturer use if she does not want to make the modification unless it actually decreases the drying time of the paint?
  (b) What alternative hypothesis should the manufacturer use if she wants to make the modification unless it actually increases the drying time of the paint?

9.7

Tests Concerning Means

Having used tests concerning means to illustrate the basic principles of hypothesis testing, let us now see how we proceed in practice. Actually, we will depart somewhat from the procedure used in the examples given earlier in this chapter.

In the wingspan example and also in the family income example we stated the test criterion in terms of values of $\bar{x}$. Now we will base it on the statistic

Statistic for large-sample test concerning mean

$$z = \frac{\bar{x} - \mu_0}{\sigma/\sqrt{n}}$$

where μ_0 is the value of the mean assumed under the null hypothesis. The reason for working with this statistic, which amounts to using standard units, is that it enables us to formulate criteria which are applicable to a great variety of problems, not just one.

We refer to the test based on the statistic above as a **large-sample test** because we are using the normal-curve approximation to the sampling distribution of the mean. Thus, the test requires that $n \geq 30$, unless the population we are sampling is normal. If that is the case, the test may be used for any value of n.

EXAMPLE   A psychologist wants to determine whether the average time it takes an adult driver to react to a certain emergency situation is really 0.56 second, as has been claimed by others. From information gathered in similar studies, she can assume that the variability of such measurements is given by a standard deviation of $\sigma = 0.082$ second. Also, she decides to base the test on a random sample of size $n = 35$ and to use the 0.05 level of significance. What will she conclude if her data yield $\bar{x} = 0.59$ second?

Solution   1. *Hypotheses*

$$H_0: \mu = 0.56 \text{ second}$$

$$H_A: \mu \neq 0.56 \text{ second}$$

where the alternative hypothesis is two-sided because the psychologist will want to reject the null hypothesis if $\mu = 0.56$ second is either too high or too low.

2. *Level of significance*

$$\alpha = 0.05$$

3. *Criterion*

Putting half of 0.05 into each tail of the sampling distribution and making use of the fact that $z_{0.025} = 1.96$, reject the null hypothesis if

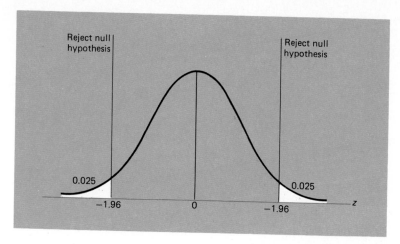

9.8

Test criterion for reaction-time example.

$z < -1.96$ or $z > 1.96$, where

$$z = \frac{\bar{x} - \mu_0}{\sigma/\sqrt{n}}$$

Otherwise, accept it or reserve judgment (see also Figure 9.8).

4. *Calculations*

Substituting $n = 35$, $\bar{x} = 0.59$, $\mu_0 = 0.56$, and $\sigma = 0.082$ into the formula for z, we get

$$z = \frac{0.59 - 0.56}{0.082/\sqrt{35}} = \frac{0.03}{0.0139} = 2.16$$

5. *Decision*

Since $z = 2.16$ exceeds 1.96, the null hypothesis must be rejected. In other words, the difference between $\bar{x} = 0.59$ and $\mu_0 = 0.56$ is too large to be attributed to chance, and this allows the psychologist to conclude that the 0.56 figure must be wrong.

In general, for tests concerning means, the dividing lines of the criteria are as shown in Figure 9.9.

1. **For the one-sided alternative $\mu < \mu_0$ the test is one-tailed, α is placed in the left-hand tail, and the dividing line is $-z_\alpha$.**

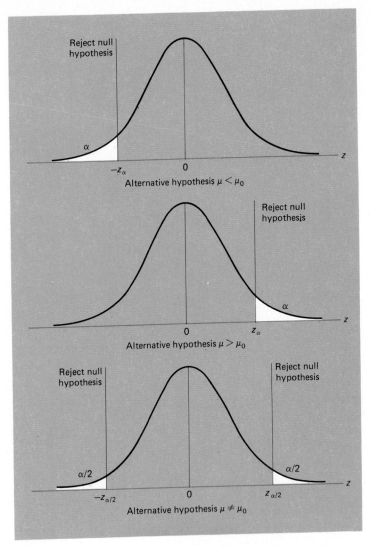

9.9

Test criteria.

2. For the one-sided alternative $\mu > \mu_0$ the test is one-tailed, α is placed in the right-hand tail, and the dividing line is z_α.
3. For the two-sided alternative $\mu \neq \mu_0$ the test is two-tailed, α is divided equally between the two tails, and the dividing lines are $-z_{\alpha/2}$ and $z_{\alpha/2}$.

If $\alpha = 0.05$, the dividing lines, or **critical values,** of the criteria are -1.645 or 1.645 for the one-sided alternatives, and -1.96 and 1.96 for the two-sided alternative; if $\alpha = 0.01$, the dividing lines of the criteria are -2.33 or 2.33 for the one-sided alternatives, and -2.575 and 2.575 for the two-sided alternative. All these values come directly from Table II (see Exercise 7.12 and the example on page 244).

In most practical situations where σ is unknown, we must substitute for it the sample standard deviation s. Again, this is permissible when the sample is large; namely, when $n \geq 30$.

EXAMPLE   A trucking firm suspects that the average lifetime of 28,000 miles claimed for certain tires is too high. To check the claim, the firm puts 40 of these tires on its trucks and gets a mean lifetime of 27,563 miles and a standard deviation of 1,348 miles. What can it conclude at the 0.01 level of significance, if it tests the null hypothesis $\mu = 28,000$ miles against an appropriate alternative?

Solution   1. *Hypotheses*

$$H_0: \mu = 28,000 \text{ miles}$$

$$H_A: \mu < 28,000 \text{ miles}$$

since the firm is interested in determining whether the tires may, perhaps, not last as long as claimed.

2. *Level of significance*

$$\alpha = 0.01$$

3. *Criterion*

Reject the null hypothesis if $z < -2.33$, where

$$z = \frac{\bar{x} - \mu_0}{\sigma/\sqrt{n}}$$

with σ replaced by s; otherwise, accept it or reserve judgment.

4. *Calculations*

Substituting $n = 40$, $\bar{x} = 27,563$, $\mu_0 = 28,000$, and $s = 1,348$ for σ into the formula for z, we get

$$z = \frac{27,563 - 28,000}{1,348/\sqrt{40}} = -2.05$$

5. *Decision*

Since $z = -2.05$ is not less than -2.33, the null hypothesis cannot be rejected. The trucking firm's suspicion is not confirmed and it may accept the claim about the tires or reserve judgment.

9.8
Tests Concerning Means (Small Samples)

When we do not know the value of the population standard deviation σ and the sample is small, $n < 30$, we assume, as on page 303, that the population we are sampling has roughly the shape of a normal distribution, and base our decision on the statistic

Statistic for
small-sample test
concerning mean

$$t = \frac{\bar{x} - \mu_0}{s/\sqrt{n}}$$

whose sampling distribution is the t distribution (see page 303) with $n - 1$ degrees of freedom. Of course, if it cannot be assumed that the population we are sampling has roughly the shape of a normal distribution, this small-sample procedure cannot be used.

The criteria we use for this test are those of Figure 9.9 on page 324 with z replaced by t and z_α and $z_{\alpha/2}$ replaced by t_α and $t_{\alpha/2}$. As was explained on page 304, for given degrees of freedom, t_α and $t_{\alpha/2}$ are values for which the area to their right under the corresponding t distribution is equal to α and $\alpha/2$. All the critical values for this **one-sample t test** may be read from Table III, with the number of degrees of freedom equal to $n - 1$.

EXAMPLE   Suppose we want to test, on the basis of a random sample of size $n = 5$, whether the fat content of a certain kind of processed meat exceeds 30%. What can we conclude at the 0.01 level of significance, if the sample values are 31.9, 30.3, 32.1, 31.7, and 30.9%?

Solution   1. *Hypotheses*

$$H_0: \mu = 30\%$$
$$H_A: \mu > 30\%$$

2. *Level of significance*

$$\alpha = 0.01$$

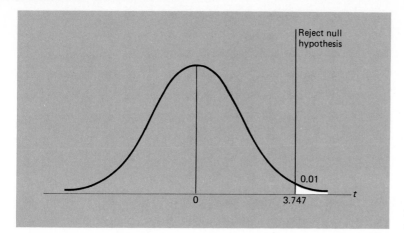

Reject null
hypothesis

0.01

0   3.747 t

9.10

*Test criterion for fat-content
example.*

3. *Criterion*

   Reject the null hypothesis if $t > 3.747$, the value of $t_{0.01}$ for $5 - 1 = 4$ degrees of freedom, where

   $$t = \frac{\bar{x} - \mu_0}{s/\sqrt{n}}$$

   (see also Figure 9.10); otherwise, accept it or reserve judgment.

4. *Calculations*

   First calculating the mean and the standard deviation of the sample, we get $\bar{x} = 31.38$ and $s = 0.756$, and substituting these values together with $n = 5$ into the formula for t, we find that

   $$t = \frac{31.38 - 30}{0.756/\sqrt{5}} = 4.08$$

5. *Decision*

   Since $t = 4.08$ exceeds 3.747, the null hypothesis must be rejected. In other words, the mean fat content of the given kind of processed meat exceeds 30%.

A computer printout of the preceding one-sample t test (with the data given as proportions instead of percentages) is shown in Figure 9.11. The $+1$ following "ALTERNATIVE" is a code for the alternative hypothesis $\mu > \mu_0$, in this case $\mu > 30\%$. "P VALUE 0.0075" means that the probability of getting a value

327

```
MTB > SET C1
DATA> .319   .303   .321   .317   .309
MTB > TTEST MU=.30, ALTERNATIVE +1, ON DATA IN C1

TEST OF MU = 0.300 VS MU G.T. 0.300

              N     MEAN     STDEV    SE MEAN      T     P VALUE
C1            5   0.31380   0.00756   0.0034     4.08    0.0075
```

9.11

Computer printout for one-sample t test.

greater than or equal to the observed value of t is 0.0075 when the null hypothesis is true. Since this probability is less than $\alpha = 0.01$, we conclude, as before, that the null hypothesis must be rejected.

EXERCISES

(Exercises 9.46 and 9.51 are practice exercises; their complete solutions are given on pages 354 and 355.)

9.46 A law student, who wants to check a professor's claim that convicted embezzlers spend on the average 12.8 months in jail, takes a random sample of 60 such cases from court files. Using his results, namely, $\bar{x} = 11.2$ months and $s = 3.5$ months, test the null hypothesis $\mu = 12.8$ months against the alternative hypothesis $\mu \neq 12.8$ months at the 0.01 level of significance.

9.47 According to the norms established for a history test, eighth graders should average 81.7 with a standard deviation of 8.5. If 100 randomly selected eighth graders from a certain school district averaged 79.6 on this test, can we conclude at the 0.05 level of significance that eighth graders from this school district can be expected to average less than the norm of 81.7 on this test?

9.48 According to specifications, the mean time required to inflate a rubber raft is 7.5 seconds. As it has been suggested that this figure might be too low, a random sample of 45 of the rafts are inflated, yielding $\bar{x} = 7.6$ seconds and $s = 0.6$ second. What can we conclude at the 0.01 level of significance?

9.49 The security department of a warehouse wants to know whether the average time required by the night watchman to walk his round is 12.0 minutes. If, in a random sample of 36 rounds, the night watchman averaged 12.3 minutes with a standard deviation of 1.2 minutes, can we reject the null hypothesis $\mu = 12.0$ minutes at the level of significance 0.05?

9.50 In a study of new sources of food, it is reported that a pound of a certain kind of fish yields on the average 2.45 ounces of FPC (fish-protein concentrate), which is used to enrich various food products. Is this figure supported by a study in which 30 samples of this kind of fish yielded on the average 2.48 ounces of FPC (per pound of fish) with a standard deviation of 0.07 ounce, if

we use

   (a) the level of significance $\alpha = 0.05$;

   (b) the level of significance $\alpha = 0.01$?

9.51 In an experiment with a new tranquilizer, the pulse rate of 12 patients was determined before they were given the tranquilizer and again five minutes later, and their pulse rate was found to be reduced on the average by 7.2 beats with a standard deviation of 1.8. At the level of significance 0.05, can we conclude that on the average this tranquilizer will reduce the pulse rate of a patient by less than 9.0 beats?

9.52 A manufacturer guarantees a certain ball bearing to have a mean outside diameter of 0.7500 inch with a standard deviation of 0.0030. If a random sample of 10 such bearings has a mean outside diameter of 0.7510, can we reject the manufacturer's guarantee with regard to the mean outside diameter at the level of significance 0.01?

9.53 A random sample of 12 graduates of a secretarial school average 72.6 words per minute with a standard deviation of 4.2 words per minute. Use the level of significance 0.05 to test an employer's claim that the school's graduates average less than 75.0 words per minute.

9.54 A soft-drink vending machine is set to dispense 6.0 ounces per cup. If the machine is tested nine times, yielding a mean cup fill of 6.2 ounces with a standard deviation of 0.15 ounce, is this evidence at the level of significance 0.05 that the machine is overfilling cups?

9.55 In ten test runs, a truck operated for 8, 10, 10, 7, 9, 12, 10, 8, 7, and 9 miles with one gallon of a certain gasoline. Is this evidence at the 0.05 level of significance that the truck is not operating at an average of 11.5 miles per gallon with this gasoline?

9.56 The yield of alfalfa from six test plots is 1.2, 2.2, 1.9, 1.1, 1.8, and 1.4 tons per acre. Test at the level of significance $\alpha = 0.05$ whether this supports the contention that the true average yield for this kind of alfalfa is less than 2.0 tons per acre.

9.57 A random sample from a company's very extensive files shows that orders for a certain piece of machinery were filled, respectively, in 12, 10, 17, 14, 13, 18, 11, and 9 days. Use the 0.01 level of significance to test the claim that on the average such orders are filled in 9.5 days. Choose the alternative hypothesis in such a way that rejection of the null hypothesis $\mu = 9.5$ days implies that it takes longer than that.

 9.58 Use a computer package to rework Exercise 9.55.

 9.59 Use a computer package to rework Exercise 9.56. What is the probability of getting a value of t less than or equal to that calculated for the given data?

 9.60 Use a computer package to rework Exercise 9.57. What is the probability of getting a value of t greater than or equal to that calculated for the given data?

329

9.9

Differences Between Means

There are many problems in which we must decide whether an observed difference between two sample means can be attributed to chance, or whether it is indicative of the fact that the two samples came from populations with unequal means. For instance, we may want to know whether there really is a difference in the mean gasoline consumption of two kinds of cars, when sample data show that one kind averaged 24.6 miles per gallon while, under the same conditions, the other kind averaged 25.7 miles per gallon. Similarly, we may want to decide on the basis of sample data whether men can perform a certain task faster than women, whether one kind of ceramic insulator is more brittle than another, whether the average diet in one country is more nutritious than that in another country, and so on.

The method we shall use to test whether an observed difference between two sample means can be attributed to chance, or whether it is statistically significant, is based on the following theory: If $\bar{x}_1$ and $\bar{x}_2$ are the means of two **independent random samples,** then the sampling distribution of the statistic $\bar{x}_1 - \bar{x}_2$ has the mean

$$\mu_1 - \mu_2$$

and the standard deviation

$$\sqrt{\frac{\sigma_1^2}{n_1} + \frac{\sigma_2^2}{n_2}}$$

where μ_1, μ_2, σ_1, and σ_2 are the means and the standard deviations of the two populations sampled. It is customary to refer to the standard deviation of this sampling distribution as the **standard error of the difference between two means.**

By "independent" samples we mean that the selection of one sample is in no way affected by the selection of the other. Thus, the theory does not apply to "before and after" kinds of comparisons, nor does it apply, say, if we want to compare the IQ's of husbands and wives. A special method for comparing the means of dependent samples is explained in Section 9.11.

To base tests of the significance between two sample means on the normal distribution, we shall have to convert to standard units, writing

$$z = \frac{\bar{x}_1 - \bar{x}_2 - (\mu_1 - \mu_2)}{\sqrt{\frac{\sigma_1^2}{n_1} + \frac{\sigma_2^2}{n_2}}}$$

where we subtracted from $\bar{x}_1 - \bar{x}_2$ the mean of its sampling distribution and then divided by the standard deviation of its sampling distribution.

Then, if we limit ourselves to large samples, $n_1 \geq 30$ and $n_2 \geq 30$, we can base the test of the null hypothesis $\mu_1 = \mu_2$ on the foregoing z statistic with $\mu_1 - \mu_2 = 0$, namely, on the statistic

Statistic for large-sample test concerning difference between two means

$$z = \frac{\bar{x}_1 - \bar{x}_2}{\sqrt{\dfrac{\sigma_1^2}{n_1} + \dfrac{\sigma_2^2}{n_2}}}$$

which has approximately the standard normal distribution.

Depending on whether the alternative hypothesis is $\mu_1 < \mu_2$, $\mu_1 > \mu_2$, or $\mu_1 \neq \mu_2$, the criteria we use for these significance tests of the difference between two means are again those of Figure 9.9 on page 324 with $\mu_1 - \mu_2$ substituted for μ and 0 substituted for μ_0. We can refer here to Figure 9.9 even though we are concerned with the sampling distribution of the difference between two means instead of the sampling distribution of the mean, because the criteria are all given in standard units.

The test we have described here is essentially a large-sample test; it is exact only when both of the populations we are sampling are normal. In most practical situations where σ_1 and σ_2 are unknown, we must make the further approximation of substituting for them the sample standard deviations s_1 and s_2.

EXAMPLE   In a study designed to test whether there is a difference between the average heights of adult females born in two different countries, random samples yielded the following results:

$$n_1 = 120 \qquad \bar{x}_1 = 62.7 \qquad s_1 = 2.50$$
$$n_2 = 150 \qquad \bar{x}_2 = 61.8 \qquad s_2 = 2.62$$

where the measurements are in inches. Use the level of significance 0.05 to test the null hypothesis that the corresponding population means are equal against the alternative hypothesis that they are not equal.

Solution   1. *Hypotheses*

$$H_0: \mu_1 = \mu_2$$
$$H_A: \mu_1 \neq \mu_2$$

2. *Level of significance*

$$\alpha = 0.05$$

3. *Criterion*
Reject the null hypothesis if $z < -1.96$ or $z > 1.96$, where

$$z = \frac{\bar{x}_1 - \bar{x}_2}{\sqrt{\dfrac{\sigma_1^2}{n_1} + \dfrac{\sigma_2^2}{n_2}}}$$

with s_1 and s_2 substituted for σ_1 and σ_2; otherwise, accept the null hypothesis or reserve judgment.

4. *Calculations*
Substituting $n_1 = 120$, $n_2 = 150$, $\bar{x}_1 = 62.7$, $\bar{x}_2 = 61.8$, $s_1 = 2.50$, and $s_2 = 2.62$ into the formula for z, we get

$$z = \frac{62.7 - 61.8}{\sqrt{\dfrac{(2.50)^2}{120} + \dfrac{(2.62)^2}{150}}} = 2.88$$

5. *Decision*
Since $z = 2.88$ exceeds 1.96, the null hypothesis must be rejected. In other words, the sample data show that there is a difference between the average heights of adult females born in the two countries.

9.10

Differences Between Means (Small Samples)

As in Section 9.8, a small-sample test of the significance of the difference between two means may be based on an appropriate t statistic. For this test, which is used when $n_1 < 30$ or $n_2 < 30$, we must assume that we have independent random samples from populations which can be approximated closely by normal distributions with the same standard deviation. Then, we can base our decision on the statistic

Statistic for small-sample test concerning difference between two means

$$t = \frac{\bar{x}_1 - \bar{x}_2}{\sqrt{\dfrac{(n_1 - 1)s_1^2 + (n_2 - 1)s_2^2}{n_1 + n_2 - 2} \cdot \left(\dfrac{1}{n_1} + \dfrac{1}{n_2}\right)}}$$

whose sampling distribution is the t distribution with $(n_1 - 1) + (n_2 - 1) = n_1 + n_2 - 2$ degrees of freedom. Of course, if we cannot assume that the two populations sampled are approximately normal and that they have the same standard deviation, this small-sample procedure cannot be used.

The criteria we use for this **two-sample t test** are again those of Figure 9.9 on page 324, with t, t_α, and $t_{\alpha/2}$ substituted for z, z_α, and $z_{\alpha/2}$, with $\mu_1 - \mu_2$ substituted for μ, and 0 substituted for μ_0.

EXAMPLE   In five games with a relatively light ball, a professional bowler scored 205, 220, 200, 210, and 201, and in five games with a somewhat heavier ball he scored 218, 204, 223, 198, and 211. At the 0.05 level of significance, can we conclude that on the average he will score higher with the heavier ball?

Solution   1. *Hypotheses*

$$H_0: \mu_1 = \mu_2$$
$$H_A: \mu_1 < \mu_2$$

2. *Level of significance*

$$\alpha = 0.05$$

3. *Criterion*

Reject the null hypothesis if $t < -1.860$, the value of $-t_{0.05}$ for $5 + 5 - 2 = 8$ degrees of freedom, where t is given by the formula above; otherwise, accept the null hypothesis or reserve judgment.

4. *Calculations*

First calculating the sample means and the sample standard deviations, we get $\bar{x}_1 = 207.20$, $\bar{x}_2 = 210.80$, $s_1 = 8.17$, and $s_2 = 10.13$, and substituting these values together with $n_1 = 5$ and $n_2 = 5$ into the formula for t, we find that

$$t = \frac{207.20 - 210.80}{\sqrt{\frac{4(8.17)^2 + 4(10.13)^2}{8} \cdot \left(\frac{1}{5} + \frac{1}{5}\right)}} = -0.62$$

5. *Decision*

Since $t = -0.62$ is not less than -1.860, the null hypothesis cannot be rejected; in other words, the difference between the two sample

333

means is so small that it may reasonably be attributed to chance. If we must reach a decision one way or the other, we conclude that the bowler performs equally well with the two balls.

Most statistical computer packages include programs for the two-sample t test. For the preceding example, we might thus have obtained the printout shown in Figure 9.12. This printout also shows a 95% confidence interval for $\mu_1 - \mu_2$, which we did not ask for, and "$P = 0.28$" means that the probability of getting a value less than or equal to the observed value of t is 0.28 when the null hypothesis is true. Since this probability exceeds 0.05, we conclude, as before, that the null hypothesis cannot be rejected.

```
MTB > SET C1
DATA> 205   220   200   210   201
MTB > SET C2
DATA> 218   204   223   198   211
MTB > POOL T, -1, C1 C2

TWOSAMPLE T FOR C1 VS C2
        N       MEAN      STDEV    SE MEAN
C1      5      207.20      8.17      3.7
C2      5      210.80     10.10      4.5

95 PCT CI FOR MU C1 - MU C2: (-17.0,  9.8)
TTEST MU C1 = MU C2 (VS LT): T=-0.62 P=0.28 DF=8.0
```

9.12

Computer printout for two-sample t test.

9.11

Differences Between Means (Paired Data)

The methods of Section 9.9 and 9.10 do not apply when the samples are not independent. For instance, they cannot be used for "before and after" kinds of comparisons, or studies of differences in IQ between husbands and wives. To handle data of this kind, we work with the (signed) differences of the paired data and test whether these differences may be looked upon as a random sample from a population which has the mean $\mu = 0$. If the sample is small, we use the one-sample t test of Section 9.8; otherwise, we use the large-sample test of Section 9.7.

EXAMPLE Use the 0.05 level of significance to test the effectiveness of an industrial safety program on the basis of the following data on the average weekly loss of labor-hours due to accidents in ten plants "before and after" the program was put into operation: 45 and 36, 73 and 60, 46 and 44, 124 and 119, 33 and 35, 57 and 51, 83 and 77, 34 and 29, 26 and 24, and 17 and 11.

Solution The differences are 9, 13, 2, 5, −2, 6, 6, 5, 2, and 6, and for these data we perform the following test:

1. *Hypotheses*

$$H_0: \mu = 0$$

$$H_A: \mu > 0$$

2. *Level of significance*

$$\alpha = 0.05$$

3. *Criterion*
Reject the null hypothesis if $t > 1.833$, the value of $t_{0.05}$ for $10 - 1 = 9$ degrees of freedom, where

$$t = \frac{\bar{x} - \mu_0}{s/\sqrt{n}}$$

Otherwise, accept the null hypothesis or reserve judgment.

4. *Calculations*
Since $\bar{x} = 5.2$ and $s = 4.08$ for the ten differences, substitution of these values together with $n = 10$ and $\mu_0 = 0$ into the formula for t yields

$$t = \frac{5.2 - 0}{4.08/\sqrt{10}} = 4.03$$

5. *Decision*
Since $t = 4.03$ exceeds 1.833, the null hypothesis must be rejected, and we have thus shown that the industrial safety program is effective.

SEC. 9.11: Differences Between Means (Paired Data)

(Exercises 9.61, 9.65, and 9.70 are practice exercises; their complete solutions are given on pages 355 and 356.)

9.61 A sample study was made of the number of business lunches that executives claim as deductible expenses per month. If 60 executives in the insurance industry averaged 9.6 such deductions with a standard deviation of 1.8 in a given month, while 50 bank executives averaged 8.4 with a standard deviation of 2.1, test at the level of significance 0.01 whether the difference between these two sample means is significant.

9.62 To compare freshmen's knowledge of current events in two community colleges, samples of 60 freshmen from each of the two community colleges were given a special test. If those from the first community college obtained an average score of 76.4 with a standard deviation of 5.0, while those from the second community college obtained an average of 71.8 with a standard deviation of 4.6, test at the 0.01 level of significance whether the difference between these two sample means is significant.

9.63 A sample study of the number of pieces of chalk the average professor uses was conducted at two universities. If 80 professors at one university averaged 12.5 pieces of chalk per month with a standard deviation of 2.4, while 60 professors at the other university averaged 11.3 pieces of chalk per month with a standard deviation of 3.3, test at the 0.05 level of significance whether the difference between the means is significant.

9.64 Suppose that we want to investigate whether males and females earn comparable wages in a certain industry. If sample data show that 60 males earn on the average $212.50 per week with a standard deviation of $15.60, while 60 females earn on the average $196.10 per week with a standard deviation of $18.20, test the null hypothesis $\mu_1 - \mu_2 = 0$ against the alternative hypothesis $\mu_1 > \mu_2$ at the level of significance 0.01.

9.65 Measurements of the heat-producing capacity of coal from two mines yielded the following results:

$$n_1 = 5 \qquad \bar{x}_1 = 8,160 \qquad s_1 = 252$$
$$n_2 = 5 \qquad \bar{x}_2 = 7,730 \qquad s_2 = 207$$

where the measurements are in millions of calories per ton. At the level of significance 0.05, can we conclude that the mean heat-producing capacity of coal from the two mines is not the same?

9.66 Fifteen randomly selected mature citrus trees of one variety have a mean height of 14.8 feet with a standard deviation of 1.3 feet, while twelve randomly selected mature citrus trees of another variety have a mean height of 13.6 feet with a standard deviation of 1.5 feet. Test at the level of significance 0.01 whether the difference between the two sample means is significant.

9.67 Twelve measurements each of the hydrogen content (in percent number of atoms) of gases collected from the eruptions of two volcanos yielded $\bar{x}_1 = 41.5$,

$\bar{x}_2 = 46.1$, $s_1 = 5.2$, and $s_2 = 6.7$. Use the level of significance $\alpha = 0.05$ to test the null hypothesis that there is no difference (with regard to hydrogen content) in the composition of the gases from the two eruptions.

9.68 The following are independent random samples of the IQ's of teenagers belonging to two different ethnic groups:

Group A: 98, 104, 101, 98, 96, 103, 99, 95, 105, 101
Group B: 105, 95, 103, 107, 100, 99, 108, 114, 107, 102

Use the level of significance $\alpha = 0.05$ to test the claim that teenagers of group A have a lower average IQ than teenagers of group B.

9.69 The following are the numbers of sales which a random sample of nine salesmen of industrial chemicals in California and a random sample of six salesmen of industrial chemicals in Oregon made over a fixed period of time:

California: 41, 47, 62, 39, 56, 64, 37, 61, 52
Oregon: 34, 63, 45, 55, 24, 43

Use the 0.01 level of significance to test whether the difference between the means of these two samples is significant.

9.70 In a study of the effectiveness of physical exercise in weight reduction, a group of 32 persons engaged in a prescribed program of physical exercise for one month showed the following results:

Weight before (pounds)	Weight after (pounds)	Weight before (pounds)	Weight after (pounds)
209	196	170	164
178	171	153	152
169	170	183	179
212	207	165	162
180	177	201	199
192	190	179	173
158	159	243	231
180	180	144	140
211	203	179	180
193	183	202	197
245	229	169	175
188	190	187	190
201	194	213	205
222	219	174	170
190	195	196	197
199	197	201	201

337

Test at the 0.01 level of significance whether the prescribed program of exercise is effective.

9.71 The following data were obtained in an experiment designed to check whether there is a systematic difference in the weights (in grams) obtained with two different scales:

	Scale I	Scale II
Rock specimen 1	12.13	12.17
Rock specimen 2	17.56	17.61
Rock specimen 3	9.33	9.35
Rock specimen 4	11.40	11.42
Rock specimen 5	28.62	28.61
Rock specimen 6	10.25	10.27
Rock specimen 7	23.37	23.42
Rock specimen 8	16.27	16.26
Rock specimen 9	12.40	12.45
Rock specimen 10	24.78	24.75

Test at the 0.01 level of significance whether the difference between the means of the weights obtained with the two scales is significant.

9.72 The following are the ratings of two supervisors of the performance of a random sample of 30 employees on a scale from 1 to 25:

Supervisor A	Supervisor B	Supervisor A	Supervisor B
25	23	24	24
23	22	24	22
21	23	25	25
22	20	20	15
15	17	16	16
22	22	19	18
17	20	21	19
18	15	17	17
25	22	23	22
21	23	19	19
20	23	19	15
17	16	20	18
16	15	18	18
23	20	24	22
21	18	23	24

Test at the 0.05 level of significance whether the difference between the means of the ratings of the two supervisors is significant.

9.73 Use a computer package to rework Exercise 9.68.

 9.74 Use a computer package to rework Exercise 9.69.

9.12

Differences Among *k* Means ✶

In the three preceding sections we developed procedures for testing the hypothesis that an observed difference between two sample means can be attributed to chance. Now we consider the problem of deciding whether observed differences among *more than two* sample means can be attributed to chance, or whether they indicate actual differences among the means of the populations sampled. For instance, we may want to decide on the basis of sample data whether there really is a difference in the effectiveness of three methods of teaching computer programming, we may want to compare the yield of four varieties of wheat, we may want to see whether there really is a difference in the average mileage obtained with five kinds of gasoline, we may want to judge whether there really is a difference in the performance of eight different hair driers, and so on.

Suppose we actually want to compare the effectiveness of three methods of teaching the programming of a certain computer—method 1, which is straight teaching-machine instruction; method 2, which involves the personal attention of an instructor and some direct experience working with the computer; and method 3, which involves the personal attention of an instructor but no work with the computer itself. Suppose, furthermore, that random samples of size four are taken from large groups of students taught by the three methods and that these students obtained the following scores in an appropriate achievement test:

$$
\begin{array}{ll}
\textit{Method 1:} & 71, 75, 65, 69 \\
\textit{Method 2:} & 90, 80, 86, 84 \\
\textit{Method 3:} & 72, 77, 76, 79
\end{array}
$$

The means of these three samples are $\bar{x}_1 = 70$, $\bar{x}_2 = 85$, and $\bar{x}_3 = 76$, and we would like to know whether the differences among them are significant or whether they can be attributed to chance.

If μ_1, μ_2, and μ_3 are the means of the three populations sampled in this example, we shall want to test the null hypothesis $\mu_1 = \mu_2 = \mu_3$ against the alternative hypothesis that these means are not all equal. This null hypothesis would be supported if the differences among the sample means, $\bar{x}_1$, $\bar{x}_2$, and $\bar{x}_3$, are small; the alternative hypothesis would be supported if at least some of the differences among the sample means are large. Thus, we need a precise measure

339

of the discrepancies among the $\bar{x}$'s, and with it a rule which tells us when the discrepancies are so large that the null hypothesis can be rejected.

Possible choices for such a measure are the standard deviation of the $\bar{x}$'s or their variance. To determine the latter, we first calculate the mean of the three $\bar{x}$'s, getting

$$\frac{70 + 85 + 76}{3} = 77$$

and then we find that

$$s_{\bar{x}}^2 = \frac{(70 - 77)^2 + (85 - 77)^2 + (76 - 77)^2}{3 - 1} = 57$$

where the subscript $\bar{x}$ serves to indicate that $s_{\bar{x}}^2$ measures the variation of the sample means.

Let us now make two assumptions which are critical to the method by which we shall continue the analysis of our problem:

1. **The populations we are sampling can be approximated closely with normal distributions.**
2. **These populations all have the same standard deviation σ.**

With reference to our example, this means that we are assuming that (1) the test scores, for each method of teaching, are values of a random variable having (at least approximately) a normal distribution, and that (2) these random variables all have the same standard deviation σ.

With these assumptions, and if the null hypothesis $\mu_1 = \mu_2 = \mu_3$ is true, we can look upon the three samples as if they came from one and the same (normal) population and, hence, upon the variance of their means, $s_{\bar{x}}^2$, as an estimate of $\sigma_{\bar{x}}^2$, the square of the standard error of the mean.

Now, since $\sigma_{\bar{x}} = \dfrac{\sigma}{\sqrt{n}}$ for samples from infinite populations, we can look upon $s_{\bar{x}}^2$ as an estimate of $\sigma_{\bar{x}}^2 = \left(\dfrac{\sigma}{\sqrt{n}}\right)^2 = \dfrac{\sigma^2}{n}$, where n is the size of each sample. Then, multiplying by n, we can look upon $n \cdot s_{\bar{x}}^2$ as an estimate of σ^2, and it is important to note that this estimate is based on the variation among the sample means. For our example, we thus have

$$n \cdot s_{\bar{x}}^2 = 4 \cdot 57 = 228$$

as an estimate of σ^2, the common variance of the three populations.

CHAP. 9: The Analysis of Measurements

If σ^2 were known, we could compare $n \cdot s_{\bar{x}}^2$ with σ^2 and reject the null hypothesis that the population means are all equal if $n \cdot s_{\bar{x}}^2$ is much larger than σ^2. However, in most practical situations σ^2 is not known and we have no choice but to estimate it on the basis of the sample data.

Since we assumed under the null hypothesis that our three samples come from identical populations, we could use any one of the sample variances, s_1^2, s_2^2, or s_3^2 as an estimate of σ^2, and we could also use their mean. Averaging, or **pooling,** the three sample variances in our example, we get

$$\frac{s_1^2 + s_2^2 + s_3^2}{3} = \frac{1}{3}\left[\frac{(71-70)^2 + (75-70)^2 + (65-70)^2 + (69-70)^2}{4-1}\right.$$

$$+ \frac{(90-85)^2 + (80-85)^2 + (86-85)^2 + (84-85)^2}{4-1}$$

$$\left. + \frac{(72-76)^2 + (77-76)^2 + (76-76)^2 + (79-76)^2}{4-1}\right]$$

$$= \frac{130}{9} \text{ (or } 14\tfrac{4}{9})$$

We now have the following two estimates of σ^2,

$$n \cdot s_{\bar{x}}^2 = 228 \quad \text{and} \quad \frac{s_1^2 + s_2^2 + s_3^2}{3} = 14\tfrac{4}{9}$$

and it should be observed that whereas the first estimate measures the variation among the sample means, the second estimate measures the variation within the three samples, namely, chance variation. If the $\bar{x}$'s are far apart and the first of these two estimates is much larger than the second, it thus stands to reason that the null hypothesis ought to be rejected. To put this comparison on a rigorous basis, we use the statistic

Statistic for
test concerning
differences
among means

$$F = \frac{\textit{variation among the samples}}{\textit{variation within the samples}}$$

where the variation among the samples is measured by $n \cdot s_{\bar{x}}^2$ and the variation within the samples is measured by the mean of the three sample variances.

If the null hypothesis is true and the assumptions we made are valid, the sampling distribution of this statistic is the **F distribution,** a theoretical distribution which depends on two parameters called the **numerator** and **denominator**

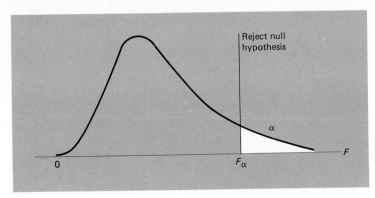

9.13

F distribution.

degrees of freedom. When the F statistic is used to compare the means of k samples of size n, the numerator and denominator degrees of freedom are, respectively, $k-1$ and $k(n-1)$.

In general, if F is close to 1 (that is, if the variation among the samples just about equals the variation within the samples and, hence, reflects only chance variation), the null hypothesis of equal population means cannot be rejected. However, if F is quite large (that is, if the variation among the samples is greater than we would expect it to be if it were due only to chance), the null hypothesis of equal population means will have to be rejected. To determine how large an F we need to reject the null hypothesis, we base our decision on the criterion of Figure 9.13, where F_α is such that the area under the curve to its right is equal to α. For $\alpha = 0.05$ and $\alpha = 0.01$, the values of F_α may be looked up in Table V at the end of the book.

Returning to our numerical example, we find that $F = \dfrac{228}{14\frac{4}{9}} = 15.8$, and since this exceeds 8.02, the value of $F_{0.01}$ for $k-1 = 3-1 = 2$ and $k(n-1) = 3(4-1) = 9$ degrees of freedom, the null hypothesis must be rejected at the 0.01 level of significance. In other words, we conclude that the differences among the sample means are too large to be attributed to chance.

EXERCISES

(Exercise 9.75 is a practice exercise; its complete solution is given on page 357.)

★ **9.75** An agronomist planted three test plots each with four varieties of wheat and obtained the following yields (in pounds per plot):

Variety A: 60, 61, 56
Variety B: 59, 52, 51
Variety C: 55, 55, 52
Variety D: 58, 58, 55

(a) Calculate $n \cdot s_{\bar{x}}^2$ for these data, the mean of the variances of the four samples, and the value of F.

(b) Use the 0.05 level of significance to test whether the differences among the four sample means can be attributed to chance.

★ 9.76 The following are the numbers of mistakes made on five occasions by three compositors setting the type for a technical report:

Compositor 1:   10, 13, 9, 11, 12
Compositor 2:   11, 13, 8, 16, 12
Compositor 3:   10, 15, 13, 11, 16

(a) Calculate $n \cdot s_{\bar{x}}^2$ for these data, the mean of the variances of the three samples, and the value of F.

(b) Use the 0.05 level of significance to test whether the differences among the three sample means can be attributed to chance.

★ 9.77 The following are the mileages which a test driver got with four gallons each of five brands of gasoline:

Brand A: 30, 25, 27, 26
Brand B: 29, 26, 29, 28
Brand C: 32, 32, 35, 37
Brand D: 29, 34, 32, 33
Brand E: 32, 26, 31, 27

(a) Calculate $n \cdot s_{\bar{x}}^2$ for these data, the mean of the variances of the five samples, and the value of F.

(b) Use the 0.01 level of significance to test whether the differences among the five sample means can be attributed to chance.

9.13
Analysis of Variance ★

Let us now look at the example of the preceding section from a different point of view, that of an **analysis of variance.** The basic idea of an analysis of variance is to express a measure of the total variation of a set of data as a sum of terms, which can be attributed to specific sources, or causes, of variation. With reference to our example, two such sources of variation would be (1) actual differences in the effectiveness of the three teaching techniques, and (2) chance, which in problems like this is referred to as the **experimental error.**

One's ability to perform an analysis of variance can be of great importance in scientific work, particularly when many factors produce certain results and we are interested in their individual contributions. For instance, we may want to see whether observed differences in cleansing action are due to the use of different detergents, differences in water temperature, differences in the hardness of the water, differences among the washing machines used in the experiment, and perhaps even differences among the instruments used to obtain the necessary readings. The method by which we analyze a complex situation like this is beyond the scope of this text, but it is a direct generalization of the work of this section.

As a measure of the total variation of kn observations consisting of k samples of size n, we shall use the **total sum of squares**[†]

$$SST = \sum_{i=1}^{k} \sum_{j=1}^{n} (x_{ij} - \bar{x}_{..})^2$$

where x_{ij} is the jth observation of the ith sample, $i = 1, 2, \ldots, k$ and $j = 1, 2, \ldots, n$, and $\bar{x}_{..}$, the mean of all the kn measurements or observations, is called the **grand mean.** Note that if we divide the total sum of squares by $kn - 1$, we get the variance of the combined data.

Letting $\bar{x}_{i.}$ denote the mean of the ith sample, $i = 1, 2, \ldots, k$, we can write the following identity, which forms the basis of a **one-way analysis of variance:**

$$SST = n \cdot \sum_{i=1}^{k} (\bar{x}_{i.} - \bar{x}_{..})^2 + \sum_{i=1}^{k} \sum_{j=1}^{n} (x_{ij} - \bar{x}_{i.})^2$$

It is customary to refer to the first term on the right-hand side, which measures the variation among the sample means, as the **treatment sum of squares** $SS(Tr)$, and to the second term, which measures the variation within the samples, as the **error sum of squares** SSE. Use of the word "treatment" is explained by the origin of many analysis-of-variance techniques in agricultural experiments where different fertilizers, for example, were regarded as different **treatments** applied to the soil. So, we shall refer to the three teaching methods on page 339 as three different treatments, and in other problems we may refer to five nationalities as five different treatments, four different levels of education as four treatments, and so on. The word "error" in "error sum of squares" pertains to the experimental error, or chance.

Thus, the identity reads $SST = SS(Tr) + SSE$, and since its proof requires a good deal of algebraic manipulation, let us merely verify it numerically.

EXAMPLE   Use the scores on page 339 (of the students taught computer programming by three different methods) to verify the identity $SST = SS(Tr) + SSE$.

[†] The use of double subscripts and double summations is explained briefly in Section 3.11.

Solution Substituting the scores, the three sample means, 70, 85, and 76, and the grand mean, 77, into the expressions for the three sums of squares, we get

$$SST = (71 - 77)^2 + (75 - 77)^2 + (65 - 77)^2 + (69 - 77)^2$$
$$+ (90 - 77)^2 + (80 - 77)^2 + (86 - 77)^2 + (84 - 77)^2$$
$$+ (72 - 77)^2 + (77 - 77)^2 + (76 - 77)^2 + (79 - 77)^2$$
$$= 586$$

$$SS(Tr) = 4[(70 - 77)^2 + (85 - 77)^2 + (76 - 77)^2]$$
$$= 456$$

and

$$SSE = (71 - 70)^2 + (75 - 70)^2 + (65 - 70)^2 + (69 - 70)^2$$
$$+ (90 - 85)^2 + (80 - 85)^2 + (86 - 85)^2 + (84 - 85)^2$$
$$+ (72 - 76)^2 + (77 - 76)^2 + (76 - 76)^2 + (79 - 76)^2$$
$$= 130$$

Since $586 = 456 + 130$, we have thus shown that $SST = SS(Tr) + SEE$ for the given data.

Examining the two terms into which the total sum of squares SST has been partitioned, we note that if we divide $SS(Tr)$ by $k - 1$ we obtain the quantity which we denoted by $n \cdot s_{\bar{x}}^2$ on page 340. Clearly,

$$\frac{SS(Tr)}{k - 1} = \frac{n \cdot \sum_{i=1}^{k} (\bar{x}_{i.} - \bar{x}_{..})^2}{k - 1} = n \cdot \left[\frac{\sum_{i=1}^{k} (\bar{x}_{i.} - \bar{x}_{..})^2}{k - 1} \right] = n \cdot s_{\bar{x}}^2$$

This quantity, which measures the variation among the samples, is called the **treatment mean square** and it is denoted by $MS(Tr)$. That is,

$$MS(Tr) = \frac{SS(Tr)}{k - 1}$$

Similarly, if we divide SSE by $k(n - 1)$ we obtain the mean of the k sample variances, for we can write

$$\frac{SSE}{k(n - 1)} = \frac{\sum_{i=1}^{k} \sum_{j=1}^{n} (x_{ij} - \bar{x}_{i.})^2}{k(n - 1)} = \frac{1}{k} \cdot \sum_{i=1}^{k} \left[\frac{\sum_{j=1}^{n} (x_{ij} - \bar{x}_{i.})^2}{n - 1} \right]$$

which equals $\frac{1}{k} \cdot [s_1^2 + s_2^2 + \cdots + s_k^2]$, the pooled variance from page 341. This quantity, which measures the variation within the samples, is called the **error mean square** and it is denoted by *MSE*. Thus,

$$MSE = \frac{SSE}{k(n-1)}$$

Since F was defined on page 341 as the ratio of these two measures of the variation among and within the samples, we can now write

Statistic for test concerning differences among means

$$F = \frac{MS(Tr)}{MSE}$$

In practice, we display the work required for the determination of F in the following kind of table, called an **analysis-of-variance table:**

Source of variation	Degrees of freedom	Sum of squares	Mean square	F
Treatments	$k-1$	$SS(Tr)$	$MS(Tr) = \dfrac{SS(Tr)}{k-1}$	$\dfrac{MS(Tr)}{MSE}$
Error	$k(n-1)$	SSE	$MSE = \dfrac{SSE}{k(n-1)}$	
Total	$kn-1$	SST		

The degrees of freedom for treatments and error are the numerator and denominator degrees of freedom referred to on page 342. Note that they are also the quantities we divide into the sums of squares to obtain the corresponding mean squares.

If we make the same assumptions as in Section 9.12, the significance test is as we described it on page 342—we reject the null hypothesis $\mu_1 = \mu_2 = \cdots = \mu_k$ against the alternative that these μ's are not all equal, if the value we get for F exceeds F_α for $k-1$ and $k(n-1)$ degrees of freedom.

EXAMPLE   Use the sums of squares obtained on page 342 to construct an analysis-of-variance table for our numerical example, and test the null hypothesis that the three methods of teaching computer programming are equally effective (against the alternative that they are not all equally effective) at the 0.01 level of significance.

Solution Copying the values of the sums of squares from page 345, we get

$$MS(Tr) = \frac{456}{2} = 228, \quad MSE = \frac{130}{9} = 14.44, \quad F = \frac{228}{14.44} = 15.8, \quad \text{and,}$$

hence,

Source of variation	Degrees of freedom	Sum of squares	Mean square	F
Treatments	2	456	228	15.8
Error	9	130	14.44	
Total	11	586		

Since $F = 15.8$ exceeds 8.02, the value of $F_{0.01}$ for $3 - 1 = 2$ and $3(4 - 1) = 9$ degrees of freedom obtained from Table V, the null hypothesis must be rejected; in other words, we conclude that the three methods of teaching computer programming are not all equally effective.

The numbers which we used in our illustration were intentionally chosen so that the calculations would be easy. In actual practice, the calculation of the sums of squares can be quite tedious unless we use the following computing formulas, in which $T_{i.}$ denotes the sum of the values in the ith sample, and $T_{..}$ denotes the grand total of all the data:

Computing formulas for sums of squares (sample sizes equal)

$$SST = \sum_{i=1}^{k} \sum_{j=1}^{n} x_{ij}^2 - \frac{1}{kn} \cdot T_{..}^2$$

$$SS(Tr) = \frac{1}{n} \cdot \sum_{i=1}^{k} T_{i.}^2 - \frac{1}{kn} \cdot T_{..}^2$$

and by subtraction

$$SSE = SST - SS(Tr)$$

Use these computing formulas to verify the sums of squares obtained on page 345.

Solution Substituting $k = 3$, $n = 4$, $T_{1.} = 280$, $T_{2.} = 340$, $T_{3.} = 304$, $T_{..} = 924$, and $\sum\sum x^2 = 71{,}734$, into the computing formulas for the three sums of squares, we get

$$SST = 71{,}734 - \frac{1}{12}(924)^2 = 586$$

$$SS(Tr) = \frac{1}{4}(280^2 + 340^2 + 304^2) - \frac{1}{12}(924)^2 = 456$$

and

$$SSE = 586 - 456 = 130$$

These values are identical with those obtained on page 345.

A computer printout of our analysis-of-variance example is shown in Figure 9.14, where the values corresponding to $SST = 586$, $SS(Tr) = 456$, and $SSE = 130$ are given in the column headed "SS." Besides the degrees of free-

9.14
Computer printout for analysis of variance.

```
MTB > SET C1
DATA> 71   75   65   69
MTB > SET C2
DATA> 90   80   86   84
MTB > SET C3
DATA> 72   77   76   79
MTB > AOVO C1-C3

ANALYSIS OF VARIANCE
SOURCE      DF        SS        MS         F
FACTOR       2      456.0     228.0     15.78
ERROR        9      130.0      14.4
TOTAL       11      586.0
                                    INDIVIDUAL 95 PCT CI'S FOR MEAN
                                    BASED ON POOLED STDEV
LEVEL       N       MEAN     STDEV   ---+---------+---------+---------+---
C1          4      70.00      4.16   (----*-----)
C2          4      85.00      4.16                     (----*-----)
C3          4      76.00      2.94            (----*-----)
                                    ---+---------+---------+---------+---
POOLED STDEV =   3.80                67.5      75.0      82.5      90.0
```

CHAP. 9: The Analysis of Measurements

dom, the sums of squares, the mean squares, and the value of F, it provides information which permits further comparisons among the population means. We shall not go into this here. Some computer programs also give the probability of getting a value greater than or equal to the observed value of F when the null hypothesis is true; for our example it is about 0.001.

The method we have discussed here applies only when the sample sizes are all equal, but minor modifications make it applicable also when the sample sizes are not all equal. If the ith sample is of size n_i, the computing formulas for the sums of squares become

Computing formulas for sums of squares (sample sizes unequal)

$$SST = \sum_{i=1}^{k} \sum_{j=1}^{n_i} x_{ij}^2 - \frac{1}{N} \cdot T_{..}^2$$

$$SS(Tr) = \sum_{i=1}^{k} \frac{T_{i.}^2}{n_i} - \frac{1}{N} \cdot T_{..}^2$$

$$SSE = SST - SS(Tr)$$

where $N = n_1 + n_2 + \cdots + n_k$. The only other change is that the total number of degrees of freedom is $N - 1$, and the degrees of freedom for treatments and error are, respectively, $k - 1$ and $N - k$.

EXAMPLE A laboratory technician wants to compare the breaking strength of three kinds of thread and originally he planned to repeat each determination six times. Not having enough time, however, he has to base his analysis on the following results (in ounces):

> *Thread 1:* 18.0, 16.4, 15.7, 19.6, 16.5, 18.2
> *Thread 2:* 21.1, 17.8, 18.6, 20.8, 17.9, 19.0
> *Thread 3:* 16.5, 17.8, 16.1

Perform an analysis of variance to test at the 0.05 level of significance whether the differences among the sample means are significant.

Solution 1. *Hypotheses*

$$H_0: \mu_1 = \mu_2 = \mu_3$$

$$H_A: \text{The } \mu\text{'s are not all equal.}$$

2. *Level of significance*

$$\alpha = 0.05$$

349

3. *Criterion*

   Reject the null hypothesis if $F > 3.89$, the value of $F_{0.05}$ for $k - 1 = 3 - 1 = 2$ and $N - k = 15 - 3 = 12$ degrees of freedom, where F is to be determined by an analysis of variance. Otherwise, accept H_0 or reserve judgment.

4. *Calculations*

   $T_{1.} = 104.4$, $T_{2.} = 115.2$, $T_{3.} = 50.4$, $T_{..} = 270.0$, and $\sum\sum x^2 = 4{,}897.46$. Then, substituting these totals together with $n_1 = 6$, $n_2 = 6$, $n_3 = 3$, and $N = 15$ into the computing formulas for the sums of squares, we get

$$SST = 4{,}897.46 - \frac{1}{15}(270.0)^2 = 37.46$$

$$SS(Tr) = \frac{104.4^2}{6} + \frac{115.2^2}{6} + \frac{50.4^2}{3} - \frac{1}{15}(270.0)^2$$

$$= 15.12$$

and

$$SSE = 37.46 - 15.12 = 22.34$$

Then, $MS(Tr) = \dfrac{15.12}{2} = 7.56$, $MSE = \dfrac{22.34}{12} = 1.86$, and $F = \dfrac{7.56}{1.86} = 4.06$. All these results are shown in the following analysis-of-variance table:

Source of variation	Degrees of freedom	Sum of squares	Mean square	F
Treatments	2	15.12	7.56	4.06
Error	12	22.34	1.86	
Total	14	37.46		

5. *Decision*

   Since $F = 4.06$ exceeds 3.89, the null hypothesis must be rejected; in other words, we conclude that there is a difference in the strength of the three kinds of thread.

(Exercises 9.78 and 9.83 are practice exercises; their complete solutions are given on pages 357 and 358.)

★ **9.78** Rework part (b) of Exercise 9.75 by performing an analysis of variance, using the computing formulas to obtain the necessary sums of squares. Compare the values of F obtained here and in part (a) of Exercise 9.75.

★ **9.79** Rework part (b) of Exercise 9.76 by performing an analysis of variance, using the computing formulas to obtain the necessary sums of squares. Compare the values of F obtained here and in part (a) of Exercise 9.76.

★ **9.80** Rework part (b) of Exercise 9.77 by performing an analysis of variance, using the computing formulas to obtain the necessary sums of squares. Compare the values of F obtained here and in part (a) of Exercise 9.77.

★ **9.81** To study the effectiveness of five different kinds of packaging, a processor of breakfast foods obtained the following data on the numbers of sales on five different days:

> *Packaging I:*   60, 52, 56, 52, 65
> *Packaging II:*   54, 64, 66, 54, 57
> *Packaging III:* 55, 66, 68, 57, 55
> *Packaging IV:*   55, 56, 70, 58, 56
> *Packaging V:*   71, 65, 60, 59, 62

Perform an analysis of variance to test at the 0.05 level of significance whether the differences among the five sample means can be attributed to chance.

★ **9.82** The following are eight consecutive weeks' earnings (in dollars) of three door-to-door cosmetics salespersons employed by a firm:

> *Salesperson A:* 309, 293, 284, 300, 306, 288, 312, 276
> *Salesperson B:* 295, 280, 299, 310, 298, 284, 293, 287
> *Salesperson C:* 311, 289, 296, 323, 287, 280, 303, 264

Perform an analysis of variance to test at the 0.01 level of significance whether the differences among the average weekly earnings of the three salespersons are significant.

★ **9.83** The following are the weight losses of certain machine parts due to friction, in milligrams, when they were used with three different lubricants:

> *Lubricant X:* 12, 11, 7, 13, 9, 11, 12, 9
> *Lubricant Y:* 8, 10, 7, 5, 6, 10, 7, 8, 11, 7, 8
> *Lubricant Z:* 9, 3, 7, 8, 4, 6, 6, 5

Perform an analysis of variance to test at the 0.01 level of significance whether the differences among the three sample means can be attributed to chance.

★ 9.84 The following are the numbers of words per minute which a secretary typed on several occasions on four different typewriters:

Typewriter C: 71, 75, 69, 77, 61, 72, 71, 78
Typewriter D: 68, 71, 74, 66, 69, 67, 70, 62
Typewriter E: 75, 70, 81, 73, 78, 72
Typewriter F: 62, 59, 71, 68, 63, 65, 72, 60, 64

Perform an analysis of variance to test at the 0.01 level of significance whether the differences among the four sample means can be attributed to chance.

★ 9.85 To study the performance of a newly designed motorboat, it was timed over a marked course under various wind and water conditions, and the following data (in minutes) were obtained:

Calm conditions: 25, 18, 15, 21
Moderate conditions: 24, 27, 24, 19, 17, 22
Choppy conditions: 22, 24, 27, 25, 30

Perform an analysis of variance to test at the 0.05 level of significance whether the differences among the three sample means are significant.

 ★ 9.86 Use a computer package to rework Exercise 9.81.

 ★ 9.87 Use a computer package to rework Exercise 9.82.

 ★ 9.88 Use a computer package to rework Exercise 9.84.

 ★ 9.89 Use a computer package to rework Exercise 9.85.

SOLUTIONS OF PRACTICE EXERCISES

9.1 The probability is 0.95 that the maximum error will be

$$E = 1.96 \cdot \frac{9.2}{\sqrt{100}} = 1.80$$

9.2 Substituting $n = 50$, $s = 0.4$, and $E = 0.1$ into the formula for E, we get

$$0.1 = z_{\alpha/2} \cdot \frac{0.4}{\sqrt{50}}$$

so that $z_{\alpha/2} = 1.77$; the corresponding entry in Table II is 0.4616, $1 - \alpha = 2(0.4616) = 0.9232$ and we can assert with 92.32% confidence that the error is at most 0.1 ounce.

9.10 Substituting $\sigma = 3.2$ and $E = 0.5$ into the formula for n, we get

$$n = \left[\frac{2.575(3.2)}{0.5}\right]^2 = 272$$

rounded up to the nearest integer.

9.14 Substituting $n = 100$, $\bar{x} = 62.7$, and $\sigma = 9.2$ into the confidence interval formula, we get

$$62.7 - 2.575 \cdot \frac{9.2}{\sqrt{100}} < \mu < 62.7 + 2.575 \cdot \frac{9.2}{\sqrt{100}}$$

$$62.7 - 2.37 < \mu < 62.7 + 2.37$$

$$60.3 < \mu < 65.1$$

9.20 (a) Substituting $\bar{x} = 2.36$, $s = 0.48$, $n = 10$, and $t_{0.005} = 3.250$ (for 9 degrees of freedom) into the small-sample confidence interval formula, we get

$$2.36 - 3.250 \cdot \frac{0.48}{\sqrt{10}} < \mu < 2.36 + 3.250 \cdot \frac{0.48}{\sqrt{10}}$$

$$1.87 < \mu < 2.85$$

(b) Substituting $s = 0.48$, $n = 10$, and $t_{0.025} = 2.262$ (for 9 degrees of freedom) into the formula for E, we get

$$E = 2.262 \cdot \frac{0.48}{\sqrt{10}} = 0.34$$

9.31 Substituting $s = 0.4$ and $n = 50$ into the large-sample confidence interval formula for σ, we get

$$\frac{0.4}{1 + \dfrac{2.575}{\sqrt{100}}} < \sigma < \frac{0.4}{1 + \dfrac{2.575}{\sqrt{100}}}$$

$$0.32 < \sigma < 0.54$$

9.36 (a) Whenever we erroneously reject a hypothesis we commit a Type I error.
(b) Whenever we erroneously accept a hypothesis we commit a Type II error.

9.40 (a) Since $\sigma_{\bar{x}} = \dfrac{0.8}{\sqrt{60}} = 0.103$, the dividing lines of the criterion, in standard units, are

$$z = \frac{13.2 - 13.4}{0.103} = -1.94 \quad \text{and} \quad z = \frac{13.6 - 13.4}{0.103} = 1.94$$

and the probability of a Type I error is $2(0.5000 - 0.4738) = 0.0524$, or approximately 0.05.

(b) The dividing lines of the criterion, in standard units, are

$$z = \frac{13.2 - 13.7}{0.103} = -4.85 \quad \text{and} \quad z = \frac{13.6 - 13.7}{0.103} = -0.97$$

and the probability of a Type II error is $0.5000 - 0.3340 = 0.1660$, or approximately 0.17.

9.44 (a) The null hypothesis $\mu = \mu_0$, where μ is the actual number of mistakes averaged by the salesperson and μ_0 is the number of mistakes averaged by all the salespersons; the alternative hypothesis $\mu > \mu_0$; the salesperson will be fired if the null hypothesis can be rejected.

(b) The null hypothesis $\mu = \mu_0$ and the alternative hypothesis $\mu < \mu_0$; the salesperson will be fired unless the null hypothesis can be rejected.

9.46 1. *Hypotheses*

$$H_0: \mu = 12.8$$

$$H_A: \mu \neq 12.8$$

2. *Level of significance*

$$\alpha = 0.01$$

3. *Criterion*
Reject the null hypothesis if $z < -2.575$ or $z > 2.575$, where

$$z = \frac{\bar{x} - \mu_0}{\sigma/\sqrt{n}}$$

with s substituted for σ. Otherwise, accept the null hypothesis or reserve judgement.

4. *Calculations*

$$z = \frac{11.2 - 12.8}{3.5/\sqrt{60}} = -3.54$$

5. *Decision*
Since $z = -3.54$ is less than -2.575, the null hypothesis must be rejected; in other words, convicted embezzlers do not spend on the average 12.8 months in jail.

9.51 1. *Hypotheses*

$$H_0: \mu = 9.0$$

$$H_A: \mu < 9.0$$

2. *Level of significance*

$$\alpha = 0.05$$

3. *Criterion*

Reject the null hypothesis if $t < -1.796$, which is the value of $-t_{0.05}$ for $12 - 1 = 11$ degrees of freedom, where

$$t = \frac{\bar{x} - \mu_0}{s/\sqrt{n}}$$

4. *Calculations*

$$t = \frac{7.2 - 9.0}{1.8/\sqrt{12}} = -3.46$$

5. *Decision*

Since $t = -3.46$ is less than -1.796, the null hypothesis must be rejected; in other words, the tranquilizer reduces the pulse rate on the average by less than 9.0 beats.

9.61 1. *Hypotheses*

$$H_0: \mu_1 = \mu_2$$

$$H_A: \mu_1 \neq \mu_2$$

2. *Level of significance*

$$\alpha = 0.01$$

3. *Criterion*

Reject the null hypothesis if $z < -2.575$ or $z > 2.575$, where

$$z = \frac{\bar{x}_1 - \bar{x}_2}{\sqrt{\dfrac{\sigma_1^2}{n_1} + \dfrac{\sigma_2^2}{n_2}}}$$

with s_1 and s_2 substituted for σ_1 and σ_2. Otherwise, state that the difference between the two sample means is not significant.

4. *Calculations*

$$z = \frac{9.6 - 8.4}{\sqrt{\frac{1.8^2}{60} + \frac{2.1^2}{50}}} = 3.18$$

5. *Decision*

Since $z = 3.18$ exceeds 2.575, the null hypothesis must be rejected; in other words, the two kinds of executives do not average equally many business lunches.

9.65 1. *Hypotheses*

$$H_0: \mu_1 = \mu_2$$

$$H_A: \mu_1 \neq \mu_2$$

2. *Level of significance*

$$\alpha = 0.05$$

3. *Criterion*

Reject the null hypothesis if $t < -2.306$ or $t > 2.306$, where 2.306 is the value of $t_{0.025}$ for $5 + 5 - 2 = 8$ degrees of freedom and t is given by the formula on page 332.

4. *Calculations*

$$t = \frac{8,160 - 7,730}{\sqrt{\frac{4(252)^2 + 4(207)^2}{5 + 5 - 2} \cdot \left(\frac{1}{5} + \frac{1}{5}\right)}} = 2.95$$

5. *Decision*

Since $t = 2.95$ exceeds 2.306, the null hypothesis must be rejected; in other words, the average heat-producing capacity of coal from the two mines is not the same.

9.70 The differences are 13, 7, -1, 5, 3, 2, -1, 0, 8, 10, 16, -2, 7, 3, -5, 2, 6, 1, 4, 3, 2, 6, 12, 4, -1, 5, -6, -3, 8, 4, -1, and 0; their mean is $\bar{x} = 3.47$, their standard deviation is $s = 5.07$, and for these data we perform the following test:

1. *Hypotheses*

$$H_0: \mu = 0$$

$$H_A: \mu > 0$$

2. *Level of significance*

$$\alpha = 0.01$$

3. *Criterion*

Reject the null hypothesis if $z > 2.33$, where

$$z = \frac{\bar{x} - \mu_0}{\sigma/\sqrt{n}}$$

with s substituted for σ; otherwise, we accept the null hypothesis or reserve judgment.

4. *Calculations*

$$z = \frac{3.47 - 0}{5.07/\sqrt{32}} = 3.87$$

5. *Decision*

Since $z = 3.87$ exceeds 2.33, the null hypothesis must be rejected; in other words, the prescribed program of exercise is effective.

9.75 (a) The four sample means are $\bar{x}_1 = 59$, $\bar{x}_2 = 54$, $\bar{x}_3 = 54$, and $\bar{x}_4 = 57$; their mean is $\dfrac{59 + 54 + 54 + 57}{4} = 56$, so that

$$s_{\bar{x}}^2 = \frac{(59 - 56)^2 + (54 - 56)^2 + (54 - 56)^2 + (57 - 56)^2}{4 - 1} = 6$$

and $n \cdot s_{\bar{x}}^2 = 3 \cdot 6 = 18$; also,

$$s_1^2 = \frac{(60 - 59)^2 + (61 - 59)^2 + (56 - 59)^2}{3 - 1} = 7$$

$s_2^2 = 19$, $s_3^2 = 3$, and $s_4^2 = 3$, so that

$$\frac{1}{4}(s_1^2 + s_2^2 + s_3^2 + s_4^2) = \frac{7 + 19 + 3 + 3}{4} = 8$$

and $F = \dfrac{18}{8} = 2.25$.

(b) Since $F = 2.25$ does not exceed 4.07, the value of $F_{0.05}$ for $k - 1 = 4 - 1 = 3$ and $k(n - 1) = 4(3 - 1) = 8$ degrees of freedom, the null hypothesis cannot be rejected; in other words, the differences among the four sample means are not significant.

9.78 1. *Hypotheses*

$$H_0: \mu_1 = \mu_2 = \mu_3 = \mu_4$$

$$H_A: \text{The } \mu\text{'s are not all equal.}$$

357

2. *Level of significance*

$$\alpha = 0.05$$

3. *Criterion*
Reject the null hypothesis if $F > 4.07$, the value of $F_{0.05}$ for $k - 1 = 4 - 1 = 3$ and $k(n - 1) = 4(3 - 1) = 8$ degrees of freedom, where F is to be determined by an analysis of variance.

4. *Calculations*
$T_{1.} = 177, T_{2.} = 162, T_{3.} = 162, T_{4.} = 171, T_{..} = 672,$ and $\sum\sum x^2 = 37{,}750.$ Substituting these values together with $k = 4$ and $n = 3$ into the computing formulas for the sums of squares, we get

$$SST = 37{,}750 - \frac{1}{12}(672)^2 = 118$$

$$SS(Tr) = \frac{1}{3}(177^2 + 162^2 + 162^2 + 171^2) - \frac{1}{12}(672)^2 = 54$$

and

$$SSE = 118 - 54 = 64$$

The remainder of the work is shown in the following analysis-of-variance table:

Source of variation	Degrees of freedom	Sum of squares	Mean square	F
Treatments	3	54	18	2.25
Error	8	64	8	
Total	11	118		

5. *Decision*
Since $F = 2.25$ does not exceed 4.07, the null hypothesis cannot be rejected; in other words, the differences among the four sample means are not significant. The value obtained here for F equals that obtained in part (a) of Exercise 9.75.

9.83 1. *Hypotheses*

$$H_0: \mu_1 = \mu_2 = \mu_3$$

$$H_A: \text{The } \mu\text{'s are not all equal.}$$

2. *Level of significance*

$$\alpha = 0.01$$

3. *Criterion*

   Reject the null hypothesis if $F > 5.61$, the value of $F_{0.01}$ for $k - 1 = 3 - 1 = 2$ and $N - k = 27 - 3 = 24$ degrees of freedom, where F is to be determined by an analysis of variance.

4. *Calculations*

   $T_1. = 84$, $T_2. = 87$, $T_3. = 48$, $T.. = 219$, and $\sum\sum x^2 = 1,947$. Substituting these values together with $n_1 = 8$, $n_2 = 11$, $n_3 = 8$, and $N = 27$ into the computing formulas for the sums of squares, we get

$$SST = 1,947 - \frac{1}{27}(219)^2 = 170.67$$

$$SS(Tr) = \frac{84^2}{8} + \frac{87^2}{11} + \frac{48^2}{8} - \frac{1}{27}(219)^2 = 81.76$$

$$SSE = 170.67 - 81.76 = 88.91$$

The remainder of the work is shown in the following analysis-of-variance table:

Source of variation	Degrees of freedom	Sum of squares	Mean square	F
Treatments	2	81.76	40.88	11.05
Error	24	88.91	3.70	
Total	26	170.67		

5. *Decision*

   Since $F = 11.05$ exceeds 5.61, the null hypothesis must be rejected; in other words, we conclude that there is a difference in the effectiveness of the lubricants in reducing the weight loss of the machine parts.

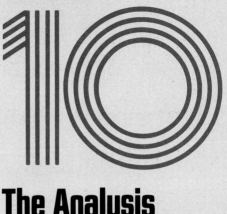

The Analysis of Count Data

Conceptually, the work of this chapter is very similar to that of Chapter 9. In problems of estimation we shall again construct confidence intervals, or determine the possible size of our error; in tests of hypotheses we shall again formulate null hypotheses and alternative hypotheses, choose a level of significance, and construct appropriate test criteria. The main difference is that we shall be concerned with other parameters—population proportions, percentages, or probabilities—and this is why we shall base our methods on

> **COUNT DATA** ('kaůnt' dāt-ə) Data obtained by counting, as contrasted to data obtained by performing measurements on continuous scales.

Such data are also referred to as **enumeration data.**

Section 10.1 deals with the estimation of proportions; Sections 10.2 and 10.3 deal with tests concerning proportions; and Sections 10.4 and 10.5 deal with tests concerning two or more proportions. In Section 10.6 we shall learn how to analyze data tallied into a two-way classification, and in Section 10.7 we shall learn how to compare observed and expected distributions.

10.1

The Estimation of Proportions

The information that is usually available for the estimation of a true proportion is a **sample proportion** $\frac{x}{n}$, where x is the number of times that an event has occurred in n trials. For instance, if 63 of 150 television viewers (interviewed in a sample survey) liked a certain new situation comedy, then $\frac{x}{n} = \frac{63}{150} = 0.42$, and we can use this figure as an estimate of the true proportion of television viewers who like the new show. Since a percentage is just a proportion multiplied by 100 and a probability may be interpreted as a proportion in the long run, we could also say that we estimate that 42% of all television viewers like the new situation comedy, or that the probability is 0.42 that any one television viewer will like the new show. We have made this point to impress upon the reader that the problem of estimating a true percentage or a true probability is essentially the same as that of estimating a true proportion.

Throughout this section it will be assumed that the situations satisfy (at least approximately) the conditions underlying the binomial distribution; that is, our information will consist of the number of successes observed in a given number of independent trials, and it will be assumed that for each trial the probability of a success—the parameter we want to estimate—has the constant value p. Thus, the sampling distribution of the counts on which our methods will be based is the binomial distribution with the mean $\mu = np$ and the standard deviation $\sigma = \sqrt{np(1 - p)}$.

We also know that this distribution can be approximated with a normal curve when np and $n(1 - p)$ are both greater than 5 (see page 253), and it follows that under these conditions

$$z = \frac{x - np}{\sqrt{np(1 - p)}}$$

has approximately the standard normal distribution. If we substitute this expression for z into the inequality $-z_{\alpha/2} < z < z_{\alpha/2}$ (as on page 302) and use some

CHAP. 10: The Analysis of Count Data

relatively simple algebra, we arrive at the inequality

$$\frac{x}{n} - z_{\alpha/2}\sqrt{\frac{p(1-p)}{n}} < p < \frac{x}{n} + z_{\alpha/2}\sqrt{\frac{p(1-p)}{n}}$$

This may look like a confidence interval formula for p and, indeed, the inequality will be satisfied with probability $1 - \alpha$ but it cannot be used in practice because the unknown parameter p appears also in $\sqrt{\frac{p(1-p)}{n}}$ to the left of the first inequality sign and to the right of the second. This quantity $\sqrt{\frac{p(1-p)}{n}}$ is called the **standard error of a proportion,** as it is, in fact, the standard deviation of the sampling distribution of a sample proportion (see Exercise 10.18). To get around this difficulty, we substitute for p in $\sqrt{\frac{p(1-p)}{n}}$ the sample proportion $\frac{x}{n}$, and we have thus arrived at the following **large-sample confidence interval for p:**

Large-sample confidence interval for p

$$\frac{x}{n} - z_{\alpha/2}\sqrt{\frac{\frac{x}{n}\left(1 - \frac{x}{n}\right)}{n}} < p < \frac{x}{n} + z_{\alpha/2}\sqrt{\frac{\frac{x}{n}\left(1 - \frac{x}{n}\right)}{n}}$$

The degree of confidence is $1 - \alpha$. As before, the confidence interval is referred to as a $(1 - \alpha)100\%$ confidence interval.

EXAMPLE   If 400 persons, constituting a random sample, are given a flu vaccine and 136 of them experienced some discomfort, construct a 95% large-sample confidence interval for the corresponding true proportion.

Solution   Substituting $n = 400$, $\frac{x}{n} = \frac{136}{400} = 0.34$, and $z_{0.025} = 1.96$ into the confidence interval formula, we get

$$0.34 - 1.96\sqrt{\frac{(0.34)(0.66)}{400}} < p < 0.34 + 1.96\sqrt{\frac{(0.34)(0.66)}{400}}$$

$$0.294 < p < 0.386$$

or, rounding to two decimals, $0.29 < p < 0.39$. To repeat what we said on page 303, the interval from 0.29 to 0.39 contains the true proportion p or it does not, and we really don't know which, but the 95% confidence implies that the interval was obtained by a method which "works 95% of the time." Note also that for $n = 400$ and p on the interval from 0.29 to 0.39, np and $n(1 - p)$ are much greater than 5, so that there can be no question about n being large enough to use the normal approximation to the binomial distribution (see page 253).

The large-sample theory which we have presented here can also be used to assess the size of the error we may be making when we use a sample proportion $\frac{x}{n}$ as a point estimate of a population proportion p. In this connection, we can assert with $(1 - \alpha)100\%$ confidence that our error is less than

Maximum error of estimate

$$E = z_{\alpha/2} \sqrt{\frac{p(1-p)}{n}} \quad \text{or approximately} \quad E = z_{\alpha/2} \sqrt{\frac{\frac{x}{n}\left(1 - \frac{x}{n}\right)}{n}}$$

The first of these formulas cannot be used in practice since p is the unknown quantity we are trying to estimate, but the second formula can be used as an approximation provided that n is sufficiently large.

EXAMPLE   With reference to the example on page 362, where 63 of 150 television viewers liked a new situation comedy, what can we say with 99% confidence about the maximum error, if we use $\frac{x}{n} = \frac{63}{150} = 0.42$ as an estimate of the true proportion of television viewers who like the new show?

Solution   Substituting $n = 150$, $\frac{x}{n} = 0.42$, and $z_{0.005} = 2.575$ into the formula for E, we get

$$E = 2.575 \sqrt{\frac{(0.42)(0.58)}{150}} = 0.10$$

rounded to two decimals.

As in the estimation of means, we can use the expression for the maximum error to determine how large a sample is needed to attain a desired degree of precision. If we want to assert with probability $1 - \alpha$ that a sample proportion will differ from the true proportion by less than E, we solve the equation

$$E = z_{\alpha/2} \sqrt{\frac{p(1 - p)}{n}}$$

for n and we get

Sample size for estimating p (with some information about p)

$$n = p(1 - p) \left[\frac{z_{\alpha/2}}{E} \right]^2$$

Since this formula involves p, it cannot be used unless we have some information about the possible values that p might assume. In that case, we substitute for p whichever of its values is closest to $\frac{1}{2}$. Without such information, we make use of the fact that $p(1 - p)$ cannot exceed $\frac{1}{4}$ (it equals $\frac{1}{4}$ for $p = \frac{1}{2}$) and use the formula

Sample size for estimating p (without information about p)

$$n = \frac{1}{4} \left[\frac{z_{\alpha/2}}{E} \right]^2$$

This may make the sample unnecessarily large, but, on the other hand, we can assert with a probability of *at least* $1 - \alpha$ that the error will be less than E.

EXAMPLE   Suppose that we want to estimate what proportion of the adult population of the United States has high blood pressure, and we want to be "99% sure" that the error of our estimate will not exceed 0.02. How large a sample will we need if
(a) we have no idea what the true proportion might be;
(b) we know that the true proportion lies on the interval from 0.05 to 0.20?

Solution   (a) Substituting $E = 0.02$ and $z_{0.005} = 2.575$ into the second formula, we get

$$n = \frac{1}{4} \left[\frac{2.575}{0.02} \right]^2 = 4,145$$

rounded up to the nearest integer.

(b) Substituting these same values together with $p = 0.20$ into the first formula, we get

$$n = (0.20)(0.80)\left[\frac{2.575}{0.02}\right]^2 = 2{,}653$$

rounded up to the nearest integer. This shows how some knowledge about the values p might take on can substantially reduce the required sample size.

The methods which we have discussed in this section are all large-sample techniques. For small samples, confidence intervals for proportions can be based on special tables, which may be found in more advanced texts.

EXERCISES

(Exercises 10.1, 10.2, 10.11, and 10.12, are practice exercises; their complete solutions are given on pages 396 and 397.)

10.1 In a sample survey, 140 of 500 persons interviewed in a large city said that they shop in the downtown area at least once a week. Construct a 99% confidence interval for the corresponding true proportion.

10.2 With reference to the preceding exercise, what can we say with 95% confidence about the maximum error, if we use the sample proportion to estimate the true proportion of persons in the given city who shop in the downtown area at least once a week?

10.3 Among 80 fish caught in a certain lake, 28 were inedible as a result of the chemical pollution of their environment. If we use the sample proportion, $\frac{28}{80} = 0.35$, to estimate the corresponding true proportion, what can we assert with 99% confidence about the maximum error?

10.4 With reference to the preceding exercise, construct a 95% confidence interval for the true proportion of fish in this lake which are inedible as a result of chemical pollution.

10.5 In a random sample of 1,200 voters interviewed nationwide, only 324 felt that the salaries of certain government officials should be raised. Construct a 95% confidence interval of the corresponding true proportion.

10.6 With reference to the preceding exercise, what can we say with 99% confidence about the maximum error, if we use the sample proportion, $\frac{324}{1{,}200} = 0.27$, as an estimate of the true proportion of voters who feel that the salaries of the government officials should be raised?

10.7 In a random sample of 250 high school seniors in a large city, 175 said that they expect to continue their education at an in-state college or university. Construct a 99% confidence interval for the corresponding true percentage.

10.8 In a random sample of 140 supposed UFO sightings, 119 could easily be explained in terms of natural phenomena. If $\frac{119}{140} \cdot 100 = 85\%$ is used as an estimate of the true percentage of UFO sightings that can easily be explained in terms of natural phenomena, what can one assert with 95% confidence about the maximum error?

10.9 In a random sample of 80 persons convicted in U.S. District Courts of narcotics charges, 36 received probation. Construct a 95% confidence interval for the probability that a person convicted in a U.S. District Court on narcotics charges will receive probation.

10.10 A random sample of 300 shoppers at various supermarkets included 207 who regularly use cents-off coupons. What can we say with 95% confidence about the maximum error if we estimate the probability that any one shopper at a supermarket will use cents-off coupons as $\frac{207}{300} = 0.69$?

10.11 Suppose that we want to estimate what proportion of all drivers exceed the 55-mph speed limit on a stretch of road between Reno and Las Vegas. How large a sample will we need to be able to assert with probability 0.95 that the error of our estimate, the sample proportion, will be less than 0.04?

10.12 Rework the preceding exercise, given that the proportion of all drivers who exceed the 55-mph speed limit on the given stretch of road is at least 0.60.

10.13 A private opinion poll is engaged by a politician to estimate what proportion of her constituents favor the decriminalization of certain narcotics violations. How large a sample will the poll have to take to be able to assert with probability 0.99 that the sample proportion will be off by less than 0.02?

10.14 Rework the preceding exercise, given that the poll has reason to believe that the proportion will not exceed 0.30.

10.15 An automobile insurance company wants to determine from a sample what proportion of its thousands of policyholders intend to buy a new car within the next twelve months. How large a sample will they need to be able to assert with a probability of at least 0.95 that the sample proportion will differ from the true proportion by less than 0.06?

10.16 With reference to the preceding exercise, how large a sample will they need to be able to assert with probability 0.95 that the sample proportion will be off by less than 0.06, if they have good reason to believe that the true proportion is somewhere between 0.10 and 0.20?

10.17 A national manufacturer wants to determine what percentage of purchases of razor blades for use by men are actually made by women. How large a sample will the manufacturer need to be able to assert with probability 0.99 that the sample percentage will be off by not more than 2.5%?

10.18 Since the proportion of successes is simply the number of successes divided by n, the mean and the standard deviation of the sampling distribution of the sample proportion may be obtained by dividing by n the mean and the standard deviation of the sampling distribution of the number of successes. Use this argument to verify the standard error formula given on page 363.

10.2

Tests Concerning Proportions

In this section we shall be concerned with tests of hypotheses which enable us to decide, on the basis of sample data, whether the true value of a proportion, percentage, or probability equals a given constant. These tests make it possible, for example, to determine whether the true proportion of tenth graders who can name the two senators of their state is 0.30, whether it is true that 10% of the answers which the IRS gives to taxpayers' telephone inquiries are in error, or whether the true probability is 0.25 that a flight from Seattle to San Francisco will be late.

Questions of this kind are usually decided on the basis of the observed number of successes in n trials, or the observed proportion of successes, and it will be assumed throughout this section that these trials are independent and that the probability of a success is the same for each trial. In other words, we shall assume that we can use the binomial distribution and that we are, in fact, testing hypotheses about the parameter p of binomial populations.

When n is small, tests concerning true proportions can be based directly on tables of binomial probabilities, as is illustrated by the following examples:

EXAMPLE   It has been claimed that at least 40% of all seniors at a large university prefer to live off campus. If 4 of 14 seniors interviewed at random prefer to live off campus, test the claim at the 0.05 level of significance.

Solution   1. *Hypotheses*

$$H_0: \; p = 0.40$$
$$H_A: \; p < 0.40$$

2. *Level of significance*

$$\alpha = 0.05$$

3. *Criterion*
Reject the null hypothesis if for $n = 14$ and $p = 0.40$ the probability of 4 or fewer successes is less than or equal to 0.05; otherwise, accept it or reserve judgment.

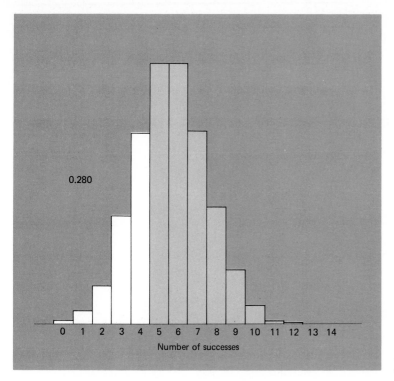

0.280

0 1 2 3 4 5 6 7 8 9 10 11 12 13 14
Number of successes

10.1

Binomial distribution with p = 0.40 and n = 14.

4. *Calculations*

Table I shows that for $n = 14$ and $p = 0.40$ the probability of 4 or fewer successes is

$$0.001 + 0.007 + 0.032 + 0.085 + 0.155 = 0.280$$

and this is the sum of the areas of the white rectangles of Figure 10.1.

5. *Decision*

Since 0.280 is greater than 0.05, the null hypothesis cannot be rejected. We accept the claim or reserve judgment pending further studies.

Note that we have departed here from the procedure which we used in the tests of Sections 9.7 through 9.13. There, we always specified the critical values, the "dividing lines" of the criteria, and then we checked where the sample value of z or t actually fell. In the preceding example, we calculated the tail probability, namely, the probability of getting a value less than or equal to the observed value (which was $x = 4$), and then we checked whether or not this probability exceeded the level of significance. We did this because, for tests based on a table of binomial probabilities, this alternative procedure can save a great deal of work. This is true, particularly, when a test is two-tailed, as in the example which follows.

EXAMPLE   It has been claimed that 60% of all shoppers can identify a highly advertised trademark. At the 0.05 level of significance, can we reject this claim, if in a random sample of 12 shoppers only three were able to identify the trademark?

Solution   1. *Hypotheses*

$$H_0: p = 0.60$$

$$H_A: p \neq 0.60$$

2. *Level of significance*

$$\alpha = 0.05$$

3. *Criterion*
Reject the null hypothesis if for $n = 12$ and $p = 0.60$ the probability of three or fewer successes, or that of three or more successes, is less than or equal to 0.025; otherwise, accept it or reserve judgment.

4. *Calculations*
Table I shows that for $n = 12$ and $p = 0.60$ the probability of three or fewer successes is

$$0.002 + 0.012 = 0.014$$

and, by subtraction, that the probability of three or more successes is

$$1 - 0.002 = 0.998$$

5. *Decision*
Since 0.014 is less than 0.025, the null hypothesis must be rejected (and we really did not have to calculate the second probability); in other words, the true percentage of shoppers who can identify the trademark is not 60%.

10.3

Tests Concerning Proportions (Large Samples)

For large n, tests concerning proportions are usually based on the normal-curve approximation to the binomial distribution. Using the same statistic which led to the large-sample confidence interval for p on page 362, we base tests of the null hypothesis $p = p_0$ on the value we obtain for

Statistic for large-sample test concerning proportion

$$z = \frac{x - np_0}{\sqrt{np_0(1 - p_0)}}$$

which has approximately the standard normal distribution.[†] The test criteria are again those of Figure 9.9 with p and p_0 substituted for μ and μ_0. For the one-sided alternative hypothesis $p < p_0$ we reject the null hypothesis if $z < -z_\alpha$, for the one-sided alternative hypothesis $p > p_0$ we reject the null hypothesis if $z > z_\alpha$, and for the two-sided alternative hypothesis $p \neq p_0$ we reject the null hypothesis if $z < -z_{\alpha/2}$ or $z > z_{\alpha/2}$.

EXAMPLE   Suppose that a nutritionist claims that at least 75% of the preschool children in a certain country have protein-deficient diets, and that a sample survey reveals that this is true for 206 preschool children in a sample of 300. Test the claim at the 0.05 level of significance.

Solution   1. *Hypotheses*

$$H_0: \ p = 0.75$$

$$H_A: \ p < 0.75$$

2. *Level of significance*

$$\alpha = 0.05$$

[†] To make the same continuity correction as on page 251, some statisticians substitute $x - \frac{1}{2}$ or $x + \frac{1}{2}$ for x into the formula for z, whichever makes z numerically smaller. However, the effect of this correction is usually negligible when n is large.

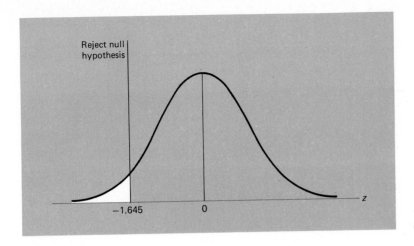

10.2

Test criterion for protein-deficiency example.

3. *Criterion*

    Reject the null hypothesis if $z < -1.645$, where

    $$z = \frac{x - np_0}{\sqrt{np_0(1 - p_0)}}$$

    (see also Figure 10.2); otherwise, accept it or reserve judgment.

4. *Calculations*

    Substituting $x = 206$, $n = 300$, and $p_0 = 0.75$ into the formula for z, we get

    $$z = \frac{206 - 300(0.75)}{\sqrt{300(0.75)(0.25)}} = -2.53$$

5. *Decision*

    Since $z = -2.53$ is less than -1.645, the null hypothesis must be rejected; in other words, we conclude that fewer than 75% of the preschool children in the given country have protein-deficient diets.

EXERCISES (Exercises 10.19 and 10.25 are practice exercises; their complete solutions are given on pages 397 and 398.)

10.19 A physicist claims that at most 10% of all persons exposed to a certain amount of radiation will feel any effects. If, in a random sample, 4 of 13 persons feel an effect, test the claim at the 0.05 level of significance.

10.20 Suppose that 5 of 12 medical students, presumably a random sample, say that they will go into private practice soon after graduation. Does this support the claim that at least 70% of all medical students go into private practice soon after graduation? Use the 0.05 level of significance.

10.21 A television critic claims that at least 80% of all viewers find the noise level of a certain commercial objectionable. If 9 out of 15 persons shown this commercial object to the noise level, what can we conclude about this claim at the 0.05 level of significance?

10.22 In a random sample of 13 undergraduate business students, 6 say that they will take advanced work in accounting. Test the claim that 20% of all undergraduate business students will take advanced work in accounting, using the alternative hypothesis $p \neq 0.20$ and the 0.01 level of significance.

10.23 A food processor wants to know whether the probability is really 0.60 that a customer will prefer a new kind of packaging to the old kind. If, in a random sample, 6 of 15 customers prefer the new kind of packaging to the old kind, test the null hypothesis $p = 0.60$ against the alternative hypothesis $p \neq 0.60$ at the 0.05 level of significance.

10.24 The manufacturer of a spot remover claims that her product removes at least 90% of all spots. If, in a random sample, the spot remover removes only 10 of 14 spots, test this claim at the 0.05 level of significance.

10.25 In a random sample of 400 automobile accidents, it was found that 128 were due at least in part to driver fatigue. Use the 0.01 level of significance to test whether this supports the claim that 35% of all automobile accidents are due at least in part to driver fatigue.

10.26 In a study of aviophobia, a psychologist claims that 30% of all women are afraid of flying. If 54 of 200 women, constituting a random sample, say that they are afraid of flying, does this refute the psychologist's claim? Use the 0.05 level of significance.

10.27 Suppose that an airline claims that only 6% of all lost luggage is never found. If 17 of 200 pieces of lost luggage are not found, test the null hypothesis $p = 0.06$ against the alternative hypothesis $p > 0.06$ at the 0.01 level of significance.

10.28 To check an ambulance service's claim that at least half its calls are life-threatening emergencies, a random sample was taken from its files, and it was found that only 63 of 150 calls were life-threatening emergencies. Test the null hypothesis $p = 0.50$ against a suitable alternative hypothesis at the 0.05 level of significance.

10.29 In a random sample of 500 cars making a left turn at a certain intersection, 169 pulled into the wrong lane. Test the null hypothesis that the actual proportion of drivers who make this mistake (at the given intersection) is 0.30 against the alternative hypothesis that this figure is too low. Use the 0.01 level of significance.

SEC. 10.3: Tests Concerning Proportions (Large Samples)

10.30 It has been claimed that 30% of all families moving away from California move to Arizona. If, in a random sample of the records of several large van lines, it is found that the belongings of 104 of 400 families moving away from California were shipped to Arizona, test the null hypothesis $p = 30$ against the alternative hypothesis $p < 0.30$ at the 0.05 level of significance.

10.4

Differences Between Proportions

There are many problems in which we must decide whether an observed difference between two sample proportions can be attributed to chance, or whether it is indicative of the fact that the corresponding true proportions are unequal. For instance, we may want to decide on the basis of sample data whether there really is a difference between the proportions of persons with and without flu shots who actually catch the disease, or we may want to check on the basis of samples whether two manufacturers of electronic equipment ship equal proportions of defectives.

The method we shall use to test whether an observed difference between two sample proportions can be attributed to chance, or whether it is statistically significant, is based on the following theory: If x_1 and x_2 are the numbers of successes obtained in n_1 trials of one kind and n_2 of another, the trials are all independent, and the corresponding probabilities of a success are, respectively, p_1 and p_2, then the sampling distribution of $\dfrac{x_1}{n_1} - \dfrac{x_2}{n_2}$ has the mean $p_1 - p_2$ and the standard deviation

$$\sqrt{\frac{p_1(1 - p_1)}{n_1} + \frac{p_2(1 - p_2)}{n_2}}$$

It is customary to refer to this standard deviation as the **standard error of the difference between two proportions.**

When we test the null hypothesis $p_1 = p_2 \, (= p)$ against an appropriate alternative hypothesis, the mean of the sampling distribution of the difference between the two sample proportions is $p_1 - p_2 = 0$, and its standard deviation can be written

$$\sqrt{p(1 - p)\left(\frac{1}{n_1} + \frac{1}{n_2}\right)}$$

where p is usually estimated by **pooling** the data and substituting for p the combined sample proportion $\hat{p} = \dfrac{x_1 + x_2}{n_1 + n_2}$, which reads "p-hat." Then, since for large samples the sampling distribution of the difference between two proportions can be approximated closely with a normal distribution, we base the test on the statistic

Statistic for test concerning difference between two proportions

$$z = \frac{\dfrac{x_1}{n_1} - \dfrac{x_2}{n_2}}{\sqrt{\hat{p}(1 - \hat{p})\left(\dfrac{1}{n_1} + \dfrac{1}{n_2}\right)}} \quad \text{with} \quad \hat{p} = \frac{x_1 + x_2}{n_1 + n_2}$$

which has approximately the standard normal distribution. The test criteria are again those of Figure 9.9 with $p_1 - p_2$ substituted for μ and 0 substituted for μ_0. For the one-sided alternative hypothesis $p_1 < p_2$ we reject the null hypothesis if $z < -z_\alpha$, for the one-sided alternative hypothesis $p_1 > p_2$ we reject the null hypothesis if $z > z_\alpha$, and for the two-sided alternative hypothesis $p_1 \neq p_2$ we reject the null hypothesis if $z < -z_{\alpha/2}$ or $z > z_{\alpha/2}$.

EXAMPLE To test the effectiveness of a new pain-relieving drug, 80 patients at a clinic were given a pill containing the drug and 80 others were given a placebo. At the 0.01 level of significance, what can we conclude about the effectiveness of the drug if in the first group 56 of the patients felt a beneficial effect while 38 of those who received the placebo felt a beneficial effect?

Solution 1. *Hypotheses*

$$H_0: p_1 = p_2$$
$$H_A: p_1 > p_2$$

2. *Level of significance*

$$\alpha = 0.01$$

3. *Criterion*

Reject the null hypothesis if $z > 2.33$, where z is given by the formula above; otherwise, accept it or reserve judgment.

4. *Calculations*

Substituting $x_1 = 56, x_2 = 38, n_1 = 80, n_2 = 80$, and $\hat{p} = \dfrac{56 + 38}{80 + 80} = 0.5875$ into the formula for z, we get

$$z = \frac{\dfrac{56}{80} - \dfrac{38}{80}}{\sqrt{(0.5875)(0.4125)\left(\dfrac{1}{80} + \dfrac{1}{80}\right)}} = 2.89$$

5. *Decision*

Since $z = 2.89$ exceeds 2.33, the null hypothesis must be rejected; in other words, we conclude that the new pain-relieving drug is effective.

EXERCISES

(Exercise 10.31 is a practice exercise; its complete solution is given on page 398.)

10.31 In random samples of 200 tractors from one assembly line and 400 tractors from another, there were, respectively, 16 tractors and 20 tractors which required extensive adjustments before they could be shipped. At the 0.05 level of significance, can we conclude that there is a difference in the quality of the work of the two assembly lines?

10.32 In a random sample of 250 persons who skipped breakfast, 102 reported that they experienced midmorning fatigue, and in a random sample of 250 persons who ate breakfast, 73 reported that they experienced midmorning fatigue. Use the 0.01 level of significance to test the null hypothesis that there is no difference between the corresponding population proportions against the alternative hypothesis that midmorning fatigue is more prevalent among persons who skip breakfast.

10.33 A study showed that 84 of 200 persons who saw a deodorant advertised during the telecast of a football game and 96 of 200 persons who saw it advertised on a variety show remembered two hours later the name of the deodorant. Use the 0.05 level of significance to test the null hypothesis that there is no difference between the corresponding true proportions.

10.34 In a true-false test, a test item is considered to be good if it discriminates between well-prepared students and poorly prepared students. What can we conclude about the merit of a test item which was answered correctly by 246 of 300 well-prepared students and by 165 of 300 poorly prepared students? Use the 0.05 level of significance.

10.35 A random sample of 150 high school students was asked whether they would turn to their father or their mother for help with a homework assignment in mathematics, and another random sample of 150 high school students was asked the same question with regard to a homework assignment in English. If 59 students in the first sample and 83 in the second sample turned to their mother for help rather than their father, test at the 0.01 level of significance whether the difference between the two sample proportions, $\frac{59}{150}$ and $\frac{83}{150}$, may be attributed to chance.

10.5

Differences Among Proportions

Following the pattern of Chapter 9, we shall now consider a test which enables us to decide whether observed differences among more than two sample proportions can be attributed to chance. For instance, if 25 of 200 brand A tires, 21 of 200 brand B tires, 32 of 200 brand C tires, and 18 of 200 brand D tires failed to last 30,000 miles, we may want to know whether the differences among the sample proportions $\frac{25}{200} = 0.125$, $\frac{21}{200} = 0.105$, $\frac{32}{200} = 0.160$, and $\frac{18}{200} = 0.090$ can be attributed to chance, or whether they are indicative of actual differences in the quality of the tires.

To illustrate the method we use to analyze this kind of data, suppose that a survey in which independent random samples of workers in three parts of the country were asked whether they feel that unemployment or inflation is a more serious problem, yielded the results shown in the following table:

	Northeast	Midwest	Southwest
Unemployment	87	73	66
Inflation	113	77	84
	200	150	150

If we let p_1, p_2, and p_3 denote the true proportions of workers in the three parts of the country who feel that unemployment is the more serious economic problem, the hypotheses we shall want to test are

$$H_0: p_1 = p_2 = p_3$$

$$H_A: \text{The } p\text{'s are not all equal.}$$

If the null hypothesis is true, we can combine the data and estimate the common proportion of workers in the three parts of the country who feel that unemployment is the more serious economic problem as

$$\hat{p} = \frac{87 + 73 + 66}{200 + 150 + 150} = \frac{226}{500} = 0.452$$

where, again, $\hat{p}$ reads "p-hat."

With this estimate we would expect $200(0.452) = 90.4$ of the 200 workers in the Northeast, $150(0.452) = 67.8$ of the 150 workers in the Midwest, and $150(0.452) = 67.8$ of the 150 workers in the Southwest to choose unemployment. Also, if we subtract these figures from the sizes of the respective samples, we find that $200 - 90.4 = 109.6$ of the 200 workers in the Northeast, $150 - 67.8 = 82.2$ of the 150 workers in the Midwest, and $150 - 67.8 = 82.2$ of the 150 workers in the Southwest would be expected to choose inflation. These results are summarized in the following table, where the **expected frequencies** are shown in parentheses below the **observed frequencies:**

	Northeast	Midwest	Southwest
Unemployment	87 (90.4)	73 (67.8)	66 (67.8)
Inflation	113 (109.6)	77 (82.2)	84 82.2)

To test the null hypothesis that the p's are all equal, we then compare the frequencies which were actually observed with the frequencies we can expect if the null hypothesis is true. Clearly, the null hypothesis should be accepted if the discrepancies between the observed and expected frequencies are small, and it should be rejected if the discrepancies between the two sets of frequencies are large.

Denoting the observed frequencies by the letter o and the expected frequencies by the letter e, we base this comparison on the following χ^2 **(chi-square) statistic:**

Statistic for test concerning differences among proportions

$$\chi^2 = \sum \frac{(o - e)^2}{e}$$

378

In words, χ^2 is the sum of the quantities obtained by dividing $(o - e)^2$ by e separately for each **cell** of the table, and for our example we get

$$\chi^2 = \frac{(87 - 90.4)^2}{90.4} + \frac{(73 - 67.8)^2}{67.8} + \frac{(66 - 67.8)^2}{67.8}$$

$$+ \frac{(113 - 109.6)^2}{109.6} + \frac{(77 - 82.2)^2}{82.2} + \frac{(84 - 82.2)^2}{82.2}$$

$$= 1.048$$

It remains to be seen whether this value is large enough to reject the null hypothesis $p_1 = p_2 = p_3$.

If the null hypothesis is true, the sampling distribution of the χ^2 statistic is approximately a theoretical distribution called the **chi-square distribution.** The parameter of this distribution, like that of the t distribution, is called the **number of degrees of freedom,** or simply the **degrees of freedom,** and it equals $k - 1$ when we compare k sample proportions.

Since the null hypothesis is to be rejected only when the value obtained for χ^2 is too large, we base our decision on the criterion of Figure 10.3, where χ_α^2 is such that the area under the curve to its right is equal to α. Values of $\chi_{0.05}^2$ and $\chi_{0.01}^2$ for 1, 2, 3, ... , and 30 degrees of freedom are given in Table IV at the end of the book.

Returning to our numerical example and supposing that the level of significance is to be 0.05, we find that $\chi_{0.05}^2 = 5.991$ for $k - 1 = 2$ degrees of freedom, and that the null hypothesis cannot be rejected since $\chi^2 = 1.048$ does not exceed 5.991. In other words, the data tend to support, rather than refute, the hypothesis that the proportion of workers who feel that unemployment is a more serious economic problem than inflation is the same in all three parts of the country.

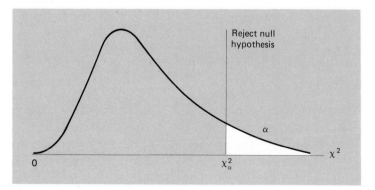

10.3

Chi-square distribution.

In general, if we want to test the null hypothesis $p_1 = p_2 = \cdots = p_k$ on the basis of random samples from k populations, we proceed as follows:

1. *Hypotheses*

$$H_0:\ p_1 = p_2 = \cdots = p_k$$
$$H_A:\ \text{The } p\text{'s are not all equal.}$$

2. *Level of significance*

$$\alpha$$

3. *Criterion*
Reject the null hypothesis if $\chi^2 > \chi_\alpha^2$ for $k - 1$ degrees of freedom, where

$$\chi^2 = \sum \frac{(o - e)^2}{e}$$

Otherwise, accept the null hypothesis or reserve judgment.

4. *Calculations*
We estimate $p_1 = p_2 = \cdots = p_k$, the common population proportion, as

$$\hat{p} = \frac{x_1 + x_2 + \cdots + x_k}{n_1 + n_2 + \cdots + n_k}$$

and then we multiply $\hat{p}$ by the respective sample sizes to obtain the expected frequencies for the first row of the table. Then we determine the expected frequencies for the second row of the table by subtracting those of the first row from the respective samples sizes, and substitute all the observed and expected frequencies into the formula for χ^2.

5. *Decision*
Depending on the value we get for χ^2, we reject the null hypothesis, or we accept it or reserve judgment.

When we calculate the expected frequencies, it is customary to round to the nearest integer or to one decimal. The entries in Table IV are given to three decimals, but there is seldom any need to carry more than two decimals in calculating the value of the χ^2 statistic. Also, the test we have been discussing is only an approximate test, and it should not be used when one (or more) of the expected frequencies is less than 5. If this is the case, we can sometimes combine two or more of the samples in such a way that none of the e's is less than 5.

The method we have discussed here can be used only to test the null hypothesis $p_1 = p_2 = \cdots = p_k$ against the alternative hypothesis that the p's are not all equal. However, in the special case where $k = 2$ we can use instead the method of Section 10.4 and test also against either of the alternative hypotheses $p_1 < p_2$ or $p_1 > p_2$. Indeed, for $k = 2$ the two methods are equivalent, as it can be shown that the χ^2 statistic equals the square of the z statistic obtained in accordance with the formula on page 375 (see Exercise 10.41).

EXERCISES

(Exercises 10.36 and 10.41 are practice exercises; their complete solutions are given on pages 399 and 400.)

10.36 On page 377 we referred to a study in which 25 of 200 brand A tires, 21 of 200 brand B tires, 32 of 200 brand C tires, and 18 of 200 brand D tires fail to last 30,000 miles. Use the 0.05 level of significance to test the null hypothesis that there is no difference in the durability of the four kinds of tires.

10.37 A market study shows that among 100 men, 100 women, and 200 children interviewed, there were, respectively, 46, 52, and 122 who did not like the flavor of a new toothpaste. Use the level of significance $\alpha = 0.05$ to test whether the differences among the corresponding sample proportions are significant.

10.38 A sample survey conducted by the dean of students of a large university showed that 203 of 240 students, 51 of 80 faculty members, and 16 of 40 staff personnel opposed late Friday afternoon classes. Use the level of significance $\alpha = 0.05$ to test the null hypothesis that there is no difference among the corresponding true proportions of persons who are opposed to late Friday afternoon classes.

10.39 In studying problems relating to its handling of reservations, an airline takes random samples of 60 of the complaints about reservations filed in each of four cities. If 49 of the complaints from city A, 54 of the complaints from city B, 41 of the complaints from city C, and 48 of the complaints from city D are about overbooking, test at the 0.05 level of significance whether the differences among the corresponding sample proportions can be attributed to chance.

10.40 The following table shows the results of a study in which samples of the members of five large unions were asked whether they are for or against a certain piece of legislation:

	Union 1	Union 2	Union 3	Union 4	Union 5
For it	74	81	69	75	91
Against it	26	19	31	25	9

Use the 0.01 level of significance to test the null hypothesis that the corresponding true proportions are all equal.

10.41 Use the method of Section 10.5 to rework Exercise 10.31, and verify that the value obtained for χ^2 equals the square of the value obtained originally for z.

10.42 Use the method of Section 10.5 to rework Exercise 10.33, and verify that the value obtained for χ^2 equals the square of the value obtained originally for z.

10.43 Use the method of Section 10.5 to rework Exercise 10.35, and verify that the value obtained for χ^2 equals the square of the value obtained originally for z.

10.6

Contingency Tables

The χ^2 statistic plays an important role in many other problems where information is obtained by counting rather than measuring. The method we shall describe here applies to two kinds of problems, which differ conceptually but are analyzed the same way.

In the first kind of problem we deal with trials permitting more than two possible outcomes. For instance, the weather can get better, remain the same, or get worse, an undergraduate can be a freshman, a sophomore, a junior, or a senior, and a movie may be rated G, PG, R, or X. In the language of Section 6.5, we could say that we are dealing with multinomial (rather than binomial) trials.

Also, in the illustration of the preceding section, each worker might have been asked whether unemployment is a more serious economic problem than inflation, whether inflation is a more serious economic problem than unemployment, or whether he or she is undecided, and this might have resulted in the following table:

	Northeast	Midwest	Southwest
Unemployment	57	53	44
Undecided	72	40	48
Inflation	71	57	58
	200	150	150

We refer to this kind of table as a 3×3 table (where 3×3 is read "3 by 3"), because it has 3 horizontal rows and 3 vertical columns; more generally, when there are r horizontal rows and c vertical columns, we refer to the table as an **$r \times c$ table**. Here, as in the table analyzed in the preceding section, the column totals, representing the sample sizes, are fixed. On the other hand, the row totals depend on the responses of the persons interviewed, and, hence, on chance.

In the second kind of problem where the method of this section applies, the column totals as well as the row totals depend on chance. To give an example, suppose that a sociologist wants to determine whether there is a relationship between the intelligence of boys who have gone through a special job-training program and their subsequent performance in their jobs, and that a sample of 400 cases taken from very extensive files yielded the following results:

		Poor	*Fair*	*Good*	
	Below average	67	64	25	*156*
IQ	*Average*	42	76	56	*174*
	Above average	10	23	37	*70*
		119	*163*	*118*	*400*

Performance

This is also a 3×3 table, and it is mainly in connection with problems like this that $r \times c$ tables are referred to as **contingency tables.** In either kind of table, the observed frequencies are referred to as the **observed cell frequencies.**

Before we demonstrate how $r \times c$ tables are analyzed, let us examine what hypotheses we want to test. In the first example we want to test whether the probabilities that a worker will choose "unemployment," "undecided," or "inflation" are the same for the three parts of the country. Formally,

H_0: For each alternative (unemployment, undecided, and inflation), the probabilities are the same for the three parts of the country

H_A: For at least one alternative, the probabilites are not the same for the three parts of the country.

In the other example we want to test whether there is a dependence (relationship) between on the job performance and intelligence, and we write

H_0: The two variables under consideration are independent.

H_A: The two variables are not independent.

To show how an $r \times c$ table is analyzed, let us refer to the second of our two examples, and let us begin by illustrating the calculation of an **expected cell frequency.**

383

EXAMPLE Assuming that intelligence and on-the-job performance are independent and using the data in the preceding table, how many of the boys
in a random sample of 400 can be expected to have below average
IQ's and perform poorly on their jobs?

Solution Under the assumption of independence, the probability of randomly
choosing a boy whose IQ is below average and whose on-the-job performance is poor is given by the product of the probability of choosing
a boy whose IQ is below average and the probability of choosing a boy
whose on-the-job performance is poor. Using the totals of the first
row and the first column to estimate these two probabilities, we get
$\frac{67 + 64 + 25}{400} = \frac{156}{400}$ for the probability of choosing a boy whose IQ is

below average, and $\frac{67 + 42 + 10}{400} = \frac{119}{400}$ for the probability of choosing

a boy whose on-the-job performance is poor. Hence, we estimate the
probability of choosing a boy whose IQ is below average and whose

on-the-job performance is poor as $\frac{156}{400} \cdot \frac{119}{400}$, and in a sample of size

400 we would expect to find

$$400 \cdot \frac{156}{400} \cdot \frac{119}{400} = \frac{156 \cdot 119}{400} = 46.4$$

boys who fit this description.

_____

In the final step of this example, $\frac{156 \cdot 119}{400}$ is just the product of the total

of the first row and the total of the first column divided by the grand total
for the entire table. Indeed, the argument which led to this result can be used
to show that in general

**The expected frequency for any cell of a contingency table may
be obtained by multiplying the total of the row to which it belongs by the total of the column to which it belongs and then
dividing by the grand total for the entire table.**

With this rule we get an expected frequency of $\frac{156 \cdot 163}{400} = 63.6$ for the second

cell of the first row, and $\frac{174 \cdot 119}{400} = 51.8$ and $\frac{174 \cdot 163}{400} = 70.9$ for the first

two cells of the second row.

384

It is not necessary to calculate all the expected cell frequencies in this way, as it can be shown that the sum of the expected frequencies for any row or column must equal the sum of the corresponding observed frequencies. Therefore, we can get some of the expected cell frequencies by subtraction from row or column totals. For instance, for our example we get

$$156 - 46.4 - 63.6 = 46.0$$

for the third cell of the first row,

$$174 - 51.8 - 70.9 = 51.3$$

for the third cell of the second row, and

$$119 - 46.4 - 51.8 = 20.8$$
$$163 - 63.6 - 70.9 = 28.5$$

and

$$118 - 46.0 - 51.3 = 20.7$$

for the three cells of the third row. These results are summarized in the following table, where the expected frequencies are shown in parentheses below the corresponding observed frequencies:

| | | Performance | | |
		Poor	Fair	Good
	Below average	67 (46.4)	64 (63.6)	25 (46.0)
IQ	Average	42 (51.8)	76 (70.9)	56 (51.3)
	Above average	10 (20.8)	23 (28.5)	37 (20.7)

From here on the work is like that of the preceding section; we calculate the χ^2 statistic according to the formula

Statistic for analysis of r × c table

$$\chi^2 = \sum \frac{(o - e)^2}{e}$$

385

with $\dfrac{(o-e)^2}{e}$ calculated separately for each cell of the table. Then we reject the null hypothesis at the level of significance α if the value obtained for χ^2 exceeds χ_α^2 for $(r-1)(c-1)$ degrees of freedom. In connection with this formula for the number of degrees of freedom, observe that after $c-1$ of the expected frequencies have been calculated for each of $r-1$ rows by means of the rule on page 384, all the other expected frequencies may be obtained by subtraction from row or column totals. For our example the number of degrees of freedom is $(3-1)(3-1)=4$, and it should be observed that after we had calculated two of the expected frequencies for each of the first two rows by means of the rule on page 384, all the others were obtained by subtraction from row or column totals.

EXAMPLE   With reference to our example, use the 0.01 level of significance to test the null hypothesis that intelligence and on-the-job performance are independent for boys who have gone through the special job-training program.

Solution   1. *Hypotheses*

H_0: Intelligence and on-the-job performance are independent.

H_A: Intelligence and on-the-job performance are not independent.

2. *Level of significance*

$$\alpha = 0.01$$

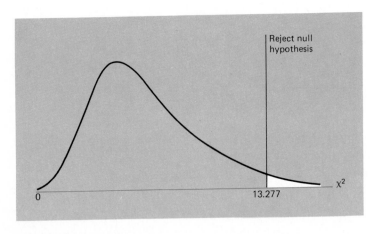

10.4

Test criterion for intelligence and on-the-job performance example.

3. *Criterion*

Reject the null hypothesis if $\chi^2 > 13.277$, the value of $\chi^2_{0.01}$ for $(3-1)(3-1) = 4$ degrees of freedom, where

$$\chi^2 = \sum \frac{(o-e)^2}{e}$$

(see also Figure 10.4); otherwise, accept it or reserve judgment.

4. *Calculations*

Copying the observed and expected cell frequencies from the table on page 385, we find that

$$\chi^2 = \frac{(67-46.4)^2}{46.4} + \frac{(64-63.6)^2}{63.6} + \frac{(25-46.0)^2}{46.0}$$

$$+ \frac{(42-51.8)^2}{51.8} + \frac{(76-70.9)^2}{70.9} + \frac{(56-51.3)^2}{51.3}$$

$$+ \frac{(10-20.8)^2}{20.8} + \frac{(23-28.5)^2}{28.5} + \frac{(37-20.7)^2}{20.7}$$

$$= 40.89$$

5. *Decision*

Since $\chi^2 = 40.89$ exceeds 13.277, the null hypothesis must be rejected; we conclude that there is a relationship between IQ and on-the-job performance.

A computer printout of the preceding chi-square analysis is shown in Figure 10.5. The difference between the values of χ^2 obtained above and in Figure 10.5 are due to rounding. Some computer programs also give the probability of getting a value greater than or equal to the observed value of χ^2 when the null hypothesis is true; for our example it is about 0.00000003.

The method which we have used here to analyze the contingency table applies also when the column totals are fixed sample sizes (as in the example on page 382) and do not depend on chance. The rule according to which we multiply the row total by the column total and then divide by the grand total has to be justified in a different way, but this is of no consequence—the expected cell frequencies are determined in exactly the same way.

EXAMPLE   Analyze the 3 × 3 table on page 382, which pertains to a study in which workers in the Northeast, Midwest, and Southwest are asked whether unemployment or inflation is the more serious economic problem, or whether they are undecided. Use the 0.01 level of significance.

387

```
MTB > READ C1 C2 C3
DATA> 67   64   25
DATA> 42   76   56
DATA> 10   23   37
MTB > CHIS C1 C2 C3

EXPECTED FREQUENCIES ARE PRINTED BELOW OBSERVED FREQUENCIES
          I  C1   I  C2   I  C3    ITOTALS
-------I-------I-------I-------I-------
    1  I   67  I   64  I   25  I    156
       I  46.4I  63.6I  46.0I
-------I-------I-------I-------I-------
    2  I   42  I   76  I   56  I    174
       I  51.8I  70.9I  51.3I
-------I-------I-------I-------I-------
    3  I   10  I   23  I   37  I     70
       I  20.8I  28.5I  20.6I
-------I-------I-------I-------I-------
TOTALS I  119 I  163 I  118 I    400

TOTAL CHI SQUARE =

          9.13 +   .00 +  9.60 +
          1.84 +   .37 +   .42 +
          5.63 +  1.07 + 12.95 +

               =   41.01

DEGREES OF FREEDOM = ( 3-1) X ( 3-1) =    4
```

10.5

Computer printout for analysis of contingency table.

Solution   1. *Hypotheses*

H_0: For each alternative (unemployment, undecided, and inflation), the probabilities are the same for the three parts of the country.

H_A: For at least one alternative, the probabilities are not the same for the three parts of the country.

2. *Level of significance*

$$\alpha = 0.01$$

3. *Criterion*

Reject the null hypothesis if $\chi^2 > 13.277$, the value of $\chi^2_{0.01}$ for $(3 - 1)(3 - 1) = 4$ degrees of freedom, where

$$\chi^2 = \sum \frac{(o - e)^2}{e}$$

Otherwise, accept the null hypothesis or reserve judgment.

4. *Calculations*

The expected frequencies for the first two cells of the first two rows are $\dfrac{154 \cdot 200}{500} = 61.6$, $\dfrac{154 \cdot 150}{500} = 46.2$, $\dfrac{160 \cdot 200}{500} = 64$, and $\dfrac{160 \cdot 150}{500} = 48$; by subtraction, those for the third cells of the first two rows are 46.2 and 48, and those for the third row are 74.4, 55.8, and 55.8. Then, substituting into the formula for χ^2, we get

$$\chi^2 = \frac{(57 - 61.6)^2}{61.6} + \frac{(53 - 46.2)^2}{46.2} + \frac{(44 - 46.2)^2}{46.2}$$
$$+ \frac{(72 - 64)^2}{64} + \frac{(40 - 48)^2}{48} + \frac{(48 - 48)^2}{48}$$
$$+ \frac{(71 - 74.4)^2}{74.4} + \frac{(57 - 55.8)^2}{55.8} + \frac{(58 - 55.8)^2}{55.8}$$
$$= 4.05$$

5. *Decision*

Since $\chi^2 = 4.05$ does not exceed 13.277, the null hypothesis cannot be rejected; the differences between the observed and expected frequencies may well be due to chance.

Since the sampling distribution of the χ^2 statistic we are using here is only approximately a chi-square distribution, it should not be used when any of the expected cell frequencies are less than 5. When there are expected cell frequencies less than 5, it may be possible to combine some of the cells, subtract 1 degree of freedom for each cell eliminated, and then perform the test just as it has been described.

EXERCISES

(Exercise 10.44 is a practice exercise; its complete solution is given on page 400.)

10.44 In a survey, 80 single persons, 120 married persons, and 100 widowed or divorced persons were asked whether they feel that friends and social life, job or primary activity, or health and physical condition contributes most to their general happiness. Use the results shown in the following table and the 0.05 level of significance to test whether the probabilities of the three alternatives are the same for persons who are single, married, or widowed or divorced.

	Single	Married	Widowed or divorced
Friends and social life	41	49	42
Job or primary activity	27	50	33
Health and physical condition	12	21	25
Total	80	120	100

10.45 Suppose that for 120 mental patients who did not receive psychotherapy and 120 mental patients who received psychotherapy a panel of psychiatrists determined after six months whether their condition had deteriorated, remained unchanged, or improved. Base on the results shown in the following table, test at the level of significance $\alpha = 0.05$ whether the therapy is effective:

	No therapy	Therapy
Deteriorated	6	11
Unchanged	65	31
Improved	49	78

10.46 The following sample data pertain to shipments received by a large firm from three different vendors:

	Vendor A	Vendor B	Vendor C
Rejected	12	8	20
Not perfect but acceptable	23	12	30
Perfect	85	60	110

Use the 0.01 level of significance to test whether the three vendors ship products of equal quality.

10.47 A sample survey, designed to show how students attending a large university get to their classes, yielded the following results:

	Freshman	Sophomore	Junior	Senior
Walk	104	87	89	72
Automobile	22	29	35	43
Bicycle	46	34	37	32
Other	28	50	39	53
Totals	200	200	200	200

Use the 0.05 level of significance to test the null hypothesis that the same proportions of freshmen, sophomores, juniors, and seniors use these means of transportation to get to class.

10.48 The following data pertain to a study in which a social scientist wants to determine whether there is a relationship between success in life (as measured by means of a questionnaire) and sense of humor (as measured by a special test):

		Sense of Humor	
		Low	Average	High
	Low	43	41	36
Success in Life	Average	107	152	81
	High	30	47	63

Use the level of significance $\alpha = 0.05$ to test the null hypothesis that there is no relationship between sense of humor and success in life.

10.49 Tests of the fidelity and selectivity of 190 radios produced the results shown in the following table:

		Fidelity		
		Low	Average	High
	Low	7	12	31
Selectivity	Average	35	59	18
	High	15	13	0

Use the 0.01 level of significance to test the null hypothesis that fidelity is independent of selectivity.

10.50 Use a computer package to rework Exercise 10.44.

10.51 Use a computer package to rework Exercise 10.45.

10.52 Use a computer package to rework Exercise 10.47.

10.53 A computer program for the analysis of a contingency table can also be used to test for differences among proportions as in Section 10.5. In that case we are, in fact, analyzing a $2 \times c$ table. Use a computer package to rework the illustration of Section 10.5, which dealt with a survey in which workers in three parts of the country were asked whether they feel that unemployment or inflation is a more serious problem.

10.54 Use a computer package to rework Exercise 10.39.

10.55 Use a computer package to rework Exercise 10.40.

10.7

Goodness of Fit

The χ^2 statistic can also be used to compare observed frequency distributions with distributions which we might expect according to theory or assumptions. We refer to such a comparison as a test of **goodness of fit.**

This kind of problem would arise, for example, if a quality control engineer takes a daily sample of ten tires coming off an assembly line, and he wants to check on the basis of the following data, the numbers of tires with imperfections observed on 200 days, whether it is true that 5% of all the tires have imperfections, namely, whether he is sampling a binomial population with $n = 10$ and $p = 0.05$:

Number with imperfections	Number of samples
0	138
1	53
2 or more	9

If the hypothesis is true, the probabilities of getting 0, 1, or 2 or more tires with imperfections are, respectively, 0.599, 0.315, and $0.075 + 0.010 + 0.001 = 0.086$ according to Table I, and hence the corresponding expected frequencies are $200(0.599) = 119.8$, $200(0.315) = 63.0$, and $200(0.086) = 17.2$.

To test whether the discrepancies between the observed frequencies and the expected frequencies can be attributed to chance, we use the same χ^2 statistic as in the preceding sections, namely,

*Statistic for test
of goodness of fit*

$$\chi^2 = \sum \frac{(o-e)^2}{e}$$

Based on the value of this statistic, we reject the null hypothesis if $\chi^2 > \chi^2_\alpha$ for $k - m - 1$ degrees of freedom, where k is the number of terms in the summation for χ^2 and m is the number of parameters that have to be estimated on the basis of the sample data.

EXAMPLE   Based on the data given above, test at the 0.05 level of significance whether the quality control engineer is sampling a binomial population with $n = 10$ and $p = 0.05$.

Solution   1. *Hypotheses*

H_0: Population is binomial with $n = 10$ and $p = 0.05$.

H_A: Population is not binomial with $n = 10$ and $p = 0.05$.

2. *Level of significance*

$$\alpha = 0.05$$

3. *Criterion*
Reject the null hypothesis if $\chi^2 > 5.991$, the value of $\chi^2_{0.05}$ for $k - m - 1 = 3 - 0 - 1 = 2$ degrees of freedom, where

$$\chi^2 = \sum \frac{(o-e)^2}{e}$$

Otherwise, accept the null hypothesis or reserve judgment.

4. *Calculations*
From page 392, the observed frequencies are 138, 53, and 9, and the expected frequencies are 119.8, 63.0, and 17.2. Substituting these

values into the formula for χ^2, we get

$$\chi^2 = \frac{(138 - 119.8)^2}{119.8} + \frac{(53 - 63.0)^2}{63.0} + \frac{(9 - 17.2)^2}{17.2}$$

$$= 8.26$$

5. *Decision*

Since $\chi^2 = 8.26$ exceeds 5.991, the null hypothesis must be rejected; we conclude that the true percentage of tires with imperfections is not 5%. (In Exercise 10.60 on page 395 the reader will find that the binomial distribution with $n = 10$ and $p = 0.04$ provides a much better fit to the data.)

In the preceding example it was assumed under the null hypothesis that $n = 10$ and $p = 0.05$, so that none of the parameters had to be estimated and the number of degrees of freedom was $k - m - 1 = 3 - 0 - 1 = 2$. However, if we want to test whether a given distribution can be approximated closely with a normal curve (see Exercise 10.61), we may have to determine the mean and the standard deviation from the sample data, and the formula for the number of degrees of freedom is $k - m - 1 = k - 2 - 1 = k - 3$.

Observe also that, as in the tests of the two preceding sections, the sampling distribution of the χ^2 statistic of a goodness-of-fit test is only approximately a chi-square distribution. So, if any of the expected frequencies is less than 5, we shall again have to combine some of the data, in this case adjacent classes of the distribution.

EXERCISES

(Exercises 10.56 and 10.61 are practice exercises; their complete solutions are given on pages 401 and 402.)

10.56 To see whether a die is balanced, it is rolled 360 times and the following results are obtained: 1 showed 57 times, 2 showed 46 times, 3 showed 68 times, 4 showed 52 times, 5 showed 72 times, and 6 showed 65 times. At the 0.05 level of significance, do these results support the hypothesis that the die is balanced?

10.57 Three coins are tossed 160 times and 0, 1, 2, and 3 heads showed 15, 54, 72, and 19 times. Using the probabilities given on page 197, namely, $\frac{1}{8}, \frac{3}{8}, \frac{3}{8}$, and $\frac{1}{8}$, test at the level of significance $\alpha = 0.01$ whether it is reasonable to suppose that the coins are balanced and randomly tossed.

10.58 Ten years' data show that in a given city there were no bank robberies in 57 months, one bank robbery in 36 months, two bank robberies in 15 months, and three or more bank robberies in 12 months. At the 0.05 level of significance, does this substantiate the claim that the probabilities of 0, 1, 2, or 3 or more bank robberies in any one month are 0.40, 0.30, 0.20, and 0.10?

10.59 The following is the distribution of the number of calls received at the switchboard of a government building during 400 five-minute intervals:

Number of calls	Frequency
0	95
1	116
2	112
3	47
4 or more	30

Use the 0.05 level of significance to test whether the number of calls received by the switchboard in a five-minute interval is a random variable having the Poisson distribution with $\lambda = 1.5$, namely, that the probabilities for 0, 1, 2, 3, or 4 or more calls are 0.22, 0.33, 0.25, 0.13, and 0.07.

10.60 If a random variable has the binomial distribution with $n = 10$ and $p = 0.04$, the probabilities of 0, 1, or 2 or more successes are 0.665, 0.277, and 0.058. Based on this information and the data on page 392, test at the 0.05 level of significance whether the quality control engineer is sampling a binomial population with $n = 10$ and $p = 0.04$.

10.61 The following is the distribution of the grades which 100 students received in a history test:

Grade	Frequency
65–69	1
70–74	10
75–79	37
80–84	36
85–89	13
90–94	2
95–99	1

As can easily be verified, the mean and the standard deviation of this distribution are $\bar{x} = 80$ and $s = 5$. To test the null hypothesis that these data constitute a random sample from a normal population with $\mu = 80$ and $\sigma = 5$, proceed with the following steps:

(a) Find the probabilities that a random variable having the normal distribution with $\mu = 80$ and $\sigma = 5$ will take on a values less than 69.5, between 69.5 and 74.5, between 74.5 and 79.5, between 79.5 and 84.5, between 84.5 and 89.5, between 89.5 and 94.5, and greater than 94.5.

(b) Multiply the probabilities obtained in part (a) by 100, the total frequency of the observed data, to find the frequencies we can expect under the null hypothesis for a random sample of size $n = 100$.

(c) Calculate χ^2 for the observed frequencies shown in the table and the expected frequencies obtained in part (b), and test at the 0.05 level of significance whether the null hypothesis can be rejected.

10.62 The following is the distribution obtained on page 16 for the amount of time that 80 college students engaged in leisure activities during a typical school week:

Hours	Frequency
10–14	8
15–19	28
20–24	27
25–29	12
30–34	4
35–39	1

As we showed on page 74, the mean and the standard deviation of this distribution are $\bar{x} = 20.7$ and $s = 5.4$. To test the null hypothesis that these data constitute a random sample from a normal population with $\mu = 20.7$ and $\sigma = 5.4$, proceed with the following steps:

(a) Find the probabilities that a random variable having the normal distribution with $\mu = 20.7$ and $\sigma = 5.4$ will take on a value less than 14.5, between 14.5 and 19.5, between 19.5 and 24.5, between 24.5 and 29.5, between 29.5 and 34.5, and greater than 34.5.

(b) Multiply the probabilities obtained in part (a) by 80, the total frequency of the observed data, to find the frequencies we can expect under the null hypothesis for a random sample of size $n = 80$.

(c) Calculate χ^2 for the observed frequencies shown in the table and the expected frequencies obtained in part (b), and test at the 0.05 level of significance whether the null hypothesis can be rejected.

SOLUTIONS OF PRACTICE EXERCISES

10.1 Substituting $n = 500$, $\dfrac{x}{n} = \dfrac{140}{500} = 0.28$, and $z_{0.005} = 2.575$ into the large-sample confidence interval formula, we get

$$0.28 - 2.575 \sqrt{\frac{(0.28)(0.72)}{500}} < p < 0.28 + 2.575 \sqrt{\frac{(0.28)(0.72)}{500}}$$

$$0.228 < p < 0.332$$

or $0.23 < p < 0.33$ rounded to two decimals.

10.2 Substituting $n = 500$, $\dfrac{x}{n} = \dfrac{140}{500} = 0.28$, and $z_{0.025} = 1.96$ into the formula for E, we get

$$E = 1.96 \sqrt{\frac{(0.28)(0.72)}{500}} = 0.039$$

We can assert with 95% confidence that the error is at most 0.039.

10.11 Substituting $E = 0.04$ and $z_{0.025} = 1.96$ into the second formula for n, we get

$$n = \frac{1}{4}\left(\frac{1.96}{0.04}\right)^2 = 601$$

rounded up to the nearest integer.

10.12 Substituting $E = 0.04$, $z_{0.025} = 1.96$, and $p = 0.60$ into the first formula for n, we get

$$n = (0.60)(0.40)\left(\frac{1.96}{0.04}\right)^2 = 577$$

rounded up to the nearest integer.

10.19 1. *Hypotheses*

$$H_0: \; p = 0.10$$

$$H_A: \; p > 0.10$$

2. *Level of significance*

$$\alpha = 0.05$$

3. *Criterion*

Reject the null hypothesis if the probability of 4 or more successes is less than or equal to 0.05; otherwise, accept it or reserve judgment.

4. *Calculations*

Table I shows that for $n = 13$ and $p = 0.10$ the probability of 4 or more successes is

$$0.028 + 0.006 + 0.001 = 0.035$$

5. *Decision*

Since 0.035 is less than 0.05, the null hypothesis must be rejected; we conclude that more than 10% of all persons exposed to the radiation will feel any effects.

10.25 1. *Hypotheses*

$$H_0: p = 0.35$$
$$H_A: p \neq 0.35$$

2. *Level of significance*

$$\alpha = 0.01$$

3. *Criterion*

Reject the null hypothesis if $z < -2.575$ or $z > 2.575$, where

$$z = \frac{x - np_0}{\sqrt{np_0(1 - p_0)}}$$

Otherwise, accept the null hypothesis or reserve judgment.

4. *Calculations*

Substituting $x = 128$, $n = 400$, and $p_0 = 0.35$ into the formula for z, we get

$$z = \frac{128 - 400(0.35)}{\sqrt{400(0.35)(0.65)}} = -1.26$$

5. *Decision*

Since $z = -1.26$ falls between -2.575 and 2.575, the null hypothesis cannot be rejected; the data do not refute the claim.

10.31 1. *Hypotheses*

$$H_0: p_1 = p_2$$
$$H_A: p_1 \neq p_2$$

2. *Level of significance*

$$\alpha = 0.05$$

3. *Criterion*

Reject the null hypothesis if $z < -1.96$ or $z > 1.96$, where z is given by the formula on page 375. Otherwise, accept the null hypothesis or reserve judgment.

4. *Calculations*

Substituting $x_1 = 16$, $x_2 = 20$, $n_1 = 200$, $n_2 = 400$, and $\hat{p} = \dfrac{16 + 20}{200 + 400} =$

0.06 into the formula for z, we get

$$z = \frac{\dfrac{16}{200} - \dfrac{20}{400}}{\sqrt{(0.06)(0.94)\left(\dfrac{1}{200} + \dfrac{1}{400}\right)}} = 1.46$$

5. *Decision*

Since $z = 1.46$ falls between -1.96 and 1.96, the null hypothesis cannot be rejected; in other words, the difference between the two sample proportions is not significant.

10.36 1. *Hypotheses*

$$H_0: p_1 = p_2 = p_3 = p_4$$

$$H_A: \text{The } p\text{'s are not all equal.}$$

2. *Level of significance*

$$\alpha = 0.05$$

3. *Criterion*

Reject the null hypothesis if $\chi^2 > 7.815$, the value of $\chi^2_{0.05}$ for $4 - 1 = 3$ degrees of freedom, where

$$\chi^2 = \sum \frac{(o - e)^2}{e}$$

Otherwise, accept it or reserve judgment.

4. *Calculations*

Since

$$\hat{p} = \frac{25 + 21 + 32 + 18}{200 + 200 + 200 + 200} = \frac{96}{800} = 0.12$$

we get $200(0.12) = 24$ for the expected frequencies for the first row and $200 - 24 = 176$ for the expected frequencies for the second row. Then, substituting into the formula for χ^2, we get

$$\chi^2 = \frac{(25 - 24)^2}{24} + \frac{(21 - 24)^2}{24} + \frac{(32 - 24)^2}{24} + \frac{(18 - 24)^2}{24}$$

$$+ \frac{(175 - 176)^2}{176} + \frac{(179 - 176)^2}{176} + \frac{(168 - 176)^2}{176} + \frac{(182 - 176)^2}{176}$$

$$= 5.21$$

5. *Decision*

Since $\chi^2 = 5.21$ does not exceed 7.815, the null hypothesis cannot be rejected; we cannot conclude that there is a real difference in the durability of the tires

10.41 1. *Hypotheses*

$$H_0: p_1 = p_2$$

$$H_A: p_1 \neq p_2$$

2. *Level of significance*

$$\alpha = 0.05$$

3. *Criterion*

Reject the null hypothesis if $\chi^2 > 3.841$, the value of $\chi^2_{0.05}$ for $2 - 1 = 1$ degrees of freedom, where

$$\chi^2 = \sum \frac{(o - e)^2}{e}$$

Otherwise, accept it or reserve judgment.

4. *Calculations*

Since $\hat{p} = \dfrac{16 + 20}{200 + 400} = 0.06$, we get $200(0.06) = 12$ and $400(0.06) = 24$ for the expected frequencies for the first row, and $200 - 12 = 188$ and $400 - 24 = 376$ for the expected frequencies for the second row. Then, substituting into the formula for χ^2, we get

$$\chi^2 = \frac{(16 - 12)^2}{12} + \frac{(20 - 24)^2}{24} + \frac{(184 - 188)^2}{188} + \frac{(380 - 376)^2}{376}$$

$$= 2.13$$

5. *Decision*

Since $\chi^2 = 2.13$ does not exceed 3.841, the null hypothesis cannot be rejected; in other words, the difference between the two sample proportions is not significant.

If we square 1.46, the value obtained for z in Exercise 10.31, we get $(1.46)^2 = 2.1316$, and, except for rounding, this equals the value which we obtained for χ^2.

10.44 1. *Hypotheses*

H_0: For each alternative, the probabilities are the same for persons who are single, married, and widowed or divorced.

H_A: For at least one alternative, the probabilities are not the same for the three kinds of persons.

CHAP. 10: The Analysis of Count Data

2. *Level of significance*

$$\alpha = 0.05$$

3. *Criterion*

   Reject the null hypothesis if $\chi^2 > 9.488$, the value of $\chi^2_{0.05}$ for $(3-1) \cdot (3-1) = 4$ degrees of freedom, where

$$\chi^2 = \sum \frac{(o-e)^2}{e}$$

   Otherwise, accept it or reserve judgment.

4. *Calculations*

   The expected frequencies for the first two cells of the first row are $\dfrac{132 \cdot 80}{300} = 35.2$ and $\dfrac{132 \cdot 120}{300} = 52.8$; those for the first two cells of the second row are $\dfrac{110 \cdot 80}{300} = 29.3$ and $\dfrac{110 \cdot 120}{300} = 44.0$; and by subtraction, those for the third cells of the first two rows are 44.0 and 36.7, and those for the third row are 15.5, 23.2, and 19.3. Thus, substitution into the formula for χ^2 yields

$$\chi^2 = \frac{(41-35.2)^2}{35.2} + \frac{(49-52.8)^2}{52.8} + \frac{(42-44.0)^2}{44.0}$$

$$+ \frac{(27-29.3)^2}{29.3} + \frac{(50-44.0)^2}{44.0} + \frac{(33-36.7)^2}{36.7}$$

$$+ \frac{(12-15.5)^2}{15.5} + \frac{(21-23.2)^2}{23.2} + \frac{(25-19.3)^2}{19.3}$$

$$= 5.37$$

5. *Decision*

   Since $\chi^2 = 5.37$ does not exceed 9.488, the null hypothesis cannot be rejected; there is no real evidence that the probabilities of the three alternatives are not the same for the three kinds of persons.

10.56 1. *Hypotheses*

$$H_0: \text{The die is balanced.}$$

$$H_A: \text{The die is not balanced.}$$

2. *Level of significance*

$$\alpha = 0.05$$

3. *Criterion*

Reject the null hypothesis if $\chi^2 > 11.070$, the value of $\chi^2_{0.05}$ for $6 - 1 = 5$ degrees of freedom, where

$$\chi^2 = \sum \frac{(o - e)^2}{e}$$

Otherwise, accept it or reserve judgment.

4. *Calculations*

The expected frequencies are all $\frac{360}{6} = 60$, so that

$$\chi^2 = \frac{(57 - 60)^2}{60} + \frac{(46 - 60)^2}{60} + \frac{(68 - 60)^2}{60}$$

$$+ \frac{(52 - 60)^2}{60} + \frac{(72 - 60)^2}{60} + \frac{(65 - 60)^2}{60}$$

$$= 8.37$$

5. *Decision*

Since $\chi^2 = 8.37$ does not exceed 11.070, the null hypothesis cannot be rejected; we conclude that there is no real evidence that the die is not balanced.

10.61 (a) In standard units the class boundaries are $z = \dfrac{69.5 - 80}{5} = -2.10$, $z = \dfrac{74.5 - 80}{5} = -1.10$, $z = \dfrac{79.5 - 80}{5} = -0.10$, $z = \dfrac{84.5 - 80}{5} = 0.90$, $z = \dfrac{89.5 - 80}{5} = 1.90$, and $z = \dfrac{94.5 - 80}{5} = 2.90$; the corresponding entries in Table II are 0.4821, 0.3643, 0.0398, 0.3159, 0.4713, and 0.4981, so that the probabilities are $0.5000 - 0.4821 = 0.0179$, $0.4821 - 0.3643 = 0.1178$, $0.3643 - 0.0398 = 0.3245$, $0.0398 + 0.3159 = 0.3557$, $0.4713 - 0.3159 = 0.1554$, $0.4981 - 0.4713 = 0.0268$, and $0.5000 - 0.4981 = 0.0019$.

(b) The expected normal curve frequencies are $100(0.0179) = 1.8$, $100(0.1178) = 11.8$, $100(0.3245) = 32.4$, $100(0.3557) = 35.6$, $100(0.1554) = 15.5$, $100(0.0268) = 2.7$, and $100(0.0019) = 0.2$. Thus, we have

o	e
$\left.\begin{array}{l}1\\10\end{array}\right\}11$	$\left.\begin{array}{l}1.8\\11.8\end{array}\right\}13.6$
37	32.4
36	35.6
$\left.\begin{array}{l}13\\2\\1\end{array}\right\}16$	$\left.\begin{array}{l}15.5\\2.7\\0.2\end{array}\right\}18.4$

where we combined some of the classes so that none of the expected frequencies is less than 5.

(c) 1. *Hypotheses*

$$H_0: \text{Data come from a normal population.}$$

$$H_A: \text{Data do not come from a normal population.}$$

2. *Level of significance*

$$\alpha = 0.05$$

3. *Criterion*

Reject the null hypothesis if $\chi^2 > 3.841$, the value of $\chi^2_{0.05}$ for $4 - 3 = 1$ degree of freedom, where

$$\chi^2 = \sum \frac{(o - e)^2}{e}$$

Otherwise, accept it or reserve judgment.

4. *Calculations*

Substituting into the formula for χ^2, we get

$$\chi^2 = \frac{(11 - 13.6)^2}{13.6} + \frac{(37 - 32.4)^2}{32.4} + \frac{(36 - 35.6)^2}{35.6} + \frac{(16 - 18.4)^2}{18.4}$$

$$= 1.47$$

5. *Decision*

Since $\chi^2 = 1.47$ does not exceed 3.841, the null hypothesis cannot be rejected; there is no real evidence to doubt that data come from a normal population.

Achievements

Having read and studied these chapters, and having worked a good portion of the exercises, you should be able to

1. Distinguish between point estimates and interval estimates.

2. Explain the difference between "probability" and "confidence."

3. Make confidence statements about the maximum error when $\bar{x}$ is used as an estimate of μ.

4. Determine the sample size needed so that a sample mean will have a desired precision.

5. Explain what is meant by "confidence interval" and "degree of confidence."

6. Construct large-sample confidence intervals for μ.

7. Construct small-sample confidence intervals for μ.

★ 8. Construct large-sample confidence intervals for σ.

9. Explain what is meant by "statistical hypothesis," "null hypothesis," and "alternative hypothesis."

10. Distinguish between Type I and Type II errors.

11. Explain what is meant by "operating characteristic curve."

12. Distinguish between simple and composite hypotheses about parameters.

13. Explain what is meant by "significance test," "level of significance," and "statistically significant."

14. List the five steps of the outline for testing a statistical hypothesis.

15. Distinguish between one-sided and two-sided alternatives, and between one-tailed and two-tailed tests.

16. Perform large-sample tests of the null hypothesis $\mu = \mu_0$.

17. Perform small-sample tests of the null hypothesis $\mu = \mu_0$.

18. Perform large-sample tests of the null hypothesis that two populations have equal means.

19. Perform small-sample tests of the null hypothesis that two populations have equal means.

20. Test the significance of the difference between the means of paired data.

★ 21. Perform a one-way analysis of variance when the sample sizes are all equal.

★ 22. Perform a one-way analysis of variance when the sample sizes are not all equal.

23. Construct large-sample confidence intervals for population proportions.

24. Make confidence statements about the maximum error when $\frac{x}{n}$ is used as an estimate of p.

25. Determine the sample size needed so that a sample proportion will have a desired precision (with and without some information about p).

26. Perform small-sample tests of the null hypothesis $p = p_0$.

27. Perform large-sample tests of the null hypothesis $p = p_0$.

28. Test for the equality of two population proportions.

29. Test for the equality of k population proportions.

30. Analyze an $r \times c$ (contingency) table.

31. Test for goodness of fit.

Checklist of Key Terms (with page references to their definitions)

REVIEW EXERCISES

R.119 To compare two kinds of baseball bats, 18 baseball players were asked to swing 20 times with each kind of bat at balls pitched by a machine, and the following are the respective numbers of balls they hit more than 300 feet: 6 and 8, 9 and 5, 4 and 4, 7 and 6, 10 and 8, 5 and 6, 9 and 7, 3 and 4, 5 and 4, 6 and 6, 12 and 9, 8 and 9, 5 and 5, 4 and 6, 9 and 6, 10 and 8, 7 and 7, and 11 and 7. Test at the 0.05 level of significance whether the difference between the means of the numbers of balls hit more than 300 feet with the two kinds of bats is significant.

R.120 A sample check reveals that 176 of 200 of a professional football team's season-ticket holders intend to renew their tickets for the next season. Construct a 95% confidence interval for the true proportion of season-ticket holders who intend to renew their tickets for the next season.

R.121 A pollster wants to determine what proportion of the population is opposed to nuclear energy. How large a sample will he need, if he wants to be able to assert with probability 0.95 that his sample proportion will be off by at most 0.03?

R.122 In a random sample of 250 retired persons, 192 stated that they prefer living in an apartment to living in a one-family home. At the 0.05 level of significance, does this refute the claim that at most 60% of all retired persons prefer living in an apartment to living in a one-family home?

R.123 Four coins are tossed 160 times and 0, 1, 2, 3, and 4 heads showed 6, 28, 72, 39, and 15 times. Making use of the fact that the probabilities for 0, 1, 2, 3, and 4 heads are $\frac{1}{16}$, $\frac{4}{16}$, $\frac{6}{16}$, $\frac{4}{16}$, and $\frac{1}{16}$, test at the 0.01 level of significance whether it is reasonable to assume that the coins are balanced and randomly tossed.

R.124 In a random sample of 90 persons having dinner by themselves at a French restaurant, 63 had wine with their dinner. If we use the sample proportion $\frac{63}{90} = 0.70$ to estimate the corresponding true proportion, what can we say with 95% confidence about the maximum error?

R.125 An efficiency expert wants to determine the average time it takes a person to buy a week's groceries at a supermarket. If preliminary studies have shown that it is reasonable to let $\sigma = 2.6$ minutes, how large a sample will she need to be able to assert with probability 0.99 that the mean of her sample will be off by at most 0.25 minute?

R.126 Thirty containers of a commercial solvent randomly selected from a large production lot have a mean weight of 24.5 pounds with a standard deviation of 0.3 pound. What can we assert with 95% confidence about the maximum error, if we use 24.5 as an estimate of the true mean weight of a container of this solvent?

R.127 The following table shows the results of a study in which samples of statisticians, economists, and mathematicians were asked to compare the merits of two statistics texts:

	Statisticians	Economists	Mathematicians
Prefer textbook A	28	44	18
Undecided	24	31	11
Prefer textbook B	48	25	71

Use the 0.01 level of significance to test the null hypothesis that in general equal proportions of statisticians, economists, and mathematicians prefer textbook A, are undecided, or prefer textbook B.

R.128 Based on the results of $n = 12$ trials, we want to test the null hypothesis $p = 0.20$ against the alternative hypothesis $p > 0.20$. If we reject the null hypothesis when the number of successes is 5 or more and otherwise we accept it, find
(a) the probability of a Type I error;
(b) the probability of a Type II error when $p = 0.30$;
(c) the probability of a Type II error when $p = 0.50$.

R.129 Five measurements of the tar content of a certain kind of cigarette yielded 14.5, 14.2, 14.4, 14.3, and 14.6 mg/cig (milligrams per cigarette). Show that the difference between the mean of this sample, $\bar{x} = 14.4$, and the average tar content claimed by the cigarette manufacturer, $\mu = 14.0$, is significant at the 0.05 level of significance.

R.130 Six guinea pigs injected with 0.5 mg of a medication took on the average 16.8 seconds to fall asleep with a standard deviation of 2.2 seconds, while six other guinea pigs injected with 1.5 mg of the medication took on the average 14.5 seconds to fall asleep with a standard deviation of 2.6 seconds. Use the 0.05 level of significance to test whether the increase in dosage decreases the mean time it takes a guinea pig to fall asleep.

★ R.131 Given that 35 one-gallon cans of a certain kind of paint covered on the average 452.2 square feet with a standard deviation of 22.8 square feet, construct a 95% confidence interval for σ.

R.132 A random sample of size $n = 50$ is taken from a population which has the standard deviation $\sigma = 5.5$ mm. If we use the mean of this sample to estimate the mean of the population, with what confidence can we assert that the error is less than 1.2 mm?

R.133 The following table shows how many times, Monday through Friday, a bus was late arriving at a given stop in 40 weeks:

Number of times bus was late	Number of weeks
0	4
1	11
2	15
3 or more	10

Use the 0.05 level of significance to test the null hypothesis that the bus is late 30% of the time, namely, the null hypothesis that the data constitute a random sample from a binomial population with $n = 5$ and $p = 0.30$.

R.134 In a study of the relationship between family size and intelligence, 40 "only children" had an average IQ of 101.5 with a standard deviation of 6.7 and 50 "firstborns" in two-child families had an average IQ of 105.9 with a standard deviation of 5.8. At the 0.01 level of significance, is the difference between these two sample means significant?

R.135 The following table shows how samples of the residents of three federally financed housing projects replied to the question whether they would continue to live there if they had the choice:

	Project 1	Project 2	Project 3
Yes	83	68	65
No	17	32	35

Test at the 0.05 level of significance whether the differences among the proportions of "yes" answers can be attributed to chance.

R.136 A cab company is considering replacing the tires on its cars with radial tires. If μ_0 is the average mileage it has been getting with its old tires, against what alternative would it test the null hypothesis $\mu = \mu_0$ if
  (a) it does not want to replace the old tires with the radial tires unless the radial tires prove to be superior;
  (b) it wants to make the change unless the radial tires actually turn out to be inferior?

R.137 In a survey conducted in a retirement community, it was found that a random sample of 10 senior citizens visited a physician on the average 6.7 times per year with a standard deviation of 1.6. Construct a 95% confidence interval for the true average number of times that a person in the population sampled visits a physician per year.

R.138 It is desired to use the mean of a random sample of size $n = 36$ to estimate the average number of days of continuous use until a new refrigerator of a certain kind will first require repairs. If it can be assumed that $\sigma = 242$ days, what can we assert with 99% confidence about the maximum error?

★ R.139 To compare four different golf-ball designs, A, B, C, and D, several balls of each kind were driven by a golf professional and the following results (distances from the tee to the point of impact in yards) were obtained:

Golf ball A: 262, 245, 237, 280, 236
Golf ball B: 244, 216, 251, 263, 214, 228
Golf ball C: 272, 265, 244, 259
Golf ball D: 250, 233, 217, 267, 258

At the level of significance 0.05, can the differences among the four sample means be attributed to chance?

R.140 In $n = 11$ trials, a certain weed killer was effective eight times. At the level of significance 0.05 test the claim that the weed killer is effective at most 50% of the time.

R.141 To compare two kinds of bumper guards, six of each kind were mounted on Buick Skylarks. Then each car was run into a concrete wall at 5 miles per hour, and the following results were obtained for the repair costs (in dollars): $\bar{x}_1 = 87.60$, $s_1 = 12.54$, $\bar{x}_2 = 107.33$, $s_2 = 11.28$. Use the level of significance $\alpha = 0.01$ to test whether the difference between the mean repair costs is significant.

R.142 A researcher wants to determine what percentage of the farm workers in a certain area are illegal aliens. How large a sample will she need if she wants to be 99% sure that her estimate will not be off by more than 4%, and she feels that the actual percentage is at most 10%?

409

R.143 Suppose we want to test the hypothesis that solar heating unit A is more efficient than solar heating unit B. Explain under what conditions we would be committing a Type I error and under what conditions we would be committing a Type II error.

R.144 A random sample of eight daily scrap records (where scrap is expressed as a percentage of material requisitioned) yielded $\bar{x} = 5.2\%$ scrap and $s = 0.8\%$ scrap. If the mean of this sample is used to estimate the mean of the population sampled, what can we assert with 95% confidence about the maximum error?

R.145 Asked about the centerfolds of a certain magazine, 56 of 400 men felt that the nudes were too explicit. At the 0.05 level of significance, does this support the claim (made by the editors of the magazine) that at most 10% of all men feel that way about the centerfolds?

R.146 During the investigation of an alleged unfair trade practice, the Federal Trade Commission takes a random sample of 49 "3-ounce" candy bars from a large shipment. If the mean and the standard deviation of their weights are, respectively, 2.94 ounces and 0.12 ounce, show that, at the level of significance 0.01, the commission has grounds upon which to proceed against the manufacturer on the unfair practice of short-weight selling.

R.147 As part of an industrial training program, some trainees are instructed by method A, which is straight teaching-machine instruction, and some are instructed by method B, which also involves the personal attention of an instructor. If random samples of size 10 are taken from large groups of trainees instructed by each of these two methods, and the scores which they obtained in an appropriate achievement test are

$$Method\ A:\quad 71, 75, 65, 69, 73, 66, 68, 71, 74, 68$$
$$Method\ B:\quad 72, 77, 84, 78, 69, 70, 77, 73, 65, 75$$

use the 0.05 level of significance to test whether method B is more effective.

★R.148 Samples of peanut butter produced by three different manufacturers are tested for aflatoxin content (ppb), with the following results:

$$Brand\ A:\quad 0.5, 6.3, 1.1, 2.7, 5.5, 4.3$$
$$Brand\ B:\quad 2.5, 1.8, 3.6, 5.2, 1.2, 0.7$$
$$Brand\ C:\quad 3.3, 1.5, 0.4, 4.8, 2.2, 1.0$$

Perform an analysis of variance to test at the 0.05 level of significance whether the differences among the three sample means are significant.

R.149 Measurements of the amount of chloroform (micrograms per liter) in 36 samples of the drinking water of a city yielded the following results: $\bar{x} = 34.8$ and $s = 4.9$. Use these data to construct a 99% confidence interval for the average amount of chloroform in the drinking water of this city.

R.150 In a study conducted at a large airport, 81 of 300 persons who had just gotten off a plane and 32 of 200 persons who were about to board a plane admitted that they were afraid of flying. At the 0.05 level of significance, test whether the difference between the corresponding sample proportions is significant.

R.151 In a French restaurant, the chef receives 26, 21, 14, 22, 18, and 20 orders for coq au vin on six different nights. Construct a 95% confidence interval for the number of orders for coq au vin the chef can expect per night.

R.152 A general achievement test is standardized so that eighth graders should average 77.2 with a standard deviation of 4.8. If 35 eighth graders from a certain school district average 79.5 on the test, does this substantiate the claim that the eighth graders in this district are above average? Use the level of significance $\alpha = 0.05$ and assume that the data constitute a random sample.

R.153 In ten test runs, a car ran for 18, 17, 21, 16, 19, 16, 19, 18, 19, and 17 miles with a gallon of a certain kind of gasoline. Construct a 95% confidence interval for the average number of miles the car will get with this gasoline.

R.154 The following table pertains to a study of the relationship between the speed of promotion and the standard of clothing of bank employees:

	Speed of Promotion		
	Slow	Average	Fast
Poorly dressed	12	8	4
Well dressed	18	25	20
Very well dressed	13	19	31

Use the 0.01 level of significance to test the null hypothesis that there is no relationship.

★ **R.155** Rework Exercise R.148 by the method of Section 9.12; that is, by calculating $n \cdot s_{\bar{x}}^2$, the mean of the three sample variances, and F, the ratio of these two quantities.

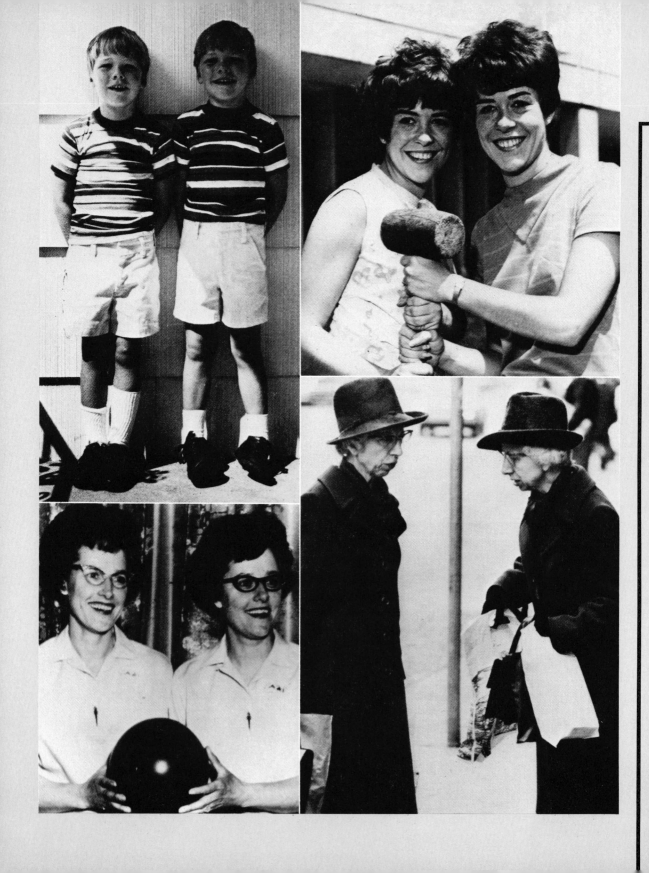

The Analysis of Paired Data

The goal of many statistical investigations is to establish relationships which make it possible to predict one variable in terms of another. For example, we might want to predict the sales of a new product in terms of the amount of money spent advertising it on television, predict family expenditures on entertainment in terms of family income, or predict a college student's grade-point average based on the number of hours he or she spent studying. As it is seldom possible to predict one quantity exactly in terms of another, we shall be concerned here with the problem of

> **REGRESSION** (ri-'gresh-ən) The relationship between the mean of a random variable and the values of one or more independent variables on which it depends.

Although we cannot predict exactly how much money a specific college graduate will earn ten years after graduation, given suitable data we can predict the average earnings of college graduates ten years after graduation. Similarly, we can predict the average yield of wheat in terms of the total rainfall in July or predict the expected grade-point average of college freshmen in terms of their SAT scores and/or their IQ's.

Of course, in some instances there may be no relationship at all, or only a very weak one, so we shall also be concerned with measuring the extent, or strength of the

> **CORRELATION** (ˌkòr-ə-'lā-shən) The relationship (association or interdependence) of the values of two or more qualitative or quantitative variables.

Problems of regression are taken up in Sections 11.1 through 11.3; then, the subject of correlation is treated in Sections 11.4 through 11.6.

413

11.1
Curve Fitting

Whenever possible, we try to express, or approximate, relationships between known quantities and quantities that are to be predicted in terms of mathematical equations. This has been very successful in the natural sciences, where it is known, for instance, that at a constant temperature the relationship between the volume, y, and the pressure, x, of a gas is given by the formula

$$y = \frac{k}{x}$$

where k is a numerical constant. Also, it has been shown that the relationship between the size of a culture of bacteria, y, and the length of time, x, it has been exposed to certain environmental conditions is given by the formula

$$y = a \cdot b^x$$

where a and b are numerical constants. More recently, equations like these have also been used to describe relationships in the behavioral sciences, the social sciences, and other fields. For instance, the first of the equations above is often used in economics to describe the relationship between price and demand, and the second has been used to describe the growth of one's vocabulary or the accumulation of wealth.

Whenever we use observed data to arrive at a mathematical equation which describes the relationship between two variables—a procedure known as **curve fitting**—we must face three kinds of problems: (1) We must decide what kind of an equation we want to use (for instance, that of a straight line or that of some other curve); (2) we must find the particular equation which is "best" in some sense; and (3) we must investigate certain questions regarding the merits of the particular equation, and of the predictions made from it. The second of these problems will be discussed in some detail in Section 11.2, and the third in Section 11.3

The first kind of problem is sometimes decided by theoretical considerations, but more often by direct inspection of the data. We plot the data on graph paper, sometimes on special graph paper with special scales, and we

414

judge visually what kind of curve best describes their overall pattern. So far as the work in this text is concerned, we shall consider only **linear equations in two unknowns,** which are of the form

$$y = a + bx$$

where a is the **y-intercept** (the value of y for $x = 0$) and b is the **slope** of the line (the change in y which accompanies an increase of one unit in x). Ordinarily, the values of a and b are estimated from the data (by the method we shall discuss in Section 11.2), and once they have been determined we can substitute a value of x into the equation and calculate the corresponding predicted value of y.

The term "linear equation" arises from the fact that the graph of $y = a + bx$ is a straight line, namely, that all pairs of values of x and y which satisfy an equation of the form $y = a + bx$ constitute points which fall on a straight line. Suppose, for instance, that

$$y = 0.23 + 4.42x$$

expresses the relationship between a certain county's annual rainfall in inches from September through August, x, and its yield of wheat in bushels per acre, y. For example, for an annual rainfall of six inches, $x = 6$ and $y = 0.23 + 4.42(6) = 26.75$ bushels of wheat per acre, and for an annual rainfall of twelve inches, $x = 12$ and $y = 0.23 + 4.42(12) = 53.27$ bushels of wheat per acre. Observe that the points $(6, 26.75)$ and $(12, 53.27)$ lie on the straight line shown in Figure 11.1, and this is true for any other points obtained in the same way.

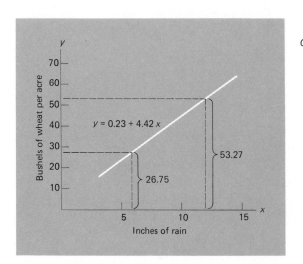

11.1

Graph of linear equation.

Since we are limiting our discussion here to linear equations, let us point out that **linear equations are useful and important not only because many relationships are actually of this form, but also because they often provide close approximations to relationships which would otherwise be difficult to describe in mathematical terms.** The problem of describing data by means of other kinds of curves, namely, the problem of **nonlinear curve fitting,** is discussed in the more theoretical texts listed in the Bibliography at the end of the book.

11.2

The Method of Least Squares

Once we have decided to fit a straight line to a given set of data, we face the second kind of problem, namely, that of finding the equation of the particular line which in some sense provides the best possible fit. To show what is involved, let us consider the following sample data obtained in a study of the relationship between the number of years that applicants for certain foreign service jobs have studied German in high school or college and the grades which they received in a proficiency test in that language:

Number of years x	Grade in test y
3	57
4	78
4	72
2	58
5	89
3	63
4	73
5	84
3	75
2	48

If we plot the points which correspond to these ten pairs of values as in Figure 11.2, we observe that, even though the points do not all fall on a straight line, the overall pattern of the relationship is reasonably well described by the white line. There is no apparent departure from linearity in the pattern of the points, so we feel justified in deciding that a straight line is a suitable description of the underlying relationship.

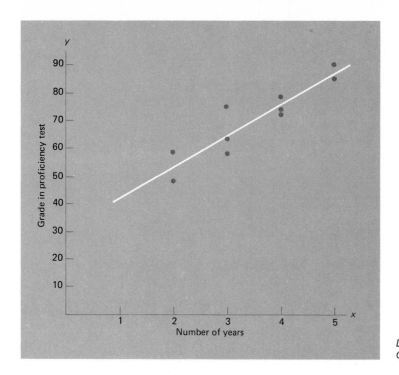

11.2

Data on number of years studied German and grade in proficiency test.

This takes care of the first kind of problem and we must now consider the second kind, namely, that of finding the equation of the line which in some sense provides the best fit to the data and which, it is hoped, will later yield the best possible predictions of *y* for given values of *x*. Logically speaking, there is no limit to the number of straight lines we can draw on a piece of paper, say, on the diagram of Figure 11.2. Some of them would fit the data so poorly that we cannot consider them seriously, but many others would come fairly close, and the problem is to find the one line which fits the data "best" in some well-defined sense. If all the points actually fell on a straight line there would be no problem, but this is an extreme case which we rarely encounter in practice. In general, we have to be satisfied with a line having certain desirable properties, short of perfection.

The criterion which, today, is used almost exclusively for defining a "best" fit dates back to the early part of the nineteenth century and the work of the French mathematician Adrien Legendre. It is called the **method of least squares,** and it requires that the sum of the squares of the vertical deviations (distances) from the points to the line be as small as possible.

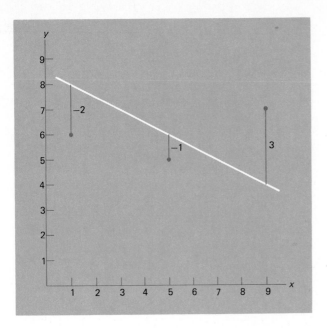

11.3

Errors made by using given line to "predict" the y values for x = 1, x=5, and x = 9.

To explain why we do this, let us refer to Figure 11.3, where we plotted points corresponding to the following data

x	y
1	6
5	5
9	7

and more or less arbitrarily "fit" a straight line; actually, we drew the line through the points (1, 8) and (9, 4).

Had we used this line to "predict" the values of y for the given values of x, namely, if we had marked $x = 1$, $x = 5$, and $x = 9$ on the horizontal scale and used the line to read off the corresponding values of y, we would have obtained 8, 6, and 4. The errors of these "predictions" are $6 - 8 = -2$, $5 - 6 = -1$, and $7 - 4 = 3$, and in Figure 11.3 they are the vertical distances from the points to the line.

The sum of these errors is $(-2) + (-1) + 3 = 0$, but this is not indicative of their size and we find ourselves in a position similar to that on page 62, which led to the definition of the standard deviation. Squaring the errors as we squared

CHAP. 11: The Analysis of Paired Data

the deviation from the mean on page 62, we find that the sum of the squares of the errors is $(-2)^2 + (-1)^2 + 3^2 = 14$.

Now let us fit another line to the same **data points** as before. The one shown in Figure 11.4 actually passes through the points (1, 5) and (9, 6), and judging by eye, it seems to provide a much better fit than the line of Figure 11.3. As can easily be verified, the errors of the "predictions," the differences between the observed values of y and the corresponding values read off the line, are now, $6 - 5 = 1$, $5 - 5.5 = -0.5$, and $7 - 6 = 1$.

The sum of these errors is $1 + (-0.5) + 1 = 1.5$, which is greater than the sum we obtained for the errors made with the line of Figure 11.3, but this is not indicative of their size. On the other hand, the sum of the squares of the errors is now $1^2 + (-0.5)^2 + 1^2 = 2.25$, and this is much less than the 14 we obtained for the line of Figure 11.3. Indeed, this is indicative of the fact that the line of Figure 11.4 provides a better fit than the line of Figure 11.3, and we define the "best-fitting" line as the one for which the sum of the squares of the errors is as small as possible. Such a line is called a **least-squares line.**

To show how a least-squares line is actually fit to a set of paired data, let us consider n pairs of numbers (x_1, y_1), (x_2, y_2), ..., and (x_n, y_n), which might represent the heights and weights of n persons, the IQ's of n mothers and sons,

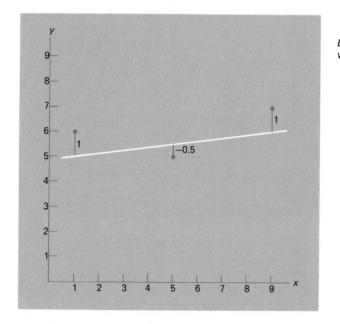

11.4

Errors made by using second line to "predict" the y values for x = 1, x = 5, and x = 9.

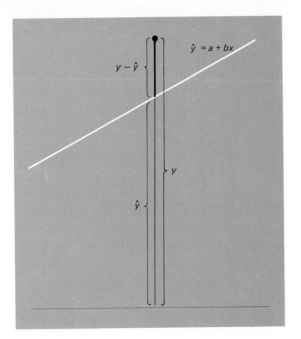

11.5

Least-squares line.

the monthly food expenditures and mortgage payments of n families, and so on. If we write the equation of the line as $\hat{y} = a + bx$, where the symbol $\hat{y}$ is used to distinguish between the observed values of y and the corresponding values $\hat{y}$ on the line, the method of least squares requires that we minimize the sum of the squares of the differences between the y's and the $\hat{y}$'s (see Figure 11.5). This means that we must find the numerical values of the constants a and b appearing in the equation $\hat{y} = a + bx$ for which

$$\sum (y - \hat{y})^2 = \sum [y - (a + bx)]^2$$

is as small as possible. As it takes calculus or fairly tedious algebra to find the expressions for a and b which minimize this sum, let us merely state the result that they are given by the solutions for a and b of the following system of linear equations:

Normal equations

$$\sum y = na + b(\sum x)$$
$$\sum xy = a(\sum x) + b(\sum x^2)$$

CHAP. 11: The Analysis of Paired Data

In these equations, called the **normal equations,** n is the number of pairs of observations, $\sum x$ and $\sum y$ are the sums of the observed x's and y's, $\sum x^2$ is the sum of the squares of the x's, and $\sum xy$ is the sum of the products obtained by multiplying each x by the corresponding y.

EXAMPLE   Fit a least-squares line to the data on page 416, which pertain to the numbers of years that certain applicants for foreign service jobs have studied German in high school or college and the grades which they received in a proficiency test in that language.

Solution   The sums needed for substitution into the normal equations are obtained by performing the calculations shown in the following table:

Number of years x	Test grade y	x^2	xy
3	57	9	171
4	78	16	312
4	72	16	288
2	58	4	116
5	89	25	445
3	63	9	189
4	73	16	292
5	84	25	420
3	75	9	225
2	48	4	96
35	697	133	2,554

(There are many desk calculators, or hand-held calculators, on which the various sums can be accumulated directly, so that there is no need to fill in all the details. Indeed, on some calculators the values of a and b can be obtained directly by recording the data and then pressing the appropriate buttons.) Substituting $\sum x = 35$, $\sum y = 697$, $\sum x^2 = 133$, $\sum xy = 2{,}554$, and $n = 10$ into the two normal equations, we get

$$697 = 10a + 35b$$

$$2{,}554 = 35a + 133b$$

and we must now solve these two simultaneous linear equations for a and b. There are several ways in which this can be done—simplest, perhaps, is the **method of elimination,** which the reader may recall from elementary algebra. Using this method, let us eliminate a by multiplying the expressions on both sides of the first normal equation by 7, multiplying the expressions on both sides of the second normal equation by 2, and then "subtracting equals from equals." We thus get

$$4{,}879 = 70a + 245b$$

$$5{,}108 = 70a + 266b$$

and by subtraction $229 = 21b$. Thus, $b = \dfrac{229}{21} = 10.90$, and if we substitute this value into the first of the two original equations, we get $697 = 10a + 35(10.90)$, $697 = 10a + 381.5$, $10a = 315.5$, and finally $a = \dfrac{315.5}{10} = 31.55$. Thus, the equation of the least-squares line is

$$\hat{y} = 31.55 + 10.90x$$

As an alternative to this procedure we can use the following formulas, which result when we symbolically solve the two normal equations for a and b:

Solutions of normal equations

$$a = \frac{(\sum y)(\sum x^2) - (\sum x)(\sum xy)}{n(\sum x^2) - (\sum x)^2}$$

$$b = \frac{n(\sum xy) - (\sum x)(\sum y)}{n(\sum x^2) - (\sum x)^2}$$

These formulas can be presented in various other forms. The way they are given here requires fewer divisions and, hence, less rounding.

EXAMPLE   Rework the preceding example, using the formulas for a and b given above.

Solution   Substituting $n = 10$ and the various sums obtained on page 421, we get

$$a = \frac{(697)(133) - (35)(2{,}554)}{10(133) - (35)^2} = \frac{3{,}311}{105} = 31.53$$

and

$$b = \frac{10(2,554) - (35)(697)}{105} = \frac{1,145}{105} = 10.90$$

The difference between the values of a obtained by the two methods, 31.55 and 31.53, is due to rounding.

There is still another way of finding a and b which is often used because of its convenience. First we calculate b by using the formula above, and then we substitute its value into the following formula:

Alternative formula for a

$$a = \frac{\sum y - b(\sum x)}{n}$$

which was obtained by solving the first normal equation for a. For our numerical example, where we had $b = 10.90$, we would thus obtain

$$a = \frac{697 - 10.90(35)}{10} = 31.55$$

A computer printout of the preceding linear regression problem is shown in Figure 11.6. The values obtained for a and b are 31.533 and 10.905, given in the column headed "COEFFICIENT." The differences between the results obtained above and in the printout are due to rounding.

It should be mentioned that there are relatively inexpensive handheld calculators which are programmed to solve linear regression problems. They yield the values of a and b after we enter the data and press the appropriate buttons.

Once we have determined the equation of a least-squares line, we can use it to make predictions.

EXAMPLE   Use the least-square line $\hat{y} = 31.53 + 10.90x$ to predict the proficiency grade of an applicant who has studied German in high school or college for two years.

423

```
MTB > NAME C2 = 'Y'
MTB > NAME Cl = 'X'
MTB > SET Cl
DATA> 3  4  4  2  5  3  4  5  3  2
MTB > SET C2
DATA> 57  78  72  58  89  63  73  84  75  48
MTB > REGR C2 1 Cl

THE REGRESSION EQUATION IS
Y = 31.5 + 10.9 X

                              ST. DEV.    T-RATIO =
COLUMN       COEFFICIENT      OF COEF.    COEF/S.D.
             31.533           6.360        4.96
X            10.905           1.744        6.25

S = 5.651

R-SQUARED = 83.0 PERCENT
R-SQUARED = 80.9 PERCENT, ADJUSTED FOR D.F.

ANALYSIS OF VARIANCE

 DUE TO      DF            SS       MS=SS/DF
REGRESSION    1         1248.6       1248.6
RESIDUAL      8          255.5         31.9
TOTAL         9         1504.1

                      Y      PRED. Y    ST.DEV.
 ROW       X          Y      VALUE      PRED. Y   RESIDUAL     ST.RES.
  9       3.00      75.00    64.25        1.99     10.75        2.03R

R DENOTES AN OBS. WITH A LARGE ST. RES.

DURBIN-WATSON STATISTIC = 2.52
```

11.6

Computer printout for linear regression.

Solution   Substituting $x = 2$ into the equation, we get

$$\hat{y} = 31.53 + 10.90(2) = 53.33$$

and this is the best prediction we can make in the least-squares sense.

When we make a prediction like this, we cannot really expect that we will always hit the answer right on the nose—in fact, we cannot possibly be right when the answer has to be a whole number, as in our example, and the answer is 53.33. With reference to this example, it would be very unreasonable to expect that every applicant who had studied German for a given number of years will get the same grade in the proficiency test; indeed, the data on page 416 show that this is not the case. Thus, to make meaningful predictions based on least-squares lines, we must look at such predictions as averages, or as mathematical expectations. Interpreted in this way, we refer to least-squares lines which enable us to read off, or calculate, expected values of y for given values of x as **regression lines.** Better yet, we refer to such lines as **estimated regression lines,** since the values of a and b are determined on the basis of sample data. Questions relating to the goodness of these estimates will be discussed in Section 11.3.

EXERCISES

(Exercise 11.1 is a practice exercise; its complete solution is given on page 449.)

11.1 The following table shows the length of time that six persons have been working at an automobile inspection station and the number of cars each of them checked between noon and 1 o'clock on a given day:

Number of weeks employed x	Cars checked y
5	16
1	15
7	19
9	23
2	14
12	21

(a) Set up the normal equations and solve them by the method of elimination, to find the equation of the least-squares line which will enable us to predict y in terms of x.

(b) Use the formulas on page 422 to check the values obtained for a and b in part (a).

(c) If a person has worked at the inspection station for 10 weeks, how many cars can we expect him or her to inspect during the given time period?

The following data pertain to the chlorine residual in a swimming pool at various times after it has been treated with chemicals:

Number of hours x	Chlorine residual (parts per million) y
2	1.8
4	1.5
6	1.4
8	1.1
10	0.9

(a) Use the formulas for *a* and *b* on page 422 to find the equation of the least-squares line from which we can predict the chlorine residual in terms of the time that has elapsed since the pool was treated with chemicals.

(b) Use the equation of the least-squares line to estimate the chlorine residual in the pool five hours after it has been treated with chemicals.

11.3 The following sample data show the demand for a product (in thousands of units) and its price (in cents) in six different market areas:

Price	Demand
19	55
23	7
21	20
15	123
16	88
18	76

(a) Fit a least-squares line from which we can predict the demand for the product in terms of its price.

(b) Estimate the demand for the product when it is priced at 20 cents.

11.4 Verify that the linear equation of the illustration on page 415 can be obtained by fitting a least-squares line to the following data:

Rainfall (inches)	Yields of wheat (bushels per acre)
12.9	62.5
7.2	28.7
11.3	52.2
18.6	80.6
8.8	41.6
10.3	44.5
15.9	71.3
13.1	54.4

11.5 The following table shows 10 years' data on a local newspaper's annual advertising volume and its annual profit before taxes:

Annual advertising volume (1,000,000 lines)	Profit before taxes (thousands of dollars)
87.4	31,338
77.0	19,745
74.0	13,703
79.3	13,004
79.7	22,887
74.9	10,966
69.0	4,834
72.3	10,467
81.6	17,685
84.5	27,445

(a) Find the equation of the least-squares line which will enable us to predict the annual profit of the newspaper from its annual advertising volume.

(b) Predict the annual profit of the newspaper for a year in which its advertising volume is 80,500,000 lines.

11.6 The following are the numbers of hours a runner has run during each of eight weeks and the corresponding times in which she ran a mile at the end of the week:

Number of hours run	Time of mile (minutes)
13	5.5
15	5.2
18	4.9
20	4.3
19	4.5
17	4.7
21	4.2
16	4.8

(a) Find the equation of the least-squares line which will allow us to predict the runner's time for the mile from the number of hours she ran that week.

(b) Use the equation obtained in part (a) to predict how fast she will run a mile at the end of a week in which she ran for 18 hours.

11.7 During its first five years of operation, a company's gross income from sales was 1.2, 1.9, 2.4, 3.3, and 3.5 million dollars. Fit a least-squares line and, assuming that the trend continues, predict this company's gross income from sales during its sixth year of operation.

11.8 The following are data on the IQ's of 25 students, the numbers of hours they studied for a certain achievement test, and their scores on the test:

IQ	Number of hours studied	Score on test
105	8	80
98	6	62
112	10	91
102	9	77
107	14	89
95	6	65
100	18	96
110	10	85
102	15	94
96	24	91
115	8	88
105	11	85
135	9	99
92	12	76
90	18	83
109	3	70
94	5	61
114	10	86
106	15	93
121	4	92
99	16	82
103	21	95
114	15	90
107	8	79
108	12	88

Use a computer package to find the least-squares line which will enable us to predict a student's score on the test in terms of his or her IQ.

11.9 With reference to the data of the preceding exercise, use a computer package to find the least-squares line which will enable us to predict a student's score on the test in terms of the number of hours he or she studied for the test.

11.10 With reference to the data of Exercise 11.8, use a computer package to find the least-squares line which will enable us to predict how many hours a student will study for the test, given his or her IQ.

11.3
Regression Analysis ⋆

In the preceding section we used a least-squares line to predict that some-one who has studied German in high school or college for two years will score 53.33 in the proficiency test, but even if we interpret the line correctly as a regression line (that is, treat predictions made from it as averages or expected values), several questions remain to be answered.

> How good are the values we found for the constants a and b in the equation $\hat{y} = a + bx$? After all, $a = 31.53$ and $b = 10.90$ are only estimates based on sample data, and if we base our work on a sample of ten other applicants for the foreign service jobs, the method of least squares would probably lead to different values of a and b.

> How good an estimate is 53.33 of the average score of persons who have had two years of German in high school or college?

In the first of these questions we said that $a = 31.53$ and $b = 10.90$ are "only estimates based on sample data," and this implies the existence of corresponding true values, usually denoted by α and β and referred to as the **regression coefficients,** and therefore of a true regression line $y = \alpha + \beta x$. To distinguish between a and b, and α and β, we refer to the former as the **estimated regression coefficients.**

To clarify the idea of a true regression line $y = \alpha + \beta x$, consider Figure 11.7, which shows the distributions of y for several values of x. In connection with our example, these curves represent the distributions of the proficiency scores of persons who have had, respectively, one, two, or three years of German in high school or college. To complete the picture, we can visualize similar curves corresponding to all other values of x within the range of values under consideration.

In **linear regression analysis** we assume that the x's are known constants, and that for each x the random variable y has a certain distribution (as pictured in Figure 11.7) with the mean $\alpha + \beta x$. In **normal regression analysis** we assume, furthermore, that these distributions are all normal distributions with the same standard deviation σ. In other words, the distributions pictured in Figure 11.7, as well as those we add mentally, are normal curves with means on the line $y = \alpha + \beta x$ and the same standard deviation σ.

Based on these assumptions, we can make all sorts of inferences about the regression coefficients α and β. They all require that we estimate σ, the common

429

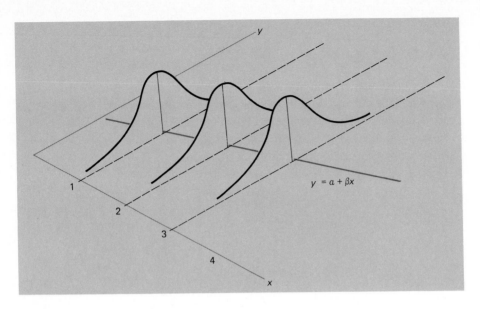

11.7

Distributions of y for given values of x.

standard deviation of the normal distributions pictured in Figure 11.6, and the estimate we shall use for this purpose is called the **standard error of estimate** and it is denoted by s_e. Its formula is

Standard error of estimate

$$s_e = \sqrt{\frac{\sum (y - \hat{y})^2}{n - 2}}$$

where, again, the y's are the observed values of y and the $\hat{y}$'s are the corresponding values on the least-squares line. Observe that s_e^2 is the sum of the squares of the vertical deviations from the points to the line in Figure 11.5 (namely, the quantity which we minimized by the method of least squares) divided by $n - 2$.

The formula above defines s_e, but in practice we calculate its value by means of the computing formula

Standard error of estimate (computing formula)

$$s_e = \sqrt{\frac{\sum y^2 - a(\sum y) - b(\sum xy)}{n - 2}}$$

where a and b are the least-squares estimates of α and β. Among the other quantities needed for substitution into this computing formula, n, $\sum y$, and $\sum xy$

CHAP. 11: The Analysis of Paired Data

are already known from the determination of a and b. So, the only new quantity we have to calculate is $\sum y^2$.

EXAMPLE   Calculate s_e for the illustration of the preceding section, where we had

Number of years x	Test grade y
3	57
4	78
4	72
2	58
5	89
3	63
4	73
5	84
3	75
2	48
35	697

and the least-squares line was found to be $\hat{y} = 31.53 + 10.90x$.

Solution   Copying $n = 10$, $\sum y = 697$, and $\sum xy = 2{,}554$ from page 421, and substituting these values together with

$$\sum y^2 = 57^2 + 78^2 + \cdots + 48^2 = 50{,}085$$

into the computing formula for s_e, we get

$$s_e = \sqrt{\frac{50{,}085 - (31.53)(697) - (10.90)(2{,}554)}{10 - 2}}$$

$$= 5.81$$

Since α is just the y-intercept of the regression line (that is, the value of y which corresponds to $x = 0$) and in many cases it has no real meaning, we shall consider here only inferences about β. They will be based on the statistic

Statistic for inferences concerning β

$$t = \frac{b - \beta}{s_e} \sqrt{\frac{n(\sum x^2) - (\sum x)^2}{n}}$$

whose sampling distribution is the t distribution with $n - 2$ degrees of freedom.

EXAMPLE   Based on the data on page 416, test the hypothesis that each additional year of German in high school or college adds another 12.5 points to the expected proficiency score of an applicant. Use the alternative hypothesis $\beta \neq 12.5$ and the level of significance 0.05.

Solution   1. *Hypotheses*

$$H_0: \beta = 12.5$$

$$H_A: \beta \neq 12.5$$

2. *Level of significance*

$$\alpha = 0.05$$

3. *Criterion*
   Reject the null hypothesis if $t < -2.306$ or $t > 2.306$, where 2.306 is the value of $t_{0.025}$ for $10 - 2 = 8$ degrees of freedom, and t is given by

$$t = \frac{b - \beta}{s_e} \sqrt{\frac{n(\sum x^2) - (\sum x)^2}{n}}$$

   Otherwise, accept the null hypothesis or reserve judgment.

4. *Calculations*
   Substituting $b = 10.90$, $\beta = 12.5$, $n = 10$, $\sum x = 35$, $\sum x^2 = 133$, and $s_e = 5.82$ (from pages 423 and 431) into the formula for t, we get

$$t = \frac{10.90 - 12.5}{5.81} \sqrt{\frac{10(133) - (35)^2}{10}}$$

$$= -0.89$$

5. *Decision*
   Since $t = -0.89$ falls on the interval from -2.306 to 2.306, the null hypothesis cannot be rejected; that is, the difference between $b = 10.90$ and $\beta = 12.5$ may reasonably be attributed to chance.

To construct confidence intervals for the regression coefficient β, we substitute for the middle term of $-t_{\alpha/2} < t < t_{\alpha/2}$ the t statistic from page 431, and relatively simple algebra yields

$$b \pm t_{\alpha/2} \cdot \frac{s_e}{\sqrt{\dfrac{n(\sum x^2) - (\sum x)^2}{n}}}$$

The degree of confidence is $1 - \alpha$ and $t_{\alpha/2}$ for $n - 2$ degrees of freedom may be read from Table III. Note that the quantity

$$\frac{s_e}{\sqrt{\dfrac{n(\sum x^2) - (\sum x)^2}{n}}}$$

by which $t_{\alpha/2}$ is multiplied, and by which $b - \beta$ is divided in the formula for the t statistic on page 431, is an estimate of the standard deviation of the sampling distribution of the least-squares estimate of β.

EXAMPLE   Based on the data on page 416, construct a 99% confidence interval for β, the expected increase in the proficiency score of an applicant for each additional year of German in high school or college.

Solution   Substituting $t_{0.005} = 3.355$ for $10 - 2 = 8$ degrees of freedom into the confidence-limits formula, as well as the values of b, n, $\sum x$, $\sum x^2$, and s_e given or obtained in the preceding example, we get

$$10.90 \pm 3.355 \cdot \frac{5.81}{\sqrt{\dfrac{10(133) - (35)^2}{10}}}$$

or 10.90 ± 6.02. Thus, the 99% confidence interval for β is

$$4.88 < \beta < 16.92$$

This interval is rather wide, and this is due to two things—the magnitude of the variation measured by s_e and the very small size of the sample.

There are many computer packages which have programs for fitting least-squares lines, and most of these programs also provide the answers for the two

433

preceding examples, or at least they greatly facilitate the calculations. The computer printout shown in Figure 11.6 on page 424 does not actually give the value of the t statistic or the confidence limits, but it tells us that s_e, referred to here as S, is 5.651, and in the column headed ST. DEV. OF COEF. in the row labeled X that the quantity

$$\frac{s_e}{\sqrt{\dfrac{n(\sum x^2) - (\sum x)^2}{n}}}$$

referred to on page 433, is 1.744. The difference between 5.651 and 5.81, the values of s_e given here and on page 431, is due to rounding. Also using $b = 10.905$ from the printout, we now get

$$t = \frac{10.905 - 12.5}{1.744} = -0.915$$

instead of $t = -0.89$ for the t statistic on page 431, and we now get

$$10.905 \pm 3.355(1.744)$$

or 10.905 ± 5.851, and hence

$$5.054 < \beta < 16.756$$

for the confidence interval for β. The differences between the results obtained here and pages 432 and 433 are all due to rounding.

To answer the second question asked on page 429, the one concerning the goodness of an estimate, or prediction, based on a least-squares equation, we use a method that is very similar to the one discussed on page 432. Basing our argument on another t statistic, we arrive at the following confidence limits for $\alpha + \beta x_0$, which is the mean of y when $x = x_0$:

Confidence limits for mean of y when $x = x_0$

$$(a + bx_0) \pm t_{\alpha/2} \cdot s_e \sqrt{\frac{1}{n} + \frac{n(x_0 - \bar{x})^2}{n(\sum x^2) - (\sum x)^2}}$$

The degree of confidence is $1 - \alpha$ and, as before, $t_{\alpha/2}$ for $n - 2$ degrees of freedom may be read from Table III.

EXAMPLE Referring again to the data on page 416, suppose that the original purpose of the study was to estimate the average proficiency score of applicants who have had two years of German in high school or college. Construct a 99% confidence interval for this mean.

Solution Copying $\sum x = 35$ and $\sum x^2 = 133$ from page 421, $a + bx_0 = 31.53 + 10.90(2) = 53.33$ from page 424, and $s_e = 5.81$ from page 431, and subsituting these values together with $n = 10$, $x_0 = 2$, $\bar{x} = \frac{35}{10} = 3.5$, and 3.355, the value of $t_{0.005}$ for $10 - 2 = 8$ degrees of freedom, into the confidence interval formula, we get

$$53.33 \pm (3.355)(5.81) \sqrt{\frac{1}{10} + \frac{10(2 - 3.5)^2}{10(133) - (35)^2}}$$

or

$$53.33 \pm 10.93$$

Hence, the 99% confidence interval for the mean proficiency score of applicants who have had two years of German in high school or college is the interval from $53.33 - 10.93 = 42.40$ to $53.33 + 10.93 = 64.26$.

EXERCISES (Exercises 11.11 and 11.16 are practice exercises; their complete solutions are given on pages 449 and 450.)

★ 11.11 With reference to Exercise 11.1 on page 425, test the null hypothesis $\beta = 1.5$ (namely, the hypothesis that each additional week on the job adds 1.5 to the number of cars a person can be expected to inspect in the given period of time) against the alternative hypothesis $\beta < 1.5$. Use the level of significance 0.05.

★ 11.12 The following table shows the assessed values and the selling prices of eight houses, constituting a random sample of all the houses sold recently in a given metropolitan area:

Assessed value (in $1,000)	Selling price (in $1,000)
40.3	63.4
72.0	118.3
32.5	55.2
44.8	74.0
27.9	48.8
51.6	81.1
80.4	123.2
58.0	92.5

Fit a least-squares line which will enable us to predict the selling price of a house in terms of its assessed value and test the null hypothesis $\beta = 1.30$ against the alternative hypothesis $\beta > 1.30$ at the level of significance 0.05.

★ 11.13 The following data show the average numbers of hours which six students studied per week and their grade-point indexes:

Hours studied x	Grade-point index y
15	2.0
28	2.7
13	1.3
20	1.9
4	0.9
10	1.7

Test the null hypothesis $\beta = 0.10$ (namely, the hypothesis that each additional hour of study per week will raise the expected grade-point index of the students by 0.10) against the alternative hypothesis $\beta < 0.10$ at the 0.01 level of significance.

★ 11.14 In a campus bar and grill, hamburger platters account for most of the food sales. The owner would like to evaluate how the price of these platters affects her weekly profit, so she experiments by varying the price during nine different weeks, with the following results:

Price	Profit
$2.00	$2,325
2.50	2,460
3.00	1,600
3.00	1,700
2.00	2,000
2.50	1,800
2.50	2,500
2.00	2,400
3.50	1,700

Fit a least-squares line to these data and construct a 95% confidence interval for the regression coefficient β.

★ 11.15 With reference to Exercise 11.3 on page 426, construct a 99% confidence interval for the regression coefficient β.

★ 11.16 With reference to Exercise 11.13, construct a 99% confidence interval for the mean grade-point index of students who study on the average 12 hours per week.

★ 11.17 With reference to Exercise 11.12, construct a 95% confidence interval for the average selling price of a house in the given metropolitan area which has an assessed value of $60,000.

★ 11.18 The following data show the advertising expenses (expressed as a percentage of total expenses) and the net operating profit (expressed as a percentage of total sales) in a random sample of five furniture stores:

Advertising expenses x	Net operating profit y
1.1	2.5
2.8	3.7
3.1	5.2
1.6	2.9
0.7	1.4

Fit a least-squares line to these data and construct a 99% confidence interval for the mean net operating profit (expressed as a percentage of total sales) when the advertising expenses are 2.0% of total expenses.

 ★ 11.19 With reference to Exercise 11.8 on page 428, use a computer package to test the null hypothesis $\beta = 0.50$ (that each unit increase in IQ adds on the average 0.50 to the score) against the alternative $\beta > 0.50$. Use the 0.05 level of significance.

 ★ 11.20 With reference to Exercise 11.8 on page 428, use a computer package to construct a 95% confidence interval for β.

 ★ 11.21 With reference to Exercise 11.9 on page 428, use a computer package to construct a 99% confidence interval for β (the average increase in the score for each additional hour studied).

 ★ 11.22 With reference to Exercise 11.10 on page 428, use a computer package to test the null hypothesis $\beta = 0$ (that a student's IQ does not affect how many hours he or she will study for the test) against the alternative $\beta \neq 0$. Use the 0.05 level of significance.

11.4

The Coefficient of Correlation

Having learned how to fit a least-squares line to a set of paired data, let us now turn to the problem of determining how well such a line actually fits the data. Of course, we can get some idea by inspecting a diagram like that of Figure 11.2, but to show how we can be more objective, let us refer back to the original data of the example dealing with the foreign service job applicants' proficiency grades in German and the number of years they have studied the language in high school or college. Copying the data from page 416, we get

Number of years x	Grade in test y
3	57
4	78
4	72
2	58
5	89
3	63
4	73
5	84
3	75
2	48

and it can be seen that there are considerable differences among the y's—the smallest value is 48 and the largest value is 89. However, it can also be seen that the grade of 48 was obtained by an applicant who has studied German for two years, while the grade of 89 was obtained by an applicant who has studied German for five years. This suggests that the differences among the grades may well be due, at least in part, to the fact that the applicants have not all studied German for the same number of years. This raises the following question:

> **Of the total variation among the y's, how much can be attributed to chance and how much can be attributed to the relationship between the two variables x and y, that is, to the fact that the observed y's correspond to different values of x?**

With reference to our example, we would thus want to know what part of the variation among the grades can be attributed to the differences in the number of years that the applicants have studied German, and what part can be attributed to all other factors (the applicants' intelligence, their health or frame of mind on the day they took the test, . . . , and their luck in guessing at some of the answers), which we combine under the general heading of chance.

A convenient measure of the total variation of the observed y's is the quantity $\sum (y - \bar{y})^2$, which is called the **total sum of squares** of the y's, and which is simply the variance of the y's multiplied by $n - 1$.

EXAMPLE   Calculate the total sum of squares for the test grades on page 416.

Solution   Since $\sum y = 697$ according to the table on page 421, we find that $\bar{y} = \frac{697}{10} = 69.7$ for these grades, and hence that

$$\sum (y - \bar{y})^2 = (57 - 69.7)^2 + (78 - 69.7)^2 + \cdots + (48 - 69.7)^2$$
$$= 1,504.10$$

This takes care of the total variation of the y's, but it remains to be seen how we might measure its two parts which are attributed, respectively, to the relationship between x and y and to chance. To this end, let us point out that if the number of years the applicants have studied German is the only thing that affects their grades in the test and the relationship is linear, then all the data points would fall on a straight line. As is apparent from Figure 11.2, this is not the case in our example, and the extent to which the points fluctuate above and below the line provides us with an indication of the size of the chance variations. Thus, chance variation is measured by the sum of the squares of the vertical deviations from the points to the line; namely, by the quantity $\sum (y - \hat{y})^2$, which is called the **residual sum of squares.** This is the quantity which we minimized on page 420 in accordance with the least-squares criterion.

EXAMPLE   Calculate the residual sum of squares for the ten test grades.

Solution   To calculate this sum of squares, we must first determine the value of $\hat{y} = a + bx = 31.53 + 10.90x$ for each of the given values of x, and we get

$$31.53 + 10.90(3) = 64.23$$
$$31.53 + 10.90(4) = 75.13$$
$$\cdots\cdots\cdots$$
$$31.53 + 10.90(2) = 53.33$$

Then,

$$\sum (y - \hat{y})^2 = (57 - 64.23)^2 + (78 - 75.13)^2 + \cdots + (48 - 53.33)^2$$
$$= 255.51$$

From the results of the two preceding examples—$\sum (y - \bar{y})^2 = 1{,}504.10$ and $\sum (y - \hat{y})^2 = 255.51$—we find that

$$\frac{\sum (y - \hat{y})^2}{\sum (y - \bar{y})^2} = \frac{255.51}{1{,}504.10} = 0.17$$

is the proportion of the total variation of the grades which can be attributed to chance, while

$$1 - \frac{\sum (y - \hat{y})^2}{\sum (y - \bar{y})^2} = 1 - 0.17 = 0.83$$

is the proportion of the total variation of the grades which can be attributed to the relationship with x, namely, to the differences in the number of years which the applicants have studied German.

If we take the square root of the last proportion (namely, the proportion of the total variation of the y's which can be attributed to the relationship with x), we obtain the statistical measure which is called the **coefficient of correlation.** It is denoted by the letter r, and its sign is chosen so that it is the same as that of the estimated regression coefficient b. Thus, for our example we get

$$r = \sqrt{0.83} = 0.91$$

rounded to two decimals.

It is of interest to note that all the quantities we have calculated so far in this section can be obtained directly from the computer printout of Figure 11.6. Under ANALYSIS OF VARIANCE, in the column headed SS, we find that the RESIDUAL sum of squares is 255.5 and that the TOTAL sum of squares is 1,504.1, where the difference between the residual sums of squares given here and above is due to rounding. Also, R-SQUARED = 83.0 PERCENT tells us that the square of the correlation coefficient is 0.83, and this agrees with the result we obtained before.

It follows from the rule for the sign of r that the correlation coefficient is positive when the least-squares line has an upward slope, namely, when the relationship between x and y is such that small values of y tend to go with small values of x and large values of y tend to go with large values of x. Also, the correlation coefficient is negative when the least-squares line has a downward slope, namely, when large values of y tend to go with small values of x and small values of y tend to go with large values of x. Geometrically, examples

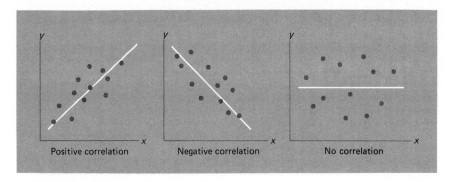

11.8
Types of correlation.

of **positive** and **negative correlations** are shown in the first two diagrams of Figures 11.8.

Observe also that since r is defined as $\pm$ the square root of a proportion, its values must lie on the interval from -1 to $+1$. When r equals -1 or $+1$, this means that 100% of the variation among the y's can be attributed to the relationship with x; in other words, $\sum(y - \hat{y})^2 = 0$ and the points must all fall on the least-squares line. When $r = 0$, this means that none of the variation among the y's can be attributed to the relationship with x, and we say that there is **no correlation.** This is pictured in the third diagram of Figure 11.8.

Our definition of r shows clearly the nature, or essence, of the coefficient of correlation, but in actual practice it is seldom used to determine its value. Instead, we use the computing formula

Computing formula for coefficient of correlation

$$r = \frac{n(\sum xy) - (\sum x)(\sum y)}{\sqrt{n(\sum x^2) - (\sum x)^2}\,\sqrt{n(\sum y^2) - (\sum y)^2}}$$

which has the added advantage that it automatically gives r the correct sign. This formula may look imposing but, with the exception of $\sum y^2$, the quantities needed for substitution are the same ones which are required to calculate the coefficients a and b of a least-squares line.

EXAMPLE   Use the computing formula to verify the value $r = 0.91$, which we obtained for the data pertaining to the proficiency grades of ten applicants for certain foreign service jobs and the number of years they have studied German in high school or college.

441

Solution   Copying $n = 10$, $\sum x = 35$, $\sum x^2 = 133$, $\sum y = 697$, and $\sum xy = 2{,}554$ from page 421, and $\sum y^2 = 50{,}085$ from page 431, we find that substitution into the computing formula for r yields

$$r = \frac{10(2{,}554) - (35)(697)}{\sqrt{10(133) - (35)^2}\sqrt{10(50{,}085) - (697)^2}}$$

$$= 0.91$$

This agrees, as it should, with the result obtained on page 440.

11.5

The Interpretation of *r*

When r equals $+1$, -1, or 0, there is no question about the interpretation of the coefficient of correlation. As we have already indicated, it is $+1$ or -1 when all the points actually fall on a straight line, and it is 0 when none of the variation among the y's can be attributed to the relationship with x, or in other words, when knowledge of x does not help in the prediction of y. In general, the definition of r tells us that the proportion of the variation of the y's which is due to the relationship with x equals r^2, or that the percentage equals $100r^2$, and this is how we interpret the strength of the relationship implied by any value of r.

For instance, if $r = 0.80$ for one set of data and $r = 0.40$ for another, it would be very misleading to say that the correlation of 0.80 is "twice as good" or "twice as strong" as the correlation of 0.40. When $r = 0.80$, then $100(0.80)^2 = 64\%$ of the variation of the y's is accounted for by the relationship with x, and when $r = 0.40$, then $100(0.40)^2 = 16$ percent of the variation of the y's is accounted for by the relationship with x. Thus, in the sense of "percentage of variation accounted for" we can say that a correlation of 0.80 is *four times as strong* as a correlation of 0.40. By the same token, a correlation of 0.60 is *nine times as strong* as a correlation of 0.20.

There are several pitfalls in the interpretation of r. First, it is often overlooked that r measures only the strength of linear relationships; second, it should be remembered that a strong correlation (a value of r close to $+1$ or -1) does not necessarily imply a cause–effect relationship.

If r is calculated indiscriminantly, for instance, for the three sets of data of Figure 11.9, we get $r = 0.75$ in each case, but it is a meaningful measure of the strength of the relationship only in the first case. In the second case there is a very strong relationship between the two variables, but it is not linear, and in the third case six of the seven points actually fall on a straight line, but the seventh

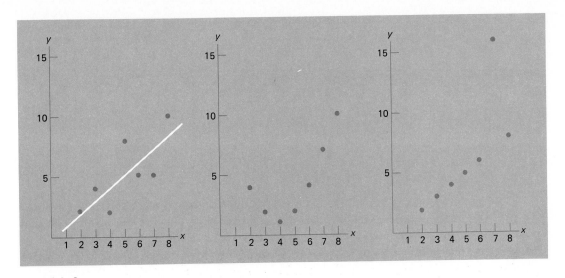

11.9

Three sets of paired data for which r = 0.75.

point is so far off that it suggests the possibility of a gross error in recording the data. Thus, before we calculate r, we should always plot the data to see whether there is reason to believe that the relationship is, in fact, linear.

The fallacy of interpreting a strong correlation as an indication of a cause–effect relationship, is best explained with a few examples. A popular illustration is the strong positive correlation that has been obtained for the annual sales of chewing gum and the incidence of crime in the United States. Obviously, one cannot conclude that the crime rate might be reduced by prohibiting the sale of chewing gum—both variables depend on the size of the population, and it is this mutual relationship with a third variable (population size) which produced the positive correlation. Another example is the strong positive correlation obtained for the number of storks seen nesting in English villages and the number of children born in the same villages. We leave it to the reader's ingenuity to explain why there might be a strong positive correlation in this case in the absence of any cause–effect relationship.

11.6

A Significance Test for *r*

It is sometimes overlooked that when r is calculated on the basis of sample data, we may get a fairly strong positive or negative correlation purely by chance, even though there is actually no relationship whatsoever between the two variables under consideration. Suppose, for instance, that we take a pair of dice, one red and one green, roll them a few times, and get the following results:

Red die x	Green die y
3	5
2	2
5	6
3	1
4	3

Calculating r for these data, we get the surprisingly high value $r = 0.66$, and this raises the question whether anything is wrong with the assumption that there is no relationship here between x and y—after all, one die does not know what the other die is doing. To answer this question, we shall have to see whether this high value of r may be attributed to chance.

When a correlation coefficient is calculated from sample data, as in the example above, the value we obtain for r is only an estimate of a corresponding parameter, the **population correlation coefficient,** which we denote by the Greek letter ρ (lowercase *rho*). What r measures for a sample, ρ measures for a population.

To test the null hypothesis of no correlation, we shall make the same assumptions as in Section 11.3. When these assumptions are met, at least within a reasonable degree of approximation, we reject the null hypothesis of no correlation if $r < -r_{\alpha/2}$ or $r > r_{\alpha/2}$, where the value of $r_{\alpha/2}$ may be obtained from Table VI for $\alpha = 0.05$ and $\alpha = 0.01$. If the null hypothesis can be rejected, we say that there is a **significant correlation;** otherwise, we say that the value of r is not statistically significant.

EXAMPLE   Use the 0.05 level of significance to test the null hypothesis of no correlation for the example in which we rolled a pair of dice five times and obtained $r = 0.66$.

Solution   1. *Hypotheses*

$$H_0: \rho = 0$$

$$H_A: \rho \neq 0$$

2. *Level of significance*

$$\alpha = 0.05$$

3. *Criterion*

Reject the null hypothesis if $r < -0.878$ or $r > 0.878$, where 0.878 is the value of $r_{0.025}$ for $n = 5$; otherwise, state that the value of r is not significant.

4. *Calculations*

$$r = 0.66$$

5. *Decision*

Since $r = 0.66$ falls on the interval from -0.878 to 0.878, the null hypothesis cannot be rejected; in other words, the correlation coefficient of 0.66 is not significant.

EXAMPLE   Use the 0.01 level of significance to test the null hypothesis of no correlation for the example in which we obtained $r = 0.91$ for the proficiency scores of ten applicants for certain foreign service jobs and the number of years they have studied German in high school or college.

Solution   1. *Hypotheses*

$$H_0\colon \rho = 0$$
$$H_A\colon \rho \neq 0$$

2. *Level of significance*

$$\alpha = 0.01$$

3. *Criterion*

Reject the null hypothesis if $r < -0.765$ or $r > 0.765$, where 0.765 is the value of $r_{0.005}$ for $n = 10$; otherwise, state that the value of r is not significant.

4. *Calculations*

$$r = 0.91$$

5. *Decision*

Since $r = 0.91$ exceeds 0.765, the null hypothesis must be rejected; in other words, there is a relationship between the two variables under consideration.

11.23 The following data pertain to a study of the effects of environmental pollution on wildlife; in particular, the effect of DDT on the thickness of the eggshells of certain birds:

DDT residues in yolk lipids (parts per million)	Thickness of eggshell (millimeters)
117	0.49
65	0.52
393	0.37
98	0.53
122	0.49

Calculate r for these data.

11.24 With reference to the preceding exercise, test the significance of the value obtained for r at the level of significance 0.01.

11.25 The following are the typing speeds and the reading speeds of nine secretaries:

Typing speed (words per minute)	Reading speed (words per minute)
60	370
56	551
52	528
63	348
70	645
58	454
44	503
79	618
62	500

Measure the extent of the relationship between typing speed and reading speed by calculating the coefficient of correlation.

11.26 With reference to the preceding exercise, test the null hypothesis of no correlation at the 0.05 level of significance.

11.27 The following table shows the percentages of the vote predicted by a poll for seven candidates for the U.S. Senate in different states, x, and the corresponding percentages of the vote which they actually received, *y*:

Poll x	Election y
42	51
34	31
59	56
41	42
53	58
40	35
55	54

Calculate *r* for these data.

11.28 Since *r* does not depend on the scales of *x* and *y*, its calculation can often be simplified by adding a suitable positive or negative number to each *x*, each *y*, or both. Rework the preceding exercise after subtracting 34 from each *x* and 31 from each *y*.

11.29 The following are the grades which 16 students received in final examinations in economics and anthropology:

Economics	Anthropology	Economics	Anthropology
51	74	45	68
68	70	73	87
72	88	93	89
97	93	66	73
55	67	20	33
73	73	91	91
95	99	74	80
74	73	80	86

(a) Calculate *r*;
(b) calculate *r* after following the suggestion of Exercise 11.28 by subtracting suitable constants from the *x*'s and the *y*'s.

11.30 With reference to the preceding exercise, test the null hypothesis $\rho = 0$ at the level of significance $\alpha = 0.05$.

11.31 With reference to Exercise 11.1 on page 425, what percentage of the variation of the numbers of cars checked can be attributed to differences in the length of time that the persons have been working at the inspection station?

11.32 With reference to Exercise 11.4 on page 426, what percentage of the variation in the yield of wheat can be attributed to differences in the amount of rain?

11.33 With reference to Exercise 11.14 on page 436, what percentage of the variation of the weekly profit can be attributed to the relationship between the weekly profit and the price of the hamburger platters?

11.34 With reference to Exercise 11.25, what percentage of the variation in typing speed can be attributed to the relationship between typing speed and reading speed?

11.35 If we calculate r for each of the following sets of data, should we be surprised if we get $r = 1$ and $r = -1$, respectively? Explain.

(a)

x	y
9	2
5	1

(b)

x	y
4	8
7	2

11.36 Check in each case whether the value of r is significant at the 0.05 level of significance:

(a) $n = 12$ and $r = -0.53$; (c) $n = 15$ and $r = -0.55$;
(b) $n = 20$ and $r = 0.58$; (d) $n = 9$ and $r = -0.61$.

11.37 Check in each case whether the value of r is significant at the 0.01 level of significance:

(a) $n = 18$ and $r = 0.62$; (c) $n = 16$ and $r = -0.58$;
(b) $n = 32$ and $r = -0.47$; (d) $n = 14$ and $r = 0.63$.

11.38 State in each case whether you would expect a positive correlation, a negative correlation, or no correlation:

(a) the ages of husbands and wives;
(b) the amount of rubber on tires and the number of miles they have been driven;
(c) income and education;
(d) shirt size and sense of humor;
(e) the number of hours that bowlers practice and their scores;
(f) hair color and one's knowledge of foreign affairs.

11.39 With reference to Exercise 11.8 on page 428, use a computer package to determine the total and residual sums of squares, and the value of r measuring the strength of the relationship between students' IQ's and their scores on the achievement test.

11.40 With reference to Exercise 11.9 on page 428, use a computer package to determine the total and residual sums of squares, and the value of r measuring the strength of the relationship between the number of hours a student studies for the test and his or her score.

11.41 With reference to Exercise 11.29, use a computer package to determine r and also the percentage of the variation of the anthropology scores that can be attributed to the relationship between students' grades in the two subjects.

11.1 (a) From the data we get $n = 6$, $\sum x = 36$, $\sum x^2 = 304$, $\sum y = 108$, and $\sum xy = 715$, so that the normal equations are

$$108 = 6a + 36b$$

$$715 = 36a + 304b$$

Multiplying the expressions on both sides of the first equation by 6 and leaving the second equation as is, we get

$$648 = 36a + 216b$$

$$715 = 36a + 304b$$

and then, by subtraction, $67 = 88b$ and $b = \frac{67}{88} = 0.7614$. Substituting this value into the first normal equation, we get $108 = 6a + 36(0.7614)$, which yields $a = 13.4316$. Thus, the equation of the least-squares line is $\hat{y} = 13.4316 + 0.7614x$.

(b) Substitution into the formulas for a and b yields

$$a = \frac{108(304) - (36)(715)}{6(304) - (36)^2} = 13.4318$$

and

$$b = \frac{6(715) - (36)(108)}{6(304) - (36)^2} = 0.7614$$

(c) Using the values of a and b obtained in part (b), we get

$$\hat{y} = 13.4318 + 0.7614(10) = 21.0458$$

or 21.05 rounded to two decimals.

11.11 1. *Hypotheses*

$$H_0: \beta = 1.5$$

$$H_A: \beta < 1.5$$

2. *Level of significance*

$$\alpha = 0.05$$

3. *Criterion*

Reject the null hypothesis if $t < -2.132$, where 2.132 is the value of $t_{0.05}$ for $6 - 2 = 4$ degrees of freedom and

$$t = \frac{b - \beta}{s_e} \sqrt{\frac{n(\sum x^2) - (\sum x)^2}{n}}$$

Otherwise, accept it or reserve judgment.

4. *Calculations*

Substituting $n = 6$, $\sum y = 108$, $\sum xy = 715$, $\sum y^2 = 2{,}008$, $a = 13.4318$, and $b = 0.7614$ into the formula for s_e, we get

$$s_e = \sqrt{\frac{2{,}008 - (13.4318)(108) - (0.7614)(715)}{6 - 2}}$$

$$= 1.800$$

Then, substituting this value together with $n = 6$, $\sum x = 36$, $\sum x^2 = 304$, $b = 0.7614$, and $\beta = 1.5$ into the formula for t, we obtain

$$t = \frac{0.7614 - 1.5}{1.800} \sqrt{\frac{6(304) - (36)^2}{6}}$$

$$= -3.85$$

5. *Decision*

Since $t = -3.85$ is less than -2.132, the null hypothesis must be rejected; in other words, the slope of the least-squares line is less than 1.5.

11.16 Copying the various quantities from the solution of Exercise 11.13 and substituting their values together with $x_0 = 12$, $\bar{x} = \frac{90}{6} = 15$, and 4.604, the value of $t_{0.005}$ for $6 - 2 = 4$ degrees of freedom, into the formula for the confidence limits, we get

$$[0.721 + (0.0686)(12)] \pm (4.604)(0.272) \sqrt{\frac{1}{6} + \frac{6(12 - 15)^2}{6(1{,}694) - (90)^2}}$$

or

$$1.544 \pm 0.550$$

rounded to three decimals. Hence, the 99% confidence interval for the mean grade-point index of students who study on the average 12 hours per week is the interval from $1.544 - 0.550 = 0.994$ to $1.544 + 0.550 = 2.094$.

11.23 Substituting $n = 5$, $\sum x = 795$, $\sum x^2 = 196{,}851$, $\sum y = 2.40$, $\sum y^2 = 1.1684$, and $\sum xy = 348.26$ into the formula for r, we get

$$r = \frac{5(348.26) - (795)(2.40)}{\sqrt{5(196{,}851) - (795)^2}\sqrt{5(1.1684) - (2.40)^2}}$$

$$= -0.98$$

11.24 1. *Hypotheses*

$$H_0 : \rho = 0$$

$$H_A : \rho \neq 0$$

2. *Level of significance*

$$\alpha = 0.01$$

3. *Criterion*
Reject the null hypothesis if $r < -0.959$ or $r > 0.959$, where 0.959 is the value of $r_{0.005}$ for $n = 5$; otherwise, state that the value of r is not significant.

4. *Calculations*
$r = -0.98$ (from the solution of Exercise 11.23).

5. *Decision*
Since $r = -0.98$ is less than -0.959, the null hypothesis must be rejected; we conclude that there is a relationship between DDT residues and eggshell thickness.

11.31 Substituting $n = 6$, $\sum x = 36$, $\sum x^2 = 304$, $\sum y = 108$, $\sum y^2 = 2{,}008$, and $\sum xy = 715$ into the formula for r, we get

$$r = \frac{6(715) - (36)(108)}{\sqrt{6(304) - (36)^2}\sqrt{6(2{,}008) - (108)^2}}$$

$$= 0.893$$

Thus, $100(0.893)^2 = 79.7\%$ of the variation of the number of cars checked can be attributed to differences in the length of time that the persons have been working at the inspection station.

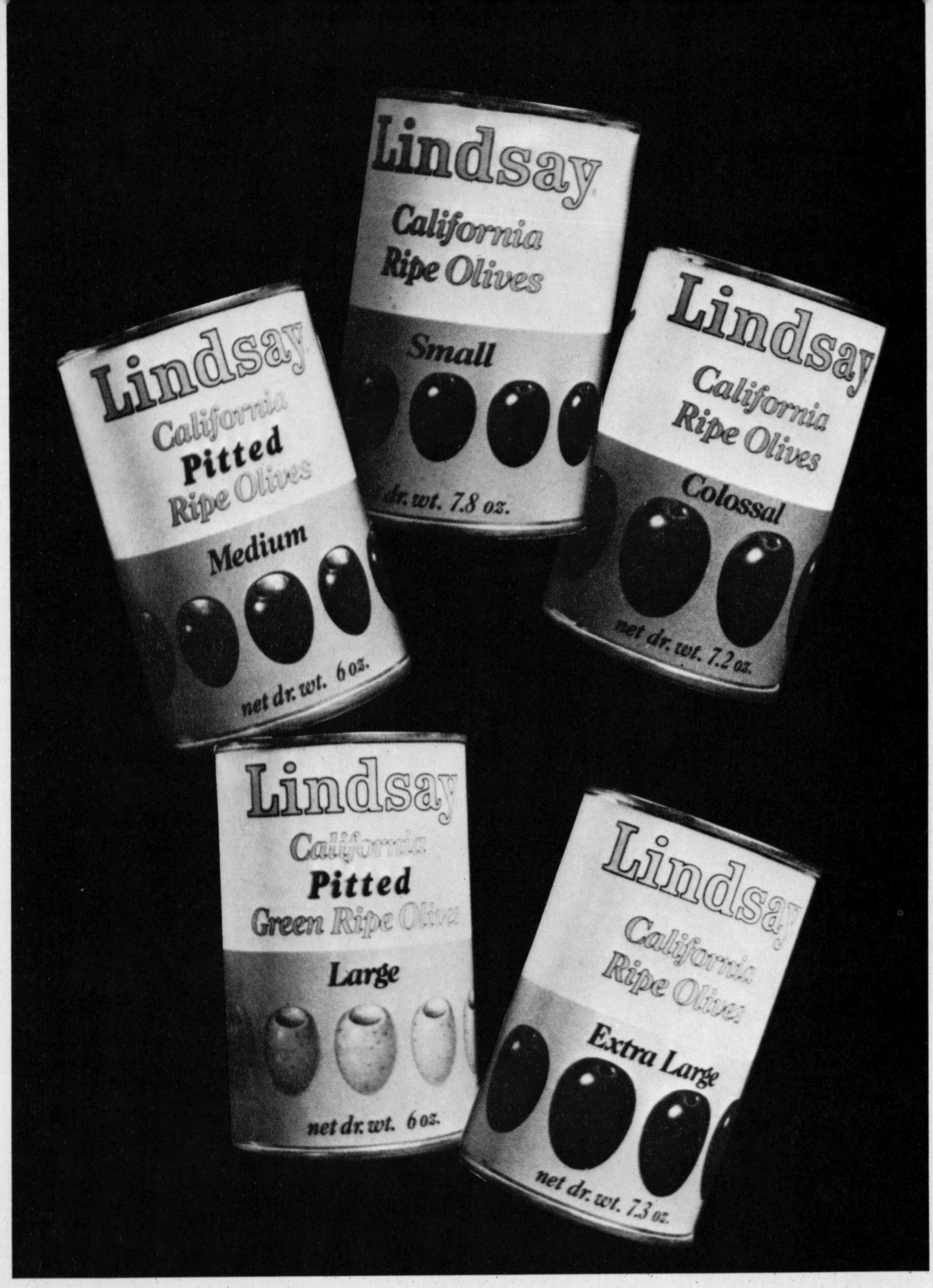
Lindsay
California
Ripe Olives
Small
dr. wt. 7.8 oz.
Lindsay
California
Pitted
Ripe Olives
Medium
net dr. wt. 6 oz.
Lindsay
California
Ripe Olives
Colossal
net dr. wt. 7.2 oz.
Lindsay
California
Pitted
Green Ripe Olives
Large
net dr. wt. 6 oz.
Lindsay
California
Ripe Olives
Extra Large
net dr. wt. 7.3 oz.

Most of the tests which we studied in the last three chapters required specific assumptions about the population, or populations, sampled. In many cases we assumed that the populations sampled can be approximated closely by normal distributions; sometimes we assumed that their standard deviations are known or are known to be equal; and sometimes we assumed that the samples are independent. Since there are many situations where these assumptions cannot be met, statisticians have developed alternative techniques based on less stringent assumptions, which have become known as

Nonparametric Tests

NONPARAMETRIC TESTS (ˌnän-ˌpar-ə-'me-trik 'tests) Statistical tests not requiring assumptions about the underlying (population) distributions.

Such tests cannot only be used under more general conditions, but they are often easier to explain and easier to understand than the standard tests which they replace. Moreover, in many nonparametric tests the computational burden is so light that they come under the heading of "quick and easy" or "shortcut" techniques. For these reasons, nonparametric tests have become quite popular, and extensive literature is devoted to their theory and application.

In Sections 12.1 through 12.3 we shall present the sign test as a nonparametric alternative to tests concerning means and differences between means for paired data; in Sections 12.4 through 12.6 we shall study methods based on rank sums, which serve as nonparametric alternatives to the two-sample t test and the one-way analysis of variance; in Sections 12.7 through 12.9 we shall learn how to test the randomness of a sample after the data have been obtained; and in Section 12.10 we shall present a nonparametric test concerning the relationship between paired data.

453

12.1

The One-Sample Sign Test

The small-sample tests concerning means and differences between means, which we studied in Sections 9.9 and 9.11, are based on the assumption that the populations sampled have roughly the shape of normal distributions. When this assumption is untenable in practice, these standard tests can be replaced by any one of several nonparametric alternatives, among them the **sign test,** which we shall study in this section and in Sections 12.2 and 12.3.

The **one-sample sign test** applies when we sample a continuous symmetrical population, so that the probability of getting a sample value less than the mean and the probability of getting a sample value greater than the mean are both $\frac{1}{2}$.

Then, to test the null hypothesis $\mu = \mu_0$ against an appropriate alternative on the basis of a random sample of size n, we replace each sample value greater than μ_0 with a plus sign and each sample value less than μ_0 with a minus sign, and test instead the null hypothesis that the total number of plus signs is a value of a random variable having the binomial distribution with $p = \frac{1}{2}$. If a sample value equals μ_0, which is not impossible or even unlikely when we deal with rounded data, it is discarded.

To perform a one-sample sign test when the sample is small, we refer directly to a table of binomial probabilities such as Table I, and proceed as in Section 10.2; when the sample is large, we use the normal approximation to the binomial distribution, as will be explained in Section 12.3.

EXAMPLE   The following data constitute a random sample of 15 measurements of the octane rating of a certain kind of gasoline:

97.5	95.2	97.3	96.0	96.8	100.3	97.4	95.3
93.2	99.1	96.1	97.6	98.2	98.5	94.9	

Use the one-sample sign test to test the null hypothesis $\mu = 98.5$ against the alternative hypothesis $\mu < 98.5$ at the 0.05 level of significance.

Solution   1. *Hypotheses*

$$H_0:\ \mu = 98.5$$

$$H_A:\ \mu < 98.5$$

2. *Level of significance*

$$\alpha = 0.05$$

3. *Criterion*

The criterion may be based on the number of plus signs or the number of minus signs. Using the number of plus signs, denoted by x, reject the null hypothesis if the probability of getting x or fewer plus signs is less than or equal to 0.05; otherwise, accept the null hypothesis or reserve judgment.

4. *Calculations*

Replacing each value greater than 98.5 with a plus sign and each value less than 98.5 with a minus sign, the 14 sample values not equal to 98.5 yield

$$- \quad - \quad - \quad - \quad - \quad + \quad - \quad - \quad - \quad + \quad - \quad - \quad - \quad -$$

Thus, $x = 2$ and Table I shows that for $n = 14$ and $p = 0.50$ the probability of $x \leq 2$ is $0.001 + 0.006 = 0.007$.

5. *Decision*

Since 0.007 is less than 0.05, the null hypothesis must be rejected; we conclude that the mean octane rating of the given kind of gasoline is less than 98.5.

12.2

The Paired-Sample Sign Test

The sign test can also be used when we deal with paired data as in Section 9.11. In such problems, each pair of sample values is replaced with a plus sign if the first value is greater than the second, with a minus sign if the first value is smaller than the second, and it is discarded when the two values are equal. Otherwise, we proceed as in Section 12.1.

EXAMPLE To determine the effectiveness of a new traffic control system, the number of accidents that occurred at ten dangerous intersections during four weeks before and four weeks after the installation of the new system was observed and the following data were obtained:

3 and 1, 4 and 2, 2 and 3, 5 and 2, 3 and 3,

2 and 0, 3 and 2, 6 and 3, 1 and 2, 1 and 0

Use the sign test to test the null hypothesis that the new traffic control system is only as effective as the old system at the 0.05 level of significance.

Solution Since one of the pairs, 3 and 3, has to be discarded, the sample size for the sign test is only $n = 9$.

1. *Hypotheses*

H_0: $\mu_1 = \mu_2$, where μ_1 and μ_2 are the mean numbers of accidents in four weeks at a dangerous intersection with the old and the new control systems.

H_A: $\mu_1 > \mu_2$

2. *Level of significance*

$$\alpha = 0.05$$

3. *Criterion*

If x is the number of plus signs, reject the null hypothesis if the probability of getting x or more plus signs is less than or equal to 0.05; otherwise, accept it or reserve judgment.

4. *Calculations*

Replacing each pair of values with a plus sign if the first value is greater than the second or with a minus sign if the first value is smaller than the second, the nine unequal sample pairs yield

$$+\ +\ -\ +\ +\ +\ +\ -\ +$$

Thus, $x = 7$ and Table I shows that for $n = 9$ and $p = 0.50$ the probability of $x \geq 7$ is $0.070 + 0.018 + 0.002 = 0.090$.

5. *Decision*

Since 0.090 exceeds 0.050, the null hypothesis cannot be rejected; in other words, we cannot conclude on the basis of this test that the new control system is more effective than the old system.

The test we have described here is only one of several nonparametric ways of analyzing such paired sample data. Another popular test used for this purpose is the Wilcoxon signed-rank test, which may be found in the books on nonparametric statistics listed in the Bibliography at the end of the book.

12.3

The Sign Test (Large Samples)

When np and $n(1 - p)$ are both greater than 5, so that we can use the normal approximation to the binomial distribution, the sign test may be based on the large-sample test of Section 10.3, namely, on the statistic

$$z = \frac{x - np_0}{\sqrt{np_0(1 - p_0)}}$$

with $p_0 = 0.50$, which has approximately the standard normal distribution.

EXAMPLE The following are measurements of the ocean depth in a certain location (in fathoms): 46.4, 48.3, 51.9, 38.8, 46.5, 45.6, 52.1, 41.0, 54.2, 44.9, 52.3, 43.6, 48.7, 42.2, and 44.9. Use the sign test at the level of significance $\alpha = 0.05$ to test the null hypothesis $\mu = 43.0$ (the previously recorded ocean depth in that location) against the alternative hypothesis $\mu \neq 43.0$.

Solution 1. *Hypotheses*

$$H_0: \mu = 43.0$$

$$H_A: \mu \neq 43.0$$

2. *Level of significance*

$$\alpha = 0.05$$

3. *Criterion*
Reject the null hypothesis if $z < -1.96$ or $z > 1.96$, where

$$z = \frac{x - np_0}{\sqrt{np_0(1 - p_0)}}$$

with $p_0 = 0.50$; otherwise, accept it or reserve judgment.

4. *Calculations*

Replacing each value greater than 43.0 with a plus sign and each value less than 43.0 with a minus sign, we get

$$+ \ + \ + \ - \ + \ + \ + \ - \ + \ + \ + \ + \ + \ - \ +$$

Thus, $x = 12$ and substitution into the formula yields

$$z = \frac{12 - 15(0.50)}{\sqrt{15(0.50)(0.50)}} = 2.32$$

5. *Decision*

Since $z = 2.32$ exceed 1.96, the null hypothesis must be rejected; in other words, the mean ocean depth at the given location is not 43.0 fathoms, as had previously been recorded.

EXERCISES

(Exercises 12.1 and 12.7 are practice exercises; their complete solutions are given on pages 483 through 484.)

12.1 On 12 occasions, Ms. Brown had to wait 3, 6, 7, 6, 4, 8, 6, 2, 8, 6, 1, and 9 minutes for the bus that takes her to work. Use the sign test based on Table I and the 0.05 level of significance to test the null hypothesis $\mu = 5$ against the alternative hypothesis $\mu \neq 5$.

12.2 Nine women buying new eyeglasses tried on, respectively, 12, 11, 14, 15, 10, 14, 11, 8, and 12 different frames. Use the sign test at the 0.05 level of significance to test the null hypothesis $\mu = 10$ against the alternative hypothesis $\mu > 10$.

12.3 In six rounds of golf at the Paradise Valley Country Club, a professional scored 71, 69, 72, 74, 71, and 72. Use the sign test at the 0.05 level of significance to test the null hypothesis $\mu = 70$ against the alternative hypothesis $\mu > 70$.

12.4 The following data are the weights (in grams) of 14 packages of a certain kind of candy: 101.0, 99.8, 100.9, 103.6, 97.1, 100.0, 102.5, 100.5, 101.0, 98.2, 100.3, 102.6, 100.0, and 100.8. Use the sign test based on Table I and the level of significance $\alpha = 0.05$ to test the null hypothesis $\mu = 100.0$ against the alternative hypothesis $\mu \neq 100.0$.

12.5 The following are the numbers of passengers carried on flights 136 and 137 between Chicago and Phoenix on 12 days: 232 and 189, 265 and 230, 249 and 236, 250 and 261, 255 and 249, 236 and 218, 270 and 258, 266 and 253, 249 and 251, 240 and 233, 257 and 254, and 239 and 249. Use the sign test based on Table I and the 0.05 level of significance to test the null hypothesis $\mu_1 = \mu_2$ against the alternative hypothesis $\mu_1 > \mu_2$.

12.6 The following are the grades which 15 students received on the midterm and final examinations in a course in European history: 66 and 73, 88 and 91, 75 and 78, 90 and 86, 63 and 69, 58 and 67, 75 and 75, 82 and 80, 73 and 76, 84 and 89, 85 and 81, 93 and 96, 70 and 76, 85 and 82, and 90 and 97. Use the sign test based on Table I and the 0.05 level of significance to test the null hypothesis $\mu_1 = \mu_2$ against the alternative hypothesis $\mu_1 < \mu_2$.

12.7 The following are the numbers of speeding tickets issued by two police officers on 20 days: 6 and 9, 11 and 13, 12 and 12, 10 and 17, 15 and 13, 7 and 11, 9 and 13, 7 and 12, 14 and 15, 11 and 13, 14 and 10, 6 and 12, 9 and 9, 12 and 14, 8 and 13, 16 and 11, 10 and 15, 12 and 14, 15 and 15, and 12 and 18. Use the sign test based on the normal approximation to the binomial distribution and the level of significance $\alpha = 0.01$ to test the null hypothesis that on the average the two police officers issue equally many speeding tickets.

12.8 The following are data on the daily sulfur oxides emission of an industrial plant (in tons):

17	15	20	29	19	18	22	25	27	9
24	20	17	6	24	14	15	23	24	26
19	23	28	19	16	22	24	17	20	13
19	10	23	18	31	13	20	17	24	14

Use the sign test and the 0.01 level of significance to test the null hypothesis $\mu = 23$ against the alternative hypothesis $\mu < 23$.

12.9 Rework Exercise 12.4 using the sign test based on the normal approximation to the binomial distribution.

12.10 The following are the miles per gallon obtained with 40 tankfuls of a certain kind of gas

24.1	25.0	24.8	24.3	24.2	25.3	24.2	23.6	24.5	24.4
24.5	23.2	24.0	23.8	23.8	25.3	24.5	24.6	24.0	25.2
25.2	24.4	24.7	24.1	24.6	24.9	24.1	25.8	24.2	24.2
24.8	24.1	25.6	24.5	25.1	24.6	24.3	25.2	24.7	23.3

Use the sign test at the 0.01 level of significance to test the null hypothesis $\mu = 24.2$ against the alternative hypothesis $\mu > 24.2$.

12.11 The following are the numbers of artifacts dug up by two archaeologists at an ancient cliff dwelling on 30 days: 1 and 0, 0 and 0, 2 and 1, 3 and 0, 1 and 2, 0 and 0, 2 and 0, 2 and 1, 3 and 1, 0 and 2, 1 and 0, 1 and 1, 4 and 2, 1 and 1, 2 and 1, 1 and 0, 3 and 2, 5 and 2, 2 and 6, 1 and 0, 3 and 2, 2 and 3, 4 and 0, 1 and 2, 3 and 1, 2 and 0, 0 and 1, 2 and 0, 4 and 1, and 2 and 0. Use the sign test at the level of significance $\alpha = 0.01$ to test the null hypothesis that the two archaeologists are equally good at finding artifacts against the alternative hypothesis that the first one is better.

12.12 Use the large-sample sign test to rework Exercise 9.70 on page 337.

12.13 The following are the numbers of employees absent from two departments of a large firm on 25 days: 4 and 3, 2 and 5, 6 and 6, 3 and 6, 1 and 4, 2 and 4, 5 and 2, 1 and 4, 3 and 4, 6 and 5, 2 and 5, 7 and 1, 4 and 6, 1 and 3, 2 and 5, 0 and 3, 6 and 5, 4 and 6, 1 and 2, 4 and 1, 2 and 4, 0 and 1, 5 and 3, 2 and 3, and 2 and 4. Use the sign test at the 0.05 level of significance to test the null hypothesis $\mu_1 = \mu_2$ against the alternative hypothesis $\mu_1 < \mu_2$.

12.4

Rank Sums: The *U* Test

In this section we shall present a nonparametric alternative to the small-sample *t* test concerning the difference between two means. It is called the **U test,** the **Wilcoxon test,** or the **Mann–Whitney test,** named after the statisticians who contributed to its development. With this test we will be able to test the null hypothesis that the two samples come from identical populations without having to assume that the populations sampled have normal distributions; in fact, the test requires only that the populations sampled are continuous to avoid ties, and in practice it does not even matter whether this assumption is satisfied.

To illustrate how the *U* test is performed, suppose that we want to compare the grain size of sand obtained from two different locations on the moon on the basis of the following diameters (in millimeters):

Location 1:   0.37, 0.70, 0.75, 0.30, 0.45, 0.16, 0.62, 0.73, 0.33
Location 2:   0.86, 0.55, 0.80, 0.42, 0.97, 0.84, 0.24, 0.51, 0.92, 0.69

The means of these two samples are 0.49 and 0.68, and their difference is large, but it remains to be seen whether it is significant.

To perform the *U* test, we first arrange the data jointly, as if they comprise one sample, in an increasing order of magnitude. For our data, we get

0.16	0.24	0.30	0.33	0.37	0.42	0.45	0.51	0.55	0.62
1	2	1	1	1	2	1	2	2	1

0.69	0.70	0.73	0.75	0.80	0.84	0.86	0.92	0.97
2	1	1	1	2	2	2	2	2

where we indicated for each value whether it came from location 1 or location 2. Assigning the data, in this order, the ranks 1, 2, 3, . . . , and 19, we find that the values of the first sample (location 1) occupy ranks 1, 3, 4, 5, 7, 10, 12, 13 and 14, while those of the second sample (location 2) occupy ranks 2, 6, 8, 9, 11, 15, 16, 17, 18, and 19. There are no ties here, but if there were, we would assign each of the tied observations the mean of the ranks which they jointly occupy. For instance, if the third and fourth values were the same, we would assign each the rank $\frac{3+4}{2} = 3.5$, and if the ninth, tenth, and eleventh values were the same, we would assign each the rank $\frac{9+10+11}{3} = 10$.

Now, if there is an appreciable difference between the means of the two populations, most of the lower ranks are likely to go to the values of one sample while most of the higher ranks are likely to go to the values of the other sample. The test of the null hypothesis that the two samples come from identical populations may thus be based on W_1, the sum of the ranks of the values of the first sample, or on W_2, the sum of the ranks of the values of the second sample. In practice, it does not matter which sample we refer to as sample 1 and which sample we refer to as sample 2, and whether we base the test on W_1 or W_2.[†] If the sample sizes are n_1 and n_2, the sum of W_1 and W_2 is simply the sum of the first $n_1 + n_2$ positive integers, which is known to equal

$$\frac{(n_1 + n_2)(n_1 + n_2 + 1)}{2}$$

This formula enables us to find W_2 if we know W_1, and vice versa. For our illustration we get

$$W_1 = 1 + 3 + 4 + 5 + 7 + 10 + 12 + 13 + 14$$
$$= 69$$

and since the sum of the first 19 positive integers is $\frac{19 \cdot 20}{2} = 190$, it follows that $W_2 = 190 - 69 = 121$. (This value may be checked by actually adding 2, 6, 8, 9, 11, 15, 16, 17, 18, and 19.)

When the use of **rank sums** was first proposed as a nonparametric alternative to the two-sample t test, the decision was based on W_1 or W_2. Nowadays,

---

[†] When the sample sizes are unequal, it is common practice to let n_1 be the smaller of the two; this is not required, however, for the work in this book.

the decision is based on either of the related statistics

U statistics

$$U_1 = n_1 n_2 + \frac{n_1(n_1 + 1)}{2} - W_1$$

or

$$U_2 = n_1 n_2 + \frac{n_2(n_2 + 1)}{2} - W_2$$

or on the statistic U, which always equals the smaller of the two. The resulting tests are equivalent to those based on W_1 or W_2, but they have the advantage that they lend themselves more readily to the construction of tables of critical values. Not only do U_1 and U_2 take on values on the interval from 0 to $n_1 n_2$—indeed, their sum is always equal to $n_1 n_2$—but their sampling distributions are symmetrical about $\frac{n_1 n_2}{2}$. The use of U, which always equals the smaller of the values of U_1 and U_2, has the added advantage that the resulting test is one-tailed, and hence easier to tabulate.

Accordingly, we test the null hypothesis that the two samples come from identical populations against the alternative hypothesis that the two populations have unequal means with the following criterion:

Reject the null hypothesis if

$$U \leq U'_\alpha$$

where U'_α is given in Table VII for $n_1 \leq 15$, $n_2 \leq 15$, and $\alpha = 0.05$ or $\alpha = 0.01$.

In the construction of Table VII, U'_α is the largest value of U for which the probability of $U \leq U'_\alpha$ is less than or equal to α, and the blank spaces indicate that the null hypothesis cannot be rejected at the given level of significance regardless of the value which we obtain for U. More extensive tables may be found in handbooks of statistical tables, but when n_1 and n_2 are both greater than 8, it is generally considered reasonable to use the large-sample test described in Section 12.5.

With reference to the grain-size data on page 460, use the U test at the 0.05 level of significance to test the null hypothesis that the two samples come from identical populations against the alternative hypothesis that the two populations have unequal means.

Solution   1. *Hypotheses*

$$H_0: \text{Populations are identical.}$$

$$H_A: \mu_1 \neq \mu_2$$

2. *Level of significance*

$$\alpha = 0.05$$

3. *Criterion*
Reject the null hypothesis if $U \leq 20$, which is the value of U'_α for $n_1 = 9$, $n_2 = 10$, and $\alpha = 0.05$; otherwise, accept it or reserve judgment.

4. *Calculations*
Having already shown that $W_1 = 69$ and $W_2 = 121$, we get

$$U_1 = 9 \cdot 10 + \frac{9 \cdot 10}{2} - 69 = 66$$

$$U_2 = 9 \cdot 10 + \frac{10 \cdot 11}{2} - 121 = 24$$

and, hence, $U = 24$. Note that $U_1 + U_2 = 66 + 24 = 90$, which equals $n_1 n_2 = 9 \cdot 10$.

5. *Decision*
Since $U = 24$ exceeds 20, the null hypothesis cannot be rejected; in other words, we cannot conclude that there is a real difference in the mean grain size of sand from the two locations on the moon.

The test which we have described here can also be used when the alternative is $\mu_1 < \mu_2$ or $\mu_1 > \mu_2$. However, since the procedure is more complicated in that case—we will have to use U_1 or U_2 instead of U—we shall discuss it only for large samples in Section 12.5.

12.5

Rank Sums: The *U* Test (Large Samples) ★

The large-sample U test may be based on either U_1 or U_2 as given on page 462, but since the resulting tests are equivalent and it does not matter which sample we denote sample 1 and which sample we denote sample 2, we shall use here the statistic U_1.

Under the null hypothesis that the two samples come from identical populations, it can be shown that the mean and the standard deviation of the sampling distribution of U_1 are[†]

Mean and standard deviation of U_1 statistic

$$\mu_{U_1} = \frac{n_1 n_2}{2}$$

and

$$\sigma_{U_1} = \sqrt{\frac{n_1 n_2 (n_1 + n_2 + 1)}{12}}$$

Furthermore, if n_1 and n_2 are both greater than 8, the sampling distribution of U_1 can be approximated closely with a normal curve. Thus, we base the test of the null hypothesis that the two samples come from identical populations on the statistic

Statistic for large-sample U test

$$z = \frac{U_1 - \mu_{U_1}}{\sigma_{U_1}}$$

which has approximately the standard normal distribution. If the alternative hypothesis is $\mu_1 \neq \mu_2$, we reject the null hypothesis for $z < -z_{\alpha/2}$ or $z > z_{\alpha/2}$; if the alternative hypothesis is $\mu_1 < \mu_2$, we reject the null hypothesis for $z > z_\alpha$ since large values of U_1 correspond to small values of W_1; and if the alternative hypothesis is $\mu_1 > \mu_2$, we reject the null hypothesis for $z < -z_\alpha$ since small values of U_1 correspond to large values of W_1.

[†] If there are ties in rank, these formulas provide only approximations, but if the number of ties is small, there is no need to make a correction.

CHAP: 12: Nonparametric Tests

EXAMPLE   The following are the weight gains (in pounds) of young turkeys, which are fed two different diets but are otherwise kept under identical conditions:

Diet 1:   16.3, 10.1, 10.7, 13.5, 14.9, 11.8, 14.3, 10.2,
          12.0, 14.7, 23.6, 15.1, 14.5, 18.4, 13.2, 14.0

Diet 2:   21.3, 23.8, 15.4, 19.6, 12.0, 13.9, 18.8, 19.2,
          15.3, 20.1, 14.8, 18.9, 20.7, 21.1, 15.8, 16.2

Use the large-sample U test at the 0.01 level of significance to test the null hypothesis that the two populations sampled are identical against the alternative hypothesis that on the average the second diet produces a greater gain in weight.

Solution   1. *Hypotheses*

H_0: Populations are identical.

H_A: $\mu_1 < \mu_2$

2. *Level of significance*

$$\alpha = 0.01$$

3. *Criterion*

Reject the null hypothesis if $z > 2.33$, where

$$z = \frac{U_1 - \mu_{U_1}}{\sigma_{U_1}}$$

Otherwise, accept it or reserve judgment.

4. *Calculations*

Arranging the data according to size, we get 10.1, 10.2, 10.7, 11.8, 12.0, 12.0, 13.2, 13.5, 13.9, 14.0, 14.3, 14.5, 14.7, 14.8, 14.9, 15.1, 15.3, 15.4, 15.8, 16.2, 16.3, 18.4, 18.8, 18.9, 19.2, 19.6, 20.1, 20.7, 21.1, 21.3, 23.6, and 23.8. Assigning the data, in this order, the ranks 1, 2, 3, ..., and 32, we find that the values of the first sample (diet 1) occupy ranks 1, 2, 3, 4, 5.5, 7, 8, 10, 11, 12, 13, 15, 16, 21, 22, and 31, while those of the second sample (diet 2) occupy ranks 5.5, 9, 14, 17, 18,

19, 20, 23, 24, 25, 26, 27, 28, 29, 30, and 32. Note that since the 5th and 6th largest values are both equal to 12.0, we assigned each the rank $\dfrac{5+6}{2} = 5.5$. Thus,

$$
\begin{aligned}
W_1 &= 1 + 2 + 3 + 4 + 5.5 + 7 + 8 + 10 + 11 + 12 \\
&\quad + 13 + 15 + 16 + 21 + 22 + 31 \\
&= 181.5
\end{aligned}
$$

and

$$
U_1 = 16 \cdot 16 + \frac{16 \cdot 17}{2} - 181.5
$$

$$
= 210.5
$$

Since $\mu_{U_1} = \dfrac{16 \cdot 16}{2} = 128$ and $\sigma_{U_1} = \sqrt{\dfrac{16 \cdot 16 \cdot 33}{12}} = 26.53$, it follows that

$$
z = \frac{210.5 - 128}{26.53} = 3.11
$$

5. *Decision*

Since $z = 3.11$ exceeds 2.33, the null hypothesis must be rejected; in other words, we conclude that on the average the second diet produces a greater gain in weight.

EXERCISES   (Exercises 12.14 and 12.19 are practice exercises; their complete solutions are given on pages 485 and 486.)

12.14 The following are figures on the number of assaults committed in a city in six weeks in the spring and in six weeks in the fall:

> *Spring:* 46, 37, 42, 48, 38, 45
> *Fall:*   35, 30, 25, 39, 28, 32

Use the U test at the level of significance 0.01 to check the claim that on the average there are equally many assaults per week in the given city in the spring and in the fall.

12.15 The following are the Rockwell hardness numbers obtained for six aluminum die castings randomly selected from production lot A and for eight aluminum die castings randomly selected from production lot B:

> *Production lot A:* 75, 56, 63, 70, 58, 74
> *Production lot B:* 63, 85, 77, 80, 86, 76, 72, 82

Use the U test at the 0.05 level of significance to check the claim that the average hardness of die castings from the two production lots is the same.

12.16 Tests made on two kinds of 9-volt batteries showed the following lifetimes (in hours) of continuous use:

> *Brand A:* 11.7, 12.0, 10.8, 11.1, 11.9, 12.9, 12.4
> *Brand B:* 11.5, 12.8, 13.5, 13.6, 11.1, 12.4, 13.3

Use the U test at the 0.05 level of significance to test whether the difference between the two sample means, 11.8 and 12.6, can be attributed to chance.

12.17 The following are the numbers of misprints counted on pages selected at random from two Sunday editions of a newspaper:

> *May 10:* 12, 6, 11, 11, 15, 7
> *May 24:* 10, 3, 6, 8, 7, 5

Use the U test at the level of significance $\alpha = 0.05$ to test the null hypothesis that the two samples come from identical populations against the alternative hypothesis that the two populations have unequal means.

12.18 The following are the numbers of minutes it took a sample of 15 men and 12 women to complete a short screening test given to job applicants at a large bank:

> *Men:* 8.8, 7.8, 6.6, 10.7, 8.9, 8.4, 6.9, 6.4, 6.3, 8.0, 8.6, 8.1, 9.1, 9.7, 9.9
> *Women:* 7.5, 8.7, 8.3, 6.2, 6.5, 7.7, 9.8, 9.6, 9.2, 10.4, 8.2, 8.5

Use the U test based on Table VII and the 0.01 level of significance to test the null hypothesis that the two samples come from identical populations against the alternative hypothesis that the two populations have unequal means.

★12.19 The following are the scores which samples of students from two minority groups obtained on a current events test:

> *Minority group 1:* 70, 62, 91, 55, 72, 94, 80, 96, 73, 44, 87, 78
> *Minority group 2:* 81, 23, 71, 30, 71, 54, 64, 93, 58, 41, 47, 56

Use the large-sample U test at the 0.05 level of significance to test whether students from the two minority groups can be expected to score equally well on this test.

★ **12.20** Use the normal approximation to the sampling distribution of U to rework the example on page 463, which dealt with the grain size of sand from two locations on the moon.

★ **12.21** Comparing two kinds of emergency flares, a consumer testing service obtained the following burning times (rounded to the nearest tenth of a minute):

> Brand X: 17.2, 18.1, 21.2, 19.3, 14.4, 21.1, 14.6, 19.1, 18.8, 15.2, 20.3, 17.5
> Brand Y: 13.6, 13.7, 11.8, 14.6, 15.2, 14.3, 22.5, 12.3, 13.5, 10.9, 14.4, 8.0

Use the large-sample U test at the 0.01 level of significance to see whether it is reasonable to say that in general brand X flares are better (last longer) than brand Y flares.

★ **12.22** Use the large-sample U test to rework Exercise 12.18.

★ **12.23** The following are the weekly food expenditures (in dollars) of families with two children chosen at random from two suburbs of a large city:

> Suburb A: 78.60, 70.50, 75.38, 86.45, 67.95, 70.78, 67.89, 72.00, 64.19, 71.15
> Suburb B: 62.63, 75.16, 55.35, 78.19, 71.72, 63.12, 75.91, 66.51, 63.76, 60.78

Use the large-sample U test at the 0.05 level of significance to check the claim that on the average such weekly food expenditures are higher in suburb A than in suburb B.

12.6
Rank Sums: The H Test ★

The **H test,** or **Kruskal–Wallis test,** is a rank-sum test which serves to test the null hypothesis that k independent random samples come from identical populations against the alternative hypothesis that the means of these populations are not all equal. Unlike the standard test which it replaces, the one-way analysis of variance of Section 9.13, it does not require the assumption that the populations sampled have, at least approximately, normal distributions.

As in the U test, the data are ranked jointly from low to high as though they constitute a single sample. Then, if R_i is the sum of the ranks assigned to the n_i values of the ith sample and $n = n_1 + n_2 + \cdots + n_k$, the H test is based on the statistic

Statistic for H test

$$H = \frac{12}{n(n+1)} \sum_{i=1}^{k} \frac{R_i^2}{n_i} - 3(n+1)$$

CHAP: 12: Nonparametric Tests

If the null hypothesis is true and each sample has at least five observations, the sampling distribution of H can be approximated closely with a chi-square distribution with $k - 1$ degrees of freedom. Consequently, we reject the null hypothesis that the populations sampled are identical and accept the alternative hypothesis that the means of these populations are not all equal, if the value we get for H exceeds χ_α^2 for $k - 1$ degrees of freedom.

EXAMPLE   The following are the final examination scores of samples of students who are taught German by three different methods (classroom instruction and language laboratory, only classroom instruction, and only self-study in language laboratory):

> *Method 1:*   94, 87, 91, 74, 87, 97
> *Method 2:*   85, 82, 79, 84, 61, 72, 80
> *Method 3:*   89, 67, 72, 76, 69

Use the H test at the 0.05 level of significance to test the null hypothesis that the three populations sampled are identical against the alternative hypothesis that their means are not all equal.

Solution   1. *Hypotheses*

H_0: Populations are identical.

H_A: The population means are not all equal.

2. *Level of significance*

$$\alpha = 0.05$$

3. *Criterion*
Reject the null hypothesis if $H > 5.991$, the value of $\chi_{0.05}^2$ for $3 - 1 = 2$ degrees of freedom, where H is calculated in accordance with the formula above. Otherwise, accept it or reserve judgment.

4. *Calculations*
Arranging the data jointly according to size, we get 61, 67, 69, 72, 72, 74, 76, 79, 80, 82, 84, 85, 87, 87, 89, 91, 94, and 97. Assigning the data, in this order, the ranks 1, 2, 3, . . . , and 18, we find that the values of the first sample occupy ranks 6, 13, 14, 16, 17, and 18, while those of the second sample occupy ranks 1, 4.5, 8, 9, 10, 11, and

12, and those of the third sample occupy ranks 2, 3, 4.5, 7, and 15. (Since the two 87's belong to the same sample, we simply assign them ranks 13 and 14.) Thus,

$$R_1 = 6 + 13 + 14 + 16 + 17 + 18 = 84$$

$$R_2 = 1 + 4.5 + 8 + 9 + 10 + 11 + 12 = 55.5$$

$$R_3 = 2 + 3 + 4.5 + 7 + 15 = 31.5$$

and it follows that

$$H = \frac{12}{18 \cdot 19} \left(\frac{84^2}{6} + \frac{55.5^2}{7} + \frac{31.5^2}{5} \right) - 3 \cdot 19$$

$$= 6.67$$

5. *Decision*

Since $H = 6.67$ exceeds 5.991, the null hypothesis must be rejected; in other words, we conclude that the three methods of teaching German are not all equally effective.

EXERCISES

(Exercise 12.24 is a practice exercise; its complete solution is given on page 487.)

★ 12.24 The following are the miles per gallon which a test driver got for six tankfuls each of three kinds of gasoline:

> *Gasoline 1:* 28, 23, 26, 31, 14, 29
> *Gasoline 2:* 21, 31, 32, 19, 27, 16
> *Gasoline 3:* 24, 17, 21, 31, 22, 18

Use the H test at the level of significance $\alpha = 0.05$ to check the claim that there is no difference in the true average mileage yield of the three kinds of gasoline.

★ 12.25 To compare four bowling balls, a professional bowler bowled five games with each ball and got the following results:

> *Bowling ball D:* 221, 232, 207, 198, 212
> *Bowling ball E:* 202, 225, 252, 218, 226
> *Bowling ball F:* 210, 205, 189, 196, 216
> *Bowling ball G:* · 229, 192, 247, 220, 208

Use the H test at the 0.05 level of significance to test the null hypothesis that the bowler performs equally well with the four bowling balls.

★ 12.26 Three groups of guinea pigs were injected, respectively, with 0.5 mg, 1.0 mg, and 1.5 mg of a tranquilizer, and the following are the numbers of seconds it took them to fall asleep:

> *0.5-mg dose:* 8.2, 10.0, 10.2, 13.7, 14.0, 7.8, 12.7, 10.9
> *1.0-mg dose:* 9.7, 13.1, 11.0, 7.5, 13.3, 12.5, 8.8, 12.9, 7.9, 10.5
> *1.5-mg dose:* 12.0, 7.2, 8.0, 9.4, 11.3, 9.0, 11.5, 8.5

Use the H test at the level of significance $\alpha = 0.01$ to test the null hypothesis that the differences in dosage have no effect on the length of time it takes guinea pigs to fall asleep.

★ 12.27 Use the H test to rework Exercise 9.82 on page 351.

12.7

Tests of Randomness: Runs

All the methods of inference which we have discussed are based on the assumption that the samples are random; yet, there are many applications where it is difficult to decide whether this assumption is justifiable. This is true, particularly, when we have little or no control over the selection of the data, as is the case, for example, when we rely on whatever records are available to make long-range predictions of the weather, when we use whatever data are available to estimate the mortality rate of a disease, or when we use sales records for past months to make predictions of a department store's sales. None of this information constitutes a random sample in the strict sense.

There are several methods of judging the randomness of a sample on the basis of the order in which the observations are obtained; they enable us to decide, after the data have been collected, whether patterns that look suspiciously nonrandom may be attributed to chance. The technique we shall describe here and in the next two sections is based on the **theory of runs.**

A **run** is a succession of identical letters (or other kinds of symbols) which is followed and preceded by different letters or no letters at all. To illustrate, consider the following arrangement of healthy, H, and diseased, D, elm trees that were planted many years ago along a country road:

$$\underline{H\,H\,H\,H}\,\underline{D\,D\,D}\,\underline{H\,H\,H\,H\,H\,H\,H}\,\underline{D\,D}\,\underline{H\,H}\,\underline{D\,D\,D\,D}$$

Using underlines to combine the letters which constitute the runs, we find that there is first a run of four H's, then a run of three D's, then a run of seven H's, then a run of two D's, then a run of two H's, and finally a run of four D's.

The **total number of runs** appearing in an arrangement of this kind is often a good indication of a possible lack of randomness. If there are too few runs we might suspect a definite grouping or clustering, or perhaps a trend; if there are too many runs, we might suspect some sort of repeated alternating, or cyclical, pattern. In the example above there seems to be a definite clustering—the diseased trees seem to come in groups—but it remains to be seen whether this is significant or whether it can be attributed to chance.

If there are n_1 letters of one kind, n_2 letters of another kind, and u runs, we base this kind of decision on the following criterion:

Reject the null hypothesis of randomness if

$$u \leq u'_{\alpha/2} \quad \text{or} \quad u \geq u_{\alpha/2}$$

where $u'_{\alpha/2}$ and $u_{\alpha/2}$ are given in Table VIII for $n_1 \leq 15$, $n_2 \leq 15$, and $\alpha = 0.05$ or $\alpha = 0.01$.

In the construction of Table VIII, $u'_{\alpha/2}$ is the largest value of u for which the probability of $u \leq u'_{\alpha/2}$ is less than or equal to $\alpha/2$, $u_{\alpha/2}$ is the smallest value of u for which the probability of $u \geq u_{\alpha/2}$ is less than or equal to $\alpha/2$, and the blank spaces indicate that the null hypothesis of randomness cannot be rejected for values in that tail of the sampling distribution of u regardless of the value which we obtain for u. More extensive tables for the **u test** may be found in handbooks of statistical tables, but when n_1 and n_2 are both at least 10, it is generally considered reasonable to use the large-sample test described in Section 12.8.

EXAMPLE   With reference to the arrangement of healthy and diseased elm trees given on page 471, use the u test at the 0.05 level of significance to test the null hypothesis of randomness against the alternative hypothesis that the arrangement is not random.

Solution   1. *Hypotheses*

H_0: Arrangement is random.

H_A: Arrangement is not random.

2. *Level of significance*

$$\alpha = 0.05$$

3. *Criterion*

   Since $n_1 = 13$, $n_2 = 9$, and $\alpha = 0.05$, we get $u'_{0.025} = 6$ and $u_{0.025} = 17$ from Table VIII; thus, the null hypothesis must be rejected if $u \leq 6$ or $u \geq 17$. Otherwise, accept it or reserve judgment.

4. *Calculations*

   $u = 6$, as can be seen from the data.

5. *Decision*

   Since $u = 6$ is less than or equal to 6, the null hypothesis must be rejected; in other words, we conclude that the arrangement of healthy and diseased elm trees is not random. Indeed, it seems that the diseased trees come in clusters.

12.8

Tests of Randomness: Runs (Large Samples) ⋆

Under the null hypothesis that n_1 letters of one kind and n_2 letters of another kind are arranged at random, it can be shown that the mean and the standard deviation of u, the total number of runs, are

Mean and standard deviation of u

$$\mu_u = \frac{2n_1 n_2}{n_1 + n_2} + 1$$

and

$$\sigma_u = \sqrt{\frac{2n_1 n_2 (2n_1 n_2 - n_1 - n_2)}{(n_1 + n_2)^2 (n_1 + n_2 - 1)}}$$

Furthermore, if neither n_1 nor n_2 is less than 10, the sampling distribution of u can be approximated closely with a normal curve. Thus, we base the test of the null hypothesis of randomness on the statistic

Statistic for large-sample u test

$$z = \frac{u - \mu_u}{\sigma_u}$$

which has approximately the standard normal distribution. If the alternative hypothesis is that the arrangement is not random, we reject the null hypothesis for $z < -z_{\alpha/2}$ or $z > z_{\alpha/2}$; if the alternative hypothesis is that there is a clustering

473

or a trend, we reject the null hypothesis for $z < -z_\alpha$; and if the alternative hypothesis is that there is an alternating, or cyclical, pattern, we reject the null hypothesis for $z > z_\alpha$.

EXAMPLE   The following is an arrangement of men, M, and women, W, lined up to purchase tickets for a rock concert:

$$M\ W\ M\ W\ M\ M\ M\ W\ M\ W\ M\ M\ M\ W\ W\ M$$

$$\text{(cont.)}\ M\ M\ M\ W\ W\ M\ W\ M\ M\ M\ W\ M\ M\ M\ W\ W$$

$$\text{(cont.)}\ W\ M\ W\ M\ M\ M\ W\ M\ W\ M\ M\ M\ M\ W\ W\ M$$

Test for randomness at the 0.05 level of significance.

Solution   1. *Hypotheses*

H_0: Arrangement is random.

H_A: Arrangement is not random.

2. *Level of significance*

$$\alpha = 0.05$$

3. *Criterion*
Reject the null hypothesis if $z < -1.96$ or $z > 1.96$, where

$$z = \frac{u - \mu_u}{\sigma_u}$$

Otherwise, accept it or reserve judgment.

4. *Calculations*
Since $n_1 = 30$, $n_2 = 18$, and $u = 27$, we get

$$\mu_u = \frac{2 \cdot 30 \cdot 18}{30 + 18} + 1 = 23.5$$

$$\sigma_u = \sqrt{\frac{2 \cdot 30 \cdot 18(2 \cdot 30 \cdot 18 - 30 - 18)}{(30 + 18)^2(30 + 18 - 1)}} = 3.21$$

and, hence,

$$z = \frac{27 - 23.5}{3.21} = 1.09$$

5. *Decision*

Since $z = 1.09$ falls between -1.96 and 1.96, the null hypothesis of randomness cannot be rejected; there is no real evidence of any lack of randomness.

12.9

Tests of Randomness: Runs Above and Below the Median

The u test is not limited to tests of the randomness of sequences of attributes, such as the H's and D's, or M's and W's, of our examples. Any sample consisting of numerical measurements or observations can be treated similarly by using the letters a and b to denote, respectively, values falling above and below the median of the sample. Numbers equal to the median are omitted. The resulting sequence of a's and b's (representing the data in their original order) can then be tested for randomness on the basis of the total number of runs of a's and b's, namely, the total number of **runs above and below the median.** Depending on the size of n_1 and n_2, we use Table VIII or the large-sample test of Section 12.8.

EXAMPLE   On 24 successive runs between two cities, a bus carried 24, 19, 32, 28, 21, 23, 26, 17, 20, 28, 30, 24, 13, 35, 26, 21, 19, 29, 27, 18, 26, 14, 21, and 23 passengers. Use the total number of runs above and below the median and the level of significance $\alpha = 0.01$, to decide whether it is reasonable to treat these data as if they constitute a random sample.

Solution   Since the median is 23.5, as can easily be verified, we get the following arrangement of values above and below the median:

$$a\ b\ a\ a\ b\ b\ a\ b\ b\ a\ a\ a\ b\ a\ a\ b\ b\ a\ a\ b\ a\ b\ b\ b$$

1. *Hypotheses*

$$H_0:\ \text{Arrangement is random.}$$

$$H_A:\ \text{Arrangement is not random.}$$

2. *Level of significance*

$$\alpha = 0.01$$

3. *Criterion*

Since $n_1 = 12$, $n_2 = 12$, and $\alpha = 0.01$, we get $u'_{0.005} = 6$ and $u_{0.005} = 20$ from Table VIII; thus, the null hypothesis must be rejected if $u \leq 6$ or $u \geq 20$; otherwise, accept it or reserve judgment.

4. *Calculations*

As can be seen from the arrangements of a's and b's above, there are $u = 14$ runs.

5. *Decision*

Since $u = 14$ falls between 6 and 20, the null hypothesis cannot be rejected; in other words, there is no real evidence to suggest that the data cannot be treated as if they constitute a random sample.

EXERCISES   (Exercises 12.28, 12.32, and 12.37 are practice exercises; their complete solutions are given on pages 487 through 489.)

12.28 The following sequence of C's and A's shows the order in which 25 cars with California or Arizona license plates crossed the Colorado river entering Arizona at Blythe, California:

$$C\ A\ A\ C\ A\ C\ C\ A\ C\ C\ C\ A\ A\ C\ A\ A\ A\ A\ C\ A\ C\ C\ A\ C\ C$$

Test for randomness at the level of significance $\alpha = 0.05$.

12.29 The following is the order in which red, R, and black, B, cards were dealt to a bridge player:

$$B\ B\ B\ R\ R\ R\ R\ R\ B\ B\ R\ R\ R$$

Test for randomness at the 0.05 level of significance.

12.30 Test at the 0.01 level of significance whether the following arrangement of defective, D, and nondefective, N, pieces coming off an assembly line may be regarded as random:

$$N\ N\ N\ N\ N\ N\ N\ D\ D\ D\ D\ N\ N\ N\ D\ D\ N\ N\ N$$

12.31 A driver buys gasoline either at a Shell station, S, or at a Chevron station, C, and the following arrangement shows where he purchased gasoline (in the given order) over a certain period of time:

$$C\ C\ C\ S\ C\ S\ C\ S\ S\ C\ C\ S\ C\ S\ C\ S\ C\ S\ S\ C\ S\ C$$

Test for randomness at the 0.05 level of significance.

★ **12.32** Representing each 0, 2, 4, 6, and 8 by the letter E and each 1, 3, 5, 7, and 9 by the letter O, check at the 0.05 level of significance whether the arrangements of the 50 digits in the first column of the random-number table on page 529 may be regarded as random.

★ **12.33** The following arrangement shows whether 50 persons interviewed consecutively in the given order are for, F, or against, A, an increase in the city sales tax:

$A\ A\ A\ A\ A\ F\ A\ A\ F\ F\ A\ A\ A\ A\ A\ A\ A\ F\ A\ A\ A\ A\ F\ F\ F$

(cont.) $A\ A\ A\ F\ A\ A\ F\ A\ A\ A\ A\ F\ F\ A\ A\ A\ A\ A\ A\ A\ A\ F\ A\ A\ A$

Test for randomness at the level of significance $\alpha = 0.05$.

★ **12.34** Use the large-sample u test to rework Exercise 12.28.

★ **12.35** To test whether a radio signal contains a message or constitutes random noise, an interval of time is subdivided into a number of very short intervals and for each of these it is determined whether the signal strength exceeds, E, or does not exceed, N, a certain level of background noise. Test at the level of significance $\alpha = 0.05$ whether the following arrangement, thus obtained, may be regarded as random, and hence that the signal contains no message and may be regarded as random noise:

$E\ N\ N\ N\ E\ N\ E\ N\ N\ N\ E\ E\ N\ N\ N\ E\ E\ N\ E\ N\ N\ N$

(cont.) $E\ E\ N\ N\ N\ N\ N\ E\ E\ N\ E\ N\ N\ E\ N\ N\ N\ E\ E\ E\ N\ N\ N$

(cont.) $E\ N\ E\ N\ N\ N\ N\ N\ E\ N$

★ **12.36** Mentally simulate fifty flips of a coin, and test at the 0.05 level of significance whether the resulting sequence of H's and T's (heads and tails) may be regarded as random.

12.37 The following are the numbers of students absent from a school on 24 consecutive days:

38	31	32	27	28	30	26	33	36	30	28	35
33	31	29	35	31	33	31	28	30	28	25	29

Test for randomness at the 0.05 level of significance.

12.38 The following are the numbers of business lunches that an insurance agent had in 30 consecutive months: 6, 7, 5, 6, 8, 6, 8, 6, 6, 4, 3, 2, 4, 4, 3, 4, 7, 5, 6, 8, 6, 6, 3, 4, 2, 5, 4, 4, 3, and 7. Discarding the three values which equal the median, 5, test for randomness at the level of significance $\alpha = 0.01$.

★ **12.39** The following are the examination grades of 40 students in the order in which they finished an examination:

75	95	77	93	89	83	69	77	92	88	62	64	91	72
76	83	50	65	84	67	63	54	58	76	70	62	65	41
63	55	32	58	61	68	54	28	35	49	82	60		

Test for randomness at the level of significance $\alpha = 0.05$.

★ **12.40** The total number of retail stores opening for business and also quitting business within the calendar years 1948–1980 in a large city were 108, 103, 109,

107, 125, 142, 147, 122, 116, 153, 144, 162, 143, 126, 145, 129, 134, 137, 143, 150, 148, 152, 125, 106, 112, 139, 132, 122, 138, 148, 155, 146, and 158. Making use of the fact that the median is 138, test at the 0.05 level of significance whether there is a significant trend.

12.10
Rank Correlation

Since the significance test for r of Section 11.6 is based on very stringent assumptions, we sometimes use a nonparametric alternative which can be applied under much more general conditions. This test of the null hypothesis of no correlation is based on the **rank-correlation coefficient,** often called **Spearman's rank-correlation coefficient** and denoted by r_S.

To calculate the rank-correlation coefficient for a given set of paired data, we first rank the x's among themselves from low to high or high to low; then we rank the y's in the same way, find the sum of the squares of the differences, d, between the ranks of the x's and the y's, and substitute into the formula

Rank-correlation coefficient

$$r_S = 1 - \frac{6(\sum d^2)}{n(n^2 - 1)}$$

where n is the number of pairs of x's and y's. When there are ties in rank, we proceed as before and assign to each of the tied observations the mean of the ranks which they jointly occupy.

EXAMPLE   The following are the numbers of hours which ten students studied for an examination and the grades which they received:

Number of hours studied x	Grade in examination y
9	56
5	44
11	79
13	72
10	70
5	54
18	94
15	85
2	33
8	65

Calculate r_S.

Solution   Ranking the x's among themselves from low to high, and also the y's, we get the ranks shown in the first two columns of the following table:

Rank of x	Rank of y	d	d²
5	4	1.0	1.00
2.5	2	0.5	0.25
7	8	−1.0	1.00
8	7	1.0	1.00
6	6	0.0	0.00
2.5	3	−0.5	0.25
10	10	0.0	0.00
9	9	0.0	0.00
1	1	0.0	0.00
4	5	−1.0	1.00
			4.50

Then, determining the d's and their squares, and substituting $n = 10$ and $\sum d^2 = 4.50$ into the formula for r_s, we get

$$r_s = 1 - \frac{6(4.50)}{10(10^2 - 1)} = 0.97$$

As can be seen from this example, r_s is easy to compute manually, and this is why it is sometimes used instead of r when no calculator is available. When there are no ties, r_s actually equals the correlation coefficient r calculated for the two sets of ranks; when ties exist there may be a small (but usually negligible) difference. Of course, by using ranks instead of the original data we lose some information, but this is usually offset by the rank-correlation coefficient's computational ease.

When we use r_s to test the null hypothesis of no correlation between two variables x and y, we do not have to make any assumptions about the nature of the populations sampled. Under the null hypothesis of no correlation—indeed, the null hypothesis that the x's and y's are randomly matched—the sampling distribution of r_s has the mean 0 and the standard deviation

$$\sigma_{rs} = \frac{1}{\sqrt{n - 1}}$$

Since this sampling distribution can be approximated with a normal distribution even for relatively small values of n, we base the test of the null hypothesis of no correlation on the statistic

$$z = \frac{r_S - 0}{1/\sqrt{n-1}} = r_S\sqrt{n-1}$$

which has approximately the standard normal distribution.

EXAMPLE   With reference to the preceding example where we had $n = 10$ and $r_S = 0.97$, test the significance of this value of r_S at the 0.01 level of significance.

Solution   1. *Hypotheses*

$$H_0: \rho = 0 \text{ (no correlation)}$$

$$H_A: \rho \neq 0$$

2. *Level of significance*

$$\alpha = 0.01$$

3. *Criterion*

Reject the null hypothesis if $z < -2.575$ or $z > 2.575$, where

$$z = r_S\sqrt{n-1}$$

Otherwise, accept it or reserve judgment.

4. *Calculations*

For $n = 10$ and $r_S = 0.97$, we get

$$z = 0.97\sqrt{10-1} = 2.91$$

5. *Decision*

Since $z = 2.91$ exceeds 2.575, the null hypothesis must be rejected; in other words, we conclude that there is a relationship between study time and examination grades in the populations sampled.

(Exercises 12.41 and 12.42 are practice exercises; their complete solutions are given on pages 489 and 490.)

12.41 Calculate r_S for the following data representing the statistics grades, x, and psychology grades, y, of a sample of 15 students:

x	y
75	70
69	64
73	70
96	89
73	84
52	70
57	66
61	63
71	68
70	83
93	91
77	79
85	73
93	84
87	85

12.42 Test at the level of significance $\alpha = 0.05$ whether the value obtained for r_S in the preceding exercise is significant.

12.43 Calculate r_S for the following sample data representing the number of minutes it took 12 mechanics to assemble a piece of machinery in the morning, x, and in the late afternoon, y:

x	y
10.8	15.1
16.6	16.8
11.1	10.9
10.3	14.2
12.0	13.8
15.1	21.5
13.7	13.2
18.5	21.1
17.3	16.4
14.2	19.3
14.8	17.4
15.3	19.0

12.44 Use the level of significance $\alpha = 0.05$ to test whether the value obtained for r_S in the preceding exercise is significant.

12.45 Ten weeks' sales of a downtown department store, x, and its suburban branch, y, are

x	y
71	49
64	31
58	24
80	68
63	30
69	40
76	62
60	22
66	35
55	16

where the units are $10,000. Calculate r_S.

12.46 Assuming that the data of the preceding exercise may be looked upon as random samples of the two stores' sales, use the level of significance $\alpha = 0.01$ to test whether the value obtained for r_S is significant.

12.47 In Exercise 11.27 we gave the percentages of the vote predicted by a poll for seven candidates for the U.S. Senate in different states, x, and the corresponding percentages of the vote which they actually received, y, as

x	y
42	51
34	31
59	56
41	42
53	58
40	35
55	54

Calculate r_S and compare it with the value of r obtained for these data in Exercise 11.27 on page 446.

12.48 If a sample of $n = 20$ pairs of data yielded $r_S = 0.41$, is this rank-correlation coefficient significant at the 0.05 level of significance?

12.49 The following table shows how a panel of nutrition experts and a panel of heads of household ranked 15 breakfast foods on their palatability:

Breakfast food	Nutrition experts	Heads of household
I	7	5
II	3	4
III	11	8
IV	9	14
V	1	2
VI	4	6
VII	10	12
VIII	8	7
IX	5	1
X	13	9
XI	12	15
XII	2	3
XIII	15	10
XIV	6	11
XV	14	13

Calculate r_S as a measure of the consistency of the two rankings.

12.50 The following are the rankings which three judges gave to the work of ten artists:

Judge A: 5,   8, 4, 2,   3, 1, 10, 7, 9, 6
Judge B: 3, 10, 1, 4,   2, 5,   6, 7, 8, 9
Judge C: 8,   5, 6, 4, 10, 2,   3, 1, 7, 9

Calculate r_S for each pair of rankings and decide
(a) which two judges are most alike in their opinions about these artists;
(b) which two judges differ the most in their opinions about these artists.

12.1 1. *Hypotheses*

$$H_0: \mu = 5$$

$$H_A: \mu \neq 5$$

2. *Level of significance*

$$\alpha = 0.05$$

3. *Criterion*

If the number of plus signs is denoted by x, reject the null hypothesis if the probability of getting x or fewer plus signs is less than or equal to 0.025, or if the probability of getting x or more plus signs is less than or equal to 0.025. Otherwise, accept it or reserve judgment.

4. *Calculations*

Replacing each value greater than 5 with a plus sign and each value less than 5 with a minus sign, we get

$$- \ + \ + \ + \ - \ + \ + \ - \ + \ + \ - \ +$$

Thus, $x = 8$ and Table I shows that for $n = 12$ and $p = 0.50$ the probability of $x \leq 8$ is

$$0.003 + 0.016 + 0.054 + 0.121 + 0.193 + 0.226 + 0.193 + 0.121 = 0.927$$

and the probability of $x \geq 8$ is

$$0.121 + 0.054 + 0.016 + 0.003 = 0.194$$

5. *Decision*

Since neither of these probabilities is less than or equal to 0.025, the null hypothesis cannot be rejected; in other words, the data do not refute the claim that on the average Ms. Brown has to wait 5 minutes for the bus that takes her to work.

12.7 Since three of the pairs (12 and 12, 9 and 9, and 15 and 15) have to be discarded, the sample size for the sign test is only $n = 17$.

1. *Hypotheses*

$$H_0: \mu_1 = \mu_2$$

$$H_A: \mu_1 \neq \mu_2$$

2. *Level of significance*

$$\alpha = 0.05$$

3. *Criterion*

Reject the null hypothesis if $z < -2.575$ *or* $z > 2.575$, where

$$z = \frac{x - np_0}{\sqrt{np_0(1 - p_0)}}$$

with $p_0 = 0.50$. Otherwise, accept it or reserve judgment.

4. *Calculations*

Replacing each positive difference with a plus sign, each negative difference with a minus sign, and discarding the three pairs of equal values, we get

$$- \ - \ - \ + \ - \ - \ - \ - \ - \ + \ - \ - \ - \ + \ - \ - \ -$$

where the number of plus signs is 3. Thus, we get

$$z = \frac{3 - 17(0.50)}{\sqrt{17(0.50)(0.50)}} = -2.67$$

5. *Decision*

Since -2.67 is less than -2.575, the null hypothesis must be rejected; in other words, we conclude that on the average the two policemen do not issue equally many speeding tickets per day.

12.14 1. *Hypotheses*

$$H_0: \text{ Populations are identical.}$$

$$H_A: \mu_1 \neq \mu_2$$

2. *Level of significance*

$$\alpha = 0.01$$

3. *Criterion*

Reject the null hypothesis if $U \leq 2$, which is the value of U'_α for $n_1 = 6$, $n_2 = 6$, and $\alpha = 0.01$; otherwise, accept it or reserve judgment.

4. *Calculations*

Arranging the data jointly according to size, we get 25, 28, 30, 32, 35, 37, 38, 39, 42, 45, 46, and 48. Assigning the data in this order the ranks 1, 2, 3, . . . , and 12, we find that the values of the first sample (Spring) occupy ranks 6, 7, 9, 10, 11, and 12, while those of the second sample (Fall) occupy ranks 1, 2, 3, 4, 5, and 8. Thus,

$$W_1 = 6 + 7 + 9 + 10 + 11 + 12 = 55$$

$$W_2 = 1 + 2 + 3 + 4 + 5 + 8 = 23$$

so that

$$U_1 = 6 \cdot 6 + \frac{6 \cdot 7}{2} - 55 = 2$$

$$U_2 = 6 \cdot 6 + \frac{6 \cdot 7}{2} - 23 = 34$$

and $U = 2$.

5. *Decision*

Since $U = 2$ is less than or equal to 2, the null hypothesis must be rejected; in other words, we conclude that on the average there are not equally many assaults per week in the spring and in the fall.

12.19 1. *Hypotheses*

$$H_0: \text{Populations are identical.}$$

$$H_A: \mu_1 \neq \mu_2$$

2. *Level of significance*

$$\alpha = 0.05$$

3. *Criterion*

Reject the null hypothesis if $z < -1.96$ or $z > 1.96$, where

$$z = \frac{U_1 - \mu_{U_1}}{\sigma_{U_1}}$$

Otherwise, accept it or reserve judgment.

4. *Calculations*

Arranging the data jointly according to size, we get 23, 30, 41, 44, 47, 54, 55, 56, 58, 62, 64, 70, 71, 71, 72, 73, 78, 80, 81, 87, 91, 93, 94, and 96. Assigning the data in this order the ranks $1, 2, 3, \ldots$, and 24, we find that the values of the first sample (minority group 1) occupy ranks 4, 7, 10, 12, 15, 16, 17, 18, 20, 21, 23, and 24, so that

$$W_1 = 4 + 7 + 10 + 12 + 15 + 16 + 17 + 18 + 20 + 21 + 23 + 24$$
$$= 187$$

and

$$U_1 = 12 \cdot 12 + \frac{12 \cdot 13}{2} - 187 = 35$$

Since $\mu_{U_1} = \dfrac{12 \cdot 12}{2} = 72$ and $\sigma_{U_1} = \sqrt{\dfrac{12 \cdot 12 \cdot 25}{12}} = 17.32$, it follows that

$$z = \frac{35 - 72}{17.32} = -2.14$$

5. *Decision*

Since $z = -2.14$ is less than -1.96, the null hypothesis must be rejected; in other words, students from the two minority groups cannot be expected to score equally well on the test.

12.24 1. *Hypotheses*

H_0: Populations are identical.

H_A: The population means are not all equal.

2. *Level of significance*

$$\alpha = 0.05$$

3. *Criterion*

Reject the null hypothesis if $H > 5.991$, which is the value of $\chi^2_{0.05}$ for $3 - 1 = 2$ degrees of freedom; otherwise, accept it or reserve judgment.

4. *Calculations*

Arranging the data jointly according to size, we get 14, 16, 17, 18, 19, 21, 21, 22, 23, 24, 26, 27, 28, 29, 31, 31, 31, and 32. Assigning the data in this order the ranks 1, 2, 3, . . . , and 18, we find that the values of the first sample occupy ranks 1, 9, 11, 13, 14, and 16, while those of the second sample occupy ranks 2, 5, 6.5, 12, 16, and 18, and those of the third sample occupy ranks 3, 4, 6.5, 8, 10, and 16. Thus,

$$R_1 = 1 + 9 + 11 + 13 + 14 + 16 = 64$$

$$R_2 = 2 + 5 + 6.5 + 12 + 16 + 18 = 59.5$$

$$R_3 = 3 + 4 + 6.5 + 8 + 10 + 16 = 47.5$$

and it follows that

$$H = \frac{12}{18 \cdot 19} \left(\frac{64^2}{6} + \frac{59.5^2}{6} + \frac{47.5^2}{6} \right) - 3 \cdot 19$$

$$= 0.85$$

5. *Decision*

Since $H = 0.85$ is less than 5.991, the null hypothesis cannot be rejected; in other words, the data tend to support the claim that there is no difference in the true average mileage yield of the three kinds of gasoline.

12.28 1. *Hypotheses*

H_0: Arrangement is random.

H_A: Arrangement is not random.

2. *Level of significance*

$$\alpha = 0.05$$

3. *Criterion*

Since $n_1 = 13$, $n_2 = 12$, and $\alpha = 0.05$, we get $u'_{0.025} = 8$ and $u_{0.025} = 19$ from Table VIII; thus, the null hypothesis must be rejected if $u \leq 8$ or $u \geq 19$.

4. *Calculations*

$$u = 15$$

5. *Decision*

Since $u = 15$ falls between 8 and 19, the null hypothesis cannot be rejected; in other words, there is no evidence to suspect that the arrangement is not random.

12.32 1. *Hypotheses*

$$H_0: \text{ Arrangement is random.}$$

$$H_A: \text{ Arrangement is not random.}$$

2. *Level of significance*

$$\alpha = 0.05$$

3. *Criterion*

Reject the null hypothesis if $z < -1.96$ or $z > 1.96$, where

$$z = \frac{u - \mu_u}{\sigma_u}$$

otherwise, accept it or reserve judgment.

4. *Calculations*

Since there are $n_1 = 28$ even digits and $n_2 = 22$ odd digits, and $u = 23$, we get

$$\mu_u = \frac{2 \cdot 28 \cdot 22}{28 + 22} + 1 = 25.64$$

$$\sigma_u = \sqrt{\frac{2 \cdot 28 \cdot 22(2 \cdot 28 \cdot 22 - 28 - 22)}{(28 + 22)^2(28 + 22 - 1)}} = 3.45$$

and, hence,

$$z = \frac{23 - 25.64}{3.45} = -0.77$$

5. *Decision*

Since $z = -0.77$ falls between -1.96 and 1.96, the null hypothesis cannot be rejected; there is no reason to suspect any lack of randomness.

12.37 1. *Hypotheses*

$$H_0: \text{Arrangement is random.}$$

$$H_A: \text{Arrangement is not random.}$$

2. *Level of significance*

$$\alpha = 0.05$$

3. *Criterion*

Since the median is 30.5, $n_1 = 12$, $n_2 = 12$, and $u'_{0.025} = 7$ and $u_{0.025} = 19$ according to Table VIII; thus, the null hypothesis must be rejected if $u \leq 7$ or $u \geq 19$. Otherwise, accept the null hypothesis or reserve judgment.

4. *Calculations*

The arrangement of values above and below the median is

$$a\ a\ a\ b\ b\ b\ b\ a\ a\ b\ b\ a\ a\ a\ b\ a\ a\ a\ a\ b\ b\ b\ b\ b$$

so that $u = 8$.

5. *Decision*

Since $u = 8$ falls between 7 and 19, the null hypothesis cannot be rejected; in other words, there is no significant deviation from randomness.

12.41 Ranking the x's among themselves, and also the y's, we get the ranks shown in the first two columns of the following table:

Rank of x	Rank of x	d	d^2
9	6	3	9
4	2	2	4
7.5	6	1.5	2.25
15	14	1	1
7.5	11.5	−4	16
1	6	−5	25
2	3	−1	1
3	1	2	4
6	4	2	4
5	10	−5	25
13.5	15	−1.5	2.25
10	9	1	1
11	8	3	9
13.5	11.5	2	4
12	13	−1	1
			108.5

SEC.12.10: Rank Correlation

Then, determining the d's and their squares, and substituting $n = 15$ and $\sum d^2 = 108.5$ into the formula for r_S, we get

$$r_S = 1 - \frac{6(108.5)}{15(15^2 - 1)} = 0.81$$

12.42 1. *Hypotheses*

$$H_0: \rho = 0 \text{ (no correlation)}$$

$$H_A: \rho \neq 0$$

2. *Level of significance*

$$\alpha = 0.05$$

3. *Criterion*
   Reject the null hypothesis if $z < -1.96$ or $z > 1.96$, where

$$z = r_S \sqrt{n - 1}$$

Otherwise, accept it or reserve judgment.

4. *Calculations*
   For $n = 15$ and $r_S = 0.81$, we get

$$z = 0.81 \sqrt{15 - 1} = 3.03$$

5. *Decision*
   Since $z = 3.03$ exceeds 1.96, the null hypothesis must be rejected; in other words, there is a relationship between students' grades in the two subjects.

Review: Chapters

Achievements

Having read and studied these chapters, and having worked a good proportion of the exercises, you should be able to

1. Explain the method of least squares.
2. Fit a least-squares line by solving the two normal equations.
3. Fit a least-squares line by using the special formulas for a and b.
4. Use the equation of a least-squares line to predict a value of y.
5. Explain what is meant by "regression line."
★ 6. State the assumptions underlying normal regression analysis.
★ 7. Calculate the standard error of estimate.
★ 8. Test hypotheses about the regression coefficient β.
★ 9. Construct confidence intervals for the regression coefficient β.
★ 10. Construct confidence limits for the mean of y when $x = x_0$.
11. Define the correlation coefficient in terms of the residual and total sums of squares.
12. Explain the difference between positive and negative correlation.

491

13. Use the computing formula to calculate r.

14. Use r to judge the strength of a linear relationship.

15. Use values of r to compare the strength of two relationships.

16. Test the significance of a value of r.

17. Explain what is meant by "nonparametric tests."

18. Perform a one-sample sign test.

19. Apply the sign test to paired data.

20. Perform a large-sample sign test.

21. Use the U test as a nonparametric alternative to the two-sample t test (concerning the difference between two means).

★ 22. Perform a large-sample U test.

★ 23. Use the H test as a nonparametric alternative to a one-way analysis of variance.

24. Test for randomness on the basis of the total number of runs.

★ 25. Perform a large-sample test of randomness based on the total number of runs.

26. Test for randomness on the basis of runs above and below the median.

27. Calculate the rank correlation coefficient.

28. Test the significance of a rank correlation coefficient.

Checklist of Key Terms (with page references to their definitions)

REVIEW
EXERCISES

R.156 If $r = 0.56$ for one set of paired data and $r = 0.97$ for another set of paired data, compare the strengths of the two relationships.

R.157 The following are the high school averages, x, and first-year-college grade-point averages, y, of ten students:

x	y
3.0	2.6
2.7	2.4
3.8	3.9
2.6	2.1
3.2	2.6
3.4	3.3
2.8	2.2
3.1	3.2
3.5	2.8
3.3	2.5

Fit a least-squares line which will enable us to predict first-year-college grade-point averages in terms of high school averages, and use it to predict the first-year-college grade-point average of a student with a high school average of 3.5.

★ **R.158** With reference to the preceding exercise, construct a 95% confidence interval for the regression coefficient β.

R.159 If a set of $n = 42$ paired data yields $r = 0.33$, test the null hypothesis of no correlation
(a) at the 0.05 level of significance;
(b) at the 0.01 level of significance.

R.160 A sample of 30 suitcases carried by an airline on transoceanic flights weighed 32, 46, 48, 27, 35, 52, 66, 41, 49, 36, 50, 44, 48, 36, 40, 35, 63, 42, 52, 40, 38, 36, 43, 41, 49, 52, 44, 60, 31, and 35 pounds. Use the sign test at the 0.01 level of significance to test the null hypothesis $\mu = 37$ pounds against the alternative hypothesis $\mu > 37$ pounds.

R.161 The following are measurements of the strength of samples of two kinds of fishing lines (in pounds):

Fishing line 1: 12.0, 11.5, 11.8, 11.3, 12.2, 11.7, 11.9
Fishing line 2: 11.1, 11.6, 11.4, 12.1, 10.5, 11.0, 10.8

Use the U test at the 0.05 level of significance to test the null hypothesis that the two samples come from identical populations against the alternative hypothesis that the two populations have unequal means.

R.162 The following sequence shows whether a certain member was present, P, or absent, A, at twenty consecutive meetings of a fraternal organization:

$$P\ P\ P\ P\ P\ P\ P\ P\ A\ P\ P\ P\ P\ P\ P\ P\ A\ A\ A\ A$$

Test for randomness at the 0.05 level of significance.

R.163 The following are the scores of twelve golfers on two consecutive Sundays: 68 and 71, 73 and 76, 70 and 73, 74 and 71, 69 and 72, 72 and 74, 67 and 70, 72 and 68, 71 and 72, 73 and 74, 68 and 69, and 70 and 72. Use the sign test at the 0.05 level of significance to test the null hypothesis $\mu_1 = \mu_2$ against the alternative hypothesis $\mu_2 > \mu_1$ (perhaps due to heavy winds on the second Sunday).

R.164 The following are the numbers of hours which ten persons (interviewed as part of a sample survey) spent reading books or magazines, x, and watching television, y, during the preceding week:

x	y
4	18
8	12
9	9
3	15
2	27
10	12
5	12
1	19
5	25
7	18

Calculate the coefficient of correlation.

R.165 With reference to the preceding exercise, test the significance of r at the 0.05 level of significance.

R.166 The following are the closing prices of a commodity on twenty consecutive trading days (in dollars): 378, 379, 379, 378, 377, 376, 374, 374, 373, 373, 374, 375, 376, 376, 376, 375, 374, 374, 373, and 374. Test for randomness at the 0.01 level of significance.

R.167 State in each case whether you would expect a positive correlation, a negative correlation, or no correlation (a zero correlation):

(a) the price of gasoline and the occupancy rate of motels;

(b) exposure to the sun and the incidence of skin cancer;

(c) shoe size and years of education;

(d) mintage figures and the values of old coins;

(e) blood pressure and eye color;

(f) baseball players' batting averages and their salaries.

R.168 The following are the numbers of minutes it took two ambulance services to reach the scenes of accidents:

Ambulance service 1:   9.3, 5.5, 13.1, 10.0, 7.6, 9.2, 11.2, 6.4, 14.0, 10.3
Ambulance service 2:   12.7, 6.6, 9.1, 4.5, 7.2, 6.4, 7.5

Use the U test at the 0.05 level of significance to test the null hypothesis that the two samples come from identical populations against the alternative hypothesis that the two populations have unequal means.

R.169 The following are the numbers of persons who attended a "singles only" dance on twelve consecutive Saturdays: 152, 188, 149, 212, 103, 146, 177, 158, 201, 175, 188, and 162. Use the sign test based on Table I and the 0.05 level of significance to test the null hypothesis $\mu = 149$ against the alternative hypothesis $\mu > 149$.

★ R.170 The following sequence shows whether a television news program had at least 12% of the viewing audience, A, or less than 12%, L, on 36 consecutive weekday evenings:

$$L \; L \; L \; L \; A \; A \; L \; L \; L \; A \; L \; L \; L \; A \; A \; A \; A \; L$$
$$\text{(cont.)} \; A \; L \; L \; L \; A \; A \; L \; L \; L \; L \; L \; A \; L \; L \; L \; L \; L \; A$$

Test for randomness at the 0.05 level of significance.

R.171 The following data were obtained in a study of the relationship between the resistance (ohms), x, and the failure time (minutes), y, of certain overloaded resistors:

x	y
33	38
41	45
48	40
46	44
35	33
30	32
50	48

Fit a least-squares line and use it to predict the failure time of such a resistor when the resistance is 40 ohms.

★ R.172 With reference to the preceding exercise, construct a 95% confidence interval for the mean failure time of such a resistor when the resistance is 40 ohms.

★ **R.173** The following are the numbers of minutes that patients had to wait in the offices of four doctors:

Doctor 1: 20, 25, 38, 31, 23
Doctor 2: 19, 12, 21, 24, 9
Doctor 3: 8, 10, 27, 25, 14
Doctor 4: 17, 25, 28, 21, 15

Use the H test at the 0.05 level of significance to test the null hypothesis that the four samples come from identical populations against the alternative hypothesis that the means of the populations are not all equal.

R.174 Calculate r for the data of Exercise R.171 and test whether it is significant at the 0.01 level of significance.

R.175 The following are the numbers of burglaries committed in two suburbs of a large city on 24 days: 20 and 17, 12 and 9, 25 and 21, 18 and 14, 12 and 15, 27 and 23, 22 and 18, 20 and 15, 18 and 18, 19 and 21, 24 and 22, 17 and 23, 20 and 11, 16 and 22, 13 and 9, 18 and 15, 20 and 17, 13 and 18, 22 and 16, 25 and 21, 20 and 13, 12 and 6, 18 and 9, and 17 and 25. Use the large-sample sign test at the 0.05 level of significance to test the null hypothesis that on the average there are equally many burglaries in the two suburbs against the alternative hypothesis that there are not equally many.

R.176 The following are the batting averages, x, and home runs hit, y, by fifteen baseball players during the first half of a baseball season:

x	y
0.252	12
0.305	6
0.299	4
0.303	15
0.285	2
0.191	2
0.283	16
0.272	6
0.310	8
0.266	10
0.215	0
0.211	3
0.272	14
0.244	6
0.320	7

Calculate the rank-correlation coefficient and check whether it is significant at the 0.05 level of significance.

R.177 Test at the 0.05 level of significance whether the following sequence of cars observing, O, and exceeding, E, the 55-mph speed limit may be regarded as random:

$$O\ O\ O\ E\ E\ O\ E\ O\ O\ O\ O\ E\ E\ E\ O\ E\ O\ O\ O\ O\ E\ O$$

★ **R.178** The following are data on the percentage kill of two kinds of insecticides used against mosquitos:

Insecticide A: 41.9, 46.9, 44.6, 43.9, 42.0, 44.0, 41.0, 43.1, 39.0, 45.2, 44.6, 42.0
Insecticide B: 45.7, 39.8, 42.8, 41.2, 45.0, 40.2, 40.2, 41.7, 37.4, 38.8, 41.7, 38.7

Use the large-sample U test at the 0.05 level of significance to test the null hypothesis that the two samples come from identical populations against the alternative hypothesis that on the average the first insecticide is more effective than the second.

R.179 Rework the preceding exercise using Table VII and the alternative hypothesis that on the average the two insecticides are not equally effective.

R.180 The following are the processing times (minutes), x, and hardness readings, y, of certain machine parts:

x	y
20	282
34	275
19	171
10	142
24	145
31	340
25	282
13	105
29	233

Fit a least-squares line and use it to predict the hardness reading of a part which has been processed for 25 minutes.

★ **R.181** With reference to the preceding exercise, test the null hypothesis $\beta = 9.5$ against the alternative hypothesis $\beta < 9.5$ at the 0.01 level of significance.

R.182 With reference to Exercise R.180, what percentage of the variation among the hardness readings can be attributed to difference in processing times?

R.183 If $r_S = 0.34$ for a set of $n = 50$ paired data, is this rank-correlation coefficient significant at the 0.01 level of significance?

Bibliography

A. PROBABILITY AND STATISTICS FOR THE LAYMAN

CAMPBELL, S. K., *Flaws and Fallacies in Statistical Thinking.* Englewood Cliffs, N.J.: Prentice-Hall, Inc., 1974.

FEDERER, W. T., *Statistics and Society.* New York: Marcel Dekker, Inc., 1973.

GARVIN, A. D., *Probability in Your Life.* Portland, Me: J. Weston Walch Publisher, 1978.

HUFF, D., *How to Lie with Statistics.* New York: W. W. Norton & Company, Inc., 1954.

HUFF, D., and GEIS, I., *How to Take a Chance.* New York: W. W. Norton & Company, Inc., 1959.

KIMBLE, G. A., *How to Use (and Misuse) Statistics.* Englewood Cliffs, N. J.: Prentice-Hall, Inc., 1978.

LARSEN, R. J., and STROUP, D. F., *Statistics in the Real World.* New York: Macmillan Publishing Co., Inc., 1976.

LEVINSON, H. C., *Chance, Luck, and Statistics.* New York: Dover Publications, Inc., 1963.

MORONEY, M. J., *Facts from Figures.* New York: Penguin Books, 1956.

MOSTELLER, F., PIETERS, R. S., KRUSKAL, W. H., RISING, G. R., LINK, R. F., CARLSON, R., and ZELINKA, M., *Statistics by Example.* Reading, Mass.: Addison-Wesley Publishing Company, Inc., 1973.

REICHMAN, W. J., *Use and Abuse of Statistics.* New York: Penguin Books, 1971.

RUNYON, R. P., *Winning with Statistics.* Reading, Mass.: Addison-Wesley Publishing Company, Inc., 1977.

TANUR, J. M. (ed.), *Statistics: A Guide to the Unknown.* San Francisco: Holden-Day, Inc., 1972.

WEAVER, W., *Lady Luck: The Theory of Probability*. New York: Dover Publications, Inc., 1982.

B. SOME BOOKS ON THE THEORY OF PROBABILITY AND STATISTICS

FREUND, J. E., and WALPOLE, R. E., *Mathematical Statistics*, 3rd ed. Englewood Cliffs, N. J.: Prentice-Hall, Inc., 1980.

GOLDBERG, S., *Probability—An Introduction*. Englewood Cliffs, N.J.: Prentice-Hall, Inc., 1960.

HODGES, J. L., and LEHMANN, E. L., *Elements of Finite Probability*. San Francisco: Holden-Day, Inc., 1965.

HOEL, P., *Introduction to Mathematical Statistics*, 4th ed. New York: John Wiley & Sons, Inc., 1971.

MENDENHALL, W., and SCHAEFFER, R. L., *Mathematical Statistics with Applications*. North Scituate, Mass.: Duxbury Press, 1973.

MOSTELLER, F., ROURKE, R. E. K., and THOMAS, G. B., *Probability with Statistical Applications*, 2nd ed. Reading, Mass.: Addison-Wesley Publishing Company, Inc., 1970.

SCHAEFFER, R. L., and MENDENHALL, W., *Introduction to Probability: Theory and Applications*. North Scituate, Mass.: Duxbury Press, 1975.

C. SOME BOOKS DEALING WITH SPECIAL TOPICS

BOX, G. E. P., HUNTER, W. G., and HUNTER, J. S., *Statistics for Experimenters*. New York: John Wiley & Sons, Inc., 1978.

CHATTERJEE, S., and PRICE, B., *Regression Analysis by Example*. New York: John Wiley & Sons, Inc., 1977.

EVERITT, B. S., *The Analysis of Contingency Tables*. New York: John Wiley & Sons, Inc., 1977.

GIBBONS, J. D., *Nonparametric Statistical Inference*. New York: McGraw-Hill Book Company, 1971.

HARTWIG, F., and DEARING, B. E., *Exploratory Data Analysis*. Beverly Hills, Calif.: Sage Publications, Inc., 1979.

LEHMANN, E. L., *Nonparametrics: Statistical Methods Based on Ranks*. San Francisco: Holden-Day, Inc., 1975.

MOSTELLER, F., and ROURKE, R. E. K., *Sturdy Statistics*. Reading, Mass.: Addison-Wesley Publishing Company, Inc., 1973.

NOETHER, G. E., *Introduction to Statistics: A Nonparametric Approach*. Boston: Houghton Mifflin Company, 1976.

WILLIAMS, W. H., *A Sampler on Sampling*. New York: John Wiley & Sons, Inc., 1978.

D. SOME GENERAL REFERENCE WORKS AND TABLES

HAUSER, P. M., and LEONARD, W. R., *Government Statistics for Business Use*, 2nd ed. New York: John Wiley & Sons, Inc., 1956.

KENDALL, M. G., and BUCKLAND, W. R., *A Dictionary of Statistical Terms*, 4th ed. London: Longman Group Ltd, 1982.

NATIONAL BUREAU OF STANDARDS, *Tables of the Binomial Probability Distribution*. Washington, D.C.: Government Printing Office, 1950.

OWEN, D. B., *Handbook of Statistical Tables*. Reading, Mass.: Addison-Wesley Publishing Company, Inc., 1962.

RAND CORPORATION, *A Million Random Digits with 100,000 Normal Deviates*. New York: The Free Press, 1955.

ROMIG, H. G., *50–100 Binomial Tables*. New York: John Wiley & Sons, Inc., 1953.

Statistical Tables

Table I   Binomial probabilities

n	x	0.05	0.1	0.2	0.3	0.4	0.5	0.6	0.7	0.8	0.9	0.95
2	0	0.902	0.810	0.640	0.490	0.360	0.250	0.160	0.090	0.040	0.010	0.002
	1	0.095	0.180	0.320	0.420	0.480	0.500	0.480	0.420	0.320	0.180	0.095
	2	0.002	0.010	0.040	0.090	0.160	0.250	0.360	0.490	0.640	0.810	0.902
3	0	0.857	0.729	0.512	0.343	0.216	0.125	0.064	0.027	0.008	0.001	
	1	0.135	0.243	0.384	0.441	0.432	0.375	0.288	0.189	0.096	0.027	0.007
	2	0.007	0.027	0.096	0.189	0.288	0.375	0.432	0.441	0.384	0.243	0.135
	3		0.001	0.008	0.027	0.064	0.125	0.216	0.343	0.512	0.729	0.857
4	0	0.815	0.656	0.410	0.240	0.130	0.062	0.026	0.008	0.002		
	1	0.171	0.292	0.410	0.412	0.346	0.250	0.154	0.076	0.026	0.004	
	2	0.014	0.049	0.154	0.265	0.346	0.375	0.346	0.265	0.154	0.049	0.014
	3		0.004	0.026	0.076	0.154	0.250	0.346	0.412	0.410	0.292	0.171
	4			0.002	0.008	0.026	0.062	0.130	0.240	0.410	0.656	0.815
5	0	0.774	0.590	0.328	0.168	0.078	0.031	0.010	0.002			
	1	0.204	0.328	0.410	0.360	0.259	0.156	0.077	0.028	0.006		
	2	0.021	0.073	0.205	0.309	0.346	0.312	0.230	0.132	0.051	0.008	0.001
	3	0.001	0.008	0.051	0.132	0.230	0.312	0.346	0.309	0.205	0.073	0.021
	4			0.006	0.028	0.077	0.156	0.259	0.360	0.410	0.328	0.204
	5				0.002	0.010	0.031	0.078	0.168	0.328	0.590	0.774
6	0	0.735	0.531	0.262	0.118	0.047	0.016	0.004	0.001			
	1	0.232	0.354	0.393	0.303	0.187	0.094	0.037	0.010	0.002		
	2	0.031	0.098	0.246	0.324	0.311	0.234	0.138	0.060	0.015	0.001	
	3	0.002	0.015	0.082	0.185	0.276	0.312	0.276	0.185	0.082	0.015	0.002
	4		0.001	0.015	0.060	0.138	0.234	0.311	0.324	0.246	0.098	0.031
	5			0.002	0.010	0.037	0.094	0.187	0.303	0.393	0.354	0.232
	6				0.001	0.004	0.016	0.047	0.118	0.262	0.531	0.735
7	0	0.698	0.478	0.210	0.082	0.028	0.008	0.002				
	1	0.257	0.372	0.367	0.247	0.131	0.055	0.017	0.004			
	2	0.041	0.124	0.275	0.318	0.261	0.164	0.077	0.025	0.004		
	3	0.004	0.023	0.115	0.227	0.290	0.273	0.194	0.097	0.029	0.003	
	4		0.003	0.029	0.097	0.194	0.273	0.290	0.227	0.115	0.023	0.004

Table I   Binomial probabilities (continued)

n	x	0.05	0.1	0.2	0.3	0.4	0.5	0.6	0.7	0.8	0.9	0.95
	5			0.004	0.025	0.077	0.164	0.261	0.318	0.275	0.124	0.041
	6				0.004	0.017	0.055	0.131	0.247	0.367	0.372	0.257
	7					0.002	0.008	0.028	0.082	0.210	0.478	0.698
8	0	0.663	0.430	0.168	0.058	0.017	0.004	0.001				
	1	0.279	0.383	0.336	0.198	0.090	0.031	0.008	0.001			
	2	0.051	0.149	0.294	0.296	0.209	0.109	0.041	0.010	0.001		
	3	0.005	0.033	0.147	0.254	0.279	0.219	0.124	0.047	0.009		
	4		0.005	0.046	0.136	0.232	0.273	0.232	0.136	0.046	0.005	
	5			0.009	0.047	0.124	0.219	0.279	0.254	0.147	0.033	0.005
	6			0.001	0.010	0.041	0.109	0.209	0.296	0.294	0.149	0.051
	7				0.001	0.008	0.031	0.090	0.198	0.336	0.383	0.279
	8					0.001	0.004	0.017	0.058	0.168	0.430	0.663
9	0	0.630	0.387	0.134	0.040	0.010	0.002					
	1	0.299	0.387	0.302	0.156	0.060	0.018	0.004				
	2	0.063	0.172	0.302	0.267	0.161	0.070	0.021	0.004			
	3	0.008	0.045	0.176	0.267	0.251	0.164	0.074	0.021	0.003		
	4	0.001	0.007	0.066	0.172	0.251	0.246	0.167	0.074	0.017	0.001	
	5		0.001	0.017	0.074	0.167	0.246	0.251	0.172	0.066	0.007	0.001
	6			0.003	0.021	0.074	0.164	0.251	0.267	0.176	0.045	0.008
	7				0.004	0.021	0.070	0.161	0.267	0.302	0.172	0.063
	8					0.004	0.018	0.060	0.156	0.302	0.387	0.299
	9						0.002	0.010	0.040	0.134	0.387	0.630
10	0	0.599	0.349	0.107	0.028	0.006	0.001					
	1	0.315	0.387	0.268	0.121	0.040	0.010	0.002				
	2	0.075	0.194	0.302	0.233	0.121	0.044	0.011	0.001			
	3	0.010	0.057	0.201	0.267	0.215	0.117	0.042	0.009	0.001		
	4	0.001	0.011	0.088	0.200	0.251	0.205	0.111	0.037	0.006		
	5		0.001	0.026	0.103	0.201	0.246	0.201	0.103	0.026	0.001	
	6			0.006	0.037	0.111	0.205	0.251	0.200	0.088	0.011	0.001
	7			0.001	0.009	0.042	0.117	0.215	0.267	0.201	0.057	0.010
	8				0.001	0.011	0.044	0.121	0.233	0.302	0.194	0.075
	9					0.002	0.010	0.040	0.121	0.268	0.387	0.315
	10						0.001	0.006	0.028	0.107	0.349	0.599

Table I   Binomial probabilities (continued)

n	x	0.05	0.1	0.2	0.3	0.4	0.5	0.6	0.7	0.8	0.9	0.95
11	0	0.569	0.314	0.086	0.020	0.004						
	1	0.329	0.384	0.236	0.093	0.027	0.005	0.001				
	2	0.087	0.213	0.295	0.200	0.089	0.027	0.005	0.001			
	3	0.014	0.071	0.221	0.257	0.177	0.081	0.023	0.004			
	4	0.001	0.016	0.111	0.220	0.236	0.161	0.070	0.017	0.002		
	5		0.002	0.039	0.132	0.221	0.226	0.147	0.057	0.010		
	6			0.010	0.057	0.147	0.226	0.221	0.132	0.039	0.002	
	7			0.002	0.017	0.070	0.161	0.236	0.220	0.111	0.016	0.001
	8				0.004	0.023	0.081	0.177	0.257	0.221	0.071	0.014
	9				0.001	0.005	0.027	0.089	0.200	0.295	0.213	0.087
	10					0.001	0.005	0.027	0.093	0.236	0.384	0.329
	11							0.004	0.020	0.086	0.314	0.569
12	0	0.540	0.282	0.069	0.014	0.002						
	1	0.341	0.377	0.206	0.071	0.017	0.003					
	2	0.099	0.230	0.283	0.168	0.064	0.016	0.002				
	3	0.017	0.085	0.236	0.240	0.142	0.054	0.012	0.001			
	4	0.002	0.021	0.133	0.231	0.213	0.121	0.042	0.008	0.001		
	5		0.004	0.053	0.158	0.227	0.193	0.101	0.029	0.003		
	6			0.016	0.079	0.177	0.226	0.177	0.079	0.016		
	7			0.003	0.029	0.101	0.193	0.227	0.158	0.053	0.004	
	8			0.001	0.008	0.042	0.121	0.213	0.231	0.133	0.021	0.002
	9				0.001	0.012	0.054	0.142	0.240	0.236	0.085	0.017
	10					0.002	0.016	0.064	0.168	0.283	0.230	0.099
	11						0.003	0.017	0.071	0.206	0.377	0.341
	12							0.002	0.014	0.069	0.282	0.540
13	0	0.513	0.254	0.055	0.010	0.001						
	1	0.351	0.367	0.179	0.054	0.011	0.002					
	2	0.111	0.245	0.268	0.139	0.045	0.010	0.001				
	3	0.021	0.100	0.246	0.218	0.111	0.035	0.006	0.001			
	4	0.003	0.028	0.154	0.234	0.184	0.087	0.024	0.003			
	5		0.006	0.069	0.180	0.221	0.157	0.066	0.014	0.001		
	6		0.001	0.023	0.103	0.197	0.209	0.131	0.044	0.006		
	7			0.006	0.044	0.131	0.209	0.197	0.103	0.023	0.001	

Table I   Binomial probabilities (continued)

n	x	0.05	0.1	0.2	0.3	0.4	0.5	0.6	0.7	0.8	0.9	0.95
	8			0.001	0.014	0.066	0.157	0.221	0.180	0.069	0.006	
	9				0.003	0.024	0.087	0.184	0.234	0.154	0.028	0.003
	10				0.001	0.006	0.035	0.111	0.218	0.246	0.100	0.021
	11					0.001	0.010	0.045	0.139	0.268	0.245	0.111
	12						0.002	0.011	0.054	0.179	0.367	0.351
	13							0.001	0.010	0.055	0.254	0.513
14	0	0.488	0.229	0.044	0.007	0.001						
	1	0.359	0.356	0.154	0.041	0.007	0.001					
	2	0.123	0.257	0.250	0.113	0.032	0.006	0.001				
	3	0.026	0.114	0.250	0.194	0.085	0.022	0.003				
	4	0.004	0.035	0.172	0.229	0.155	0.061	0.014	0.001			
	5		0.008	0.086	0.196	0.207	0.122	0.041	0.007			
	6		0.001	0.032	0.126	0.207	0.183	0.092	0.023	0.002		
	7			0.009	0.062	0.157	0.209	0.157	0.062	0.009		
	8			0.002	0.023	0.092	0.183	0.207	0.126	0.032	0.001	
	9				0.007	0.041	0.122	0.207	0.196	0.086	0.008	
	10				0.001	0.014	0.061	0.155	0.229	0.172	0.035	0.004
	11					0.003	0.022	0.085	0.194	0.250	0.114	0.026
	12					0.001	0.006	0.032	0.113	0.250	0.257	0.123
	13						0.001	0.007	0.041	0.154	0.356	0.359
	14							0.001	0.007	0.044	0.229	0.488
15	0	0.463	0.206	0.035	0.005							
	1	0.366	0.343	0.132	0.031	0.005						
	2	0.135	0.267	0.231	0.092	0.022	0.003					
	3	0.031	0.129	0.250	0.170	0.063	0.014	0.002				
	4	0.005	0.043	0.188	0.219	0.127	0.042	0.007	0.001			
	5	0.001	0.010	0.103	0.206	0.186	0.092	0.024	0.003			
	6		0.002	0.043	0.147	0.207	0.153	0.061	0.012	0.001		
	7			0.014	0.081	0.177	0.196	0.118	0.035	0.003		
	8			0.003	0.035	0.118	0.196	0.177	0.081	0.014		
	9			0.001	0.012	0.061	0.153	0.207	0.147	0.043	0.002	
	10				0.003	0.024	0.092	0.186	0.206	0.103	0.010	0.001
	11				0.001	0.007	0.042	0.127	0.219	0.188	0.043	0.005
	12					0.002	0.014	0.063	0.170	0.250	0.129	0.031
	13						0.003	0.022	0.092	0.231	0.267	0.135
	14							0.005	0.031	0.132	0.343	0.366
	15								0.005	0.035	0.206	0.463

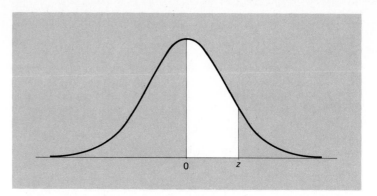

The entries in Table II are the probabilities that a random variable having the standard normal distribution will take on a value between 0 and z, they are given by the area of the white region under the curve in the figure shown above.

Table II   Normal-curve areas

z	.00	.01	.02	.03	.04	.05	.06	.07	.08	.09
0.0	.0000	.0040	.0080	.0120	.0160	.0199	.0239	.0279	.0319	.0359
0.1	.0398	.0438	.0478	.0517	.0557	.0596	.0636	.0675	.0714	.0753
0.2	.0793	.0832	.0871	.0910	.0948	.0987	.1026	.1064	.1103	.1141
0.3	.1179	.1217	.1255	.1293	.1331	.1368	.1406	.1443	.1480	.1517
0.4	.1554	.1591	.1628	.1664	.1700	.1736	.1772	.1808	.1844	.1879
0.5	.1915	.1950	.1985	.2019	.2054	.2088	.2123	.2157	.2190	.2224
0.6	.2257	.2291	.2324	.2357	.2389	.2422	.2454	.2486	.2517	.2549
0.7	.2580	.2611	.2642	.2673	.2704	.2734	.2764	.2794	.2823	.2852
0.8	.2881	.2910	.2939	.2967	.2995	.3023	.3051	.3078	.3106	.3133
0.9	.3159	.3186	.3212	.3238	.3264	.3289	.3315	.3340	.3365	.3389
1.0	.3413	.3438	.3461	.3485	.3508	.3531	.3554	.3577	.3599	.3621
1.1	.3643	.3665	.3686	.3708	.3729	.3749	.3770	.3790	.3810	.3830
1.2	.3849	.3869	.3888	.3907	.3925	.3944	.3962	.3980	.3997	.4015
1.3	.4032	.4049	.4066	.4082	.4099	.4115	.4131	.4147	.4162	.4177
1.4	.4192	.4207	.4222	.4236	.4251	.4265	.4279	.4292	.4306	.4319
1.5	.4332	.4345	.4357	.4370	.4382	.4394	.4406	.4418	.4429	.4441
1.6	.4452	.4463	.4474	.4484	.4495	.4505	.4515	.4525	.4535	.4545
1.7	.4554	.4564	.4573	.4582	.4591	.4599	.4608	.4616	.4625	.4633
1.8	.4641	.4649	.4656	.4664	.4671	.4678	.4686	.4693	.4699	.4706
1.9	.4713	.4719	.4726	.4732	.4738	.4744	.4750	.4756	.4761	.4767
2.0	.4772	.4778	.4783	.4788	.4793	.4798	.4803	.4808	.4812	.4817
2.1	.4821	.4826	.4830	.4834	.4838	.4842	.4846	.4850	.4854	.4857
2.2	.4861	.4864	.4868	.4871	.4875	.4878	.4881	.4884	.4887	.4890
2.3	.4893	.4896	.4898	.4901	.4904	.4906	.4909	.4911	.4913	.4916
2.4	.4918	.4920	.4922	.4925	.4927	.4929	.4931	.4932	.4934	.4936
2.5	.4938	.4940	.4941	.4943	.4945	.4946	.4948	.4949	.4951	.4952
2.6	.4953	.4955	.4956	.4957	.4959	.4960	.4961	.4962	.4963	.4964
2.7	.4965	.4966	.4967	.4968	.4969	.4970	.4971	.4972	.4973	.4974
2.8	.4974	.4975	.4976	.4977	.4977	.4978	.4979	.4979	.4980	.4981
2.9	.4981	.4982	.4982	.4983	.4984	.4984	.4985	.4985	.4986	.4986
3.0	.4987	.4987	.4987	.4988	.4988	.4989	.4989	.4989	.4990	.4990

Also, for $z = 4.0$, 5.0, and 6.0, the areas are 0.49997, 0.4999997, and 0.499999999.

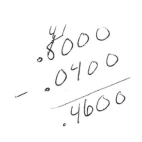

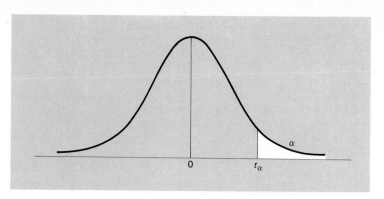

The entries in Table III are values for which the area to their right under the t distribution with given degrees of freedom (the white area in the figure shown above) is equal to α.

Table III Values of $t^{\dagger}$

d.f.	$t_{.050}$	$t_{.025}$	$t_{.010}$	$t_{.005}$	d.f.
1	6.314	12.706	31.821	63.657	1
2	2.920	4.303	6.965	9.925	2
3	2.353	3.182	4.541	5.841	3
4	2.132	2.776	3.747	4.604	4
5	2.015	2.571	3.365	4.032	5
6	1.943	2.447	3.143	3.707	6
7	1.895	2.365	2.998	3.499	7
8	1.860	2.306	2.896	3.355	8
9	1.833	2.262	2.821	3.250	9
10	1.812	2.228	2.764	3.169	10
11	1.796	2.201	2.718	3.106	11
12	1.782	2.179	2.681	3.055	12
13	1.771	2.160	2.650	3.012	13
14	1.761	2.145	2.624	2.977	14
15	1.753	2.131	2.602	2.947	15
16	1.746	2.120	2.583	2.921	16
17	1.740	2.110	2.567	2.898	17
18	1.734	2.101	2.552	2.878	18
19	1.729	2.093	2.539	2.861	19
20	1.725	2.086	2.528	2.845	20
21	1.721	2.080	2.518	2.831	21
22	1.717	2.074	2.508	2.819	22
23	1.714	2.069	2.500	2.807	23
24	1.711	2.064	2.492	2.797	24
25	1.708	2.060	2.485	2.787	25
26	1.706	2.056	2.479	2.779	26
27	1.703	2.052	2.473	2.771	27
28	1.701	2.048	2.467	2.763	28
29	1.699	2.045	2.462	2.756	29
inf.	1.645	1.960	2.326	2.576	inf.

† Richard A. Johnson and Dean W. Wichern, *Applied Multivariate Statistical Analysis*, © 1982, p. 582. Adapted by permission of Prentice-Hall, Inc., Englewood Cliffs, N.J.

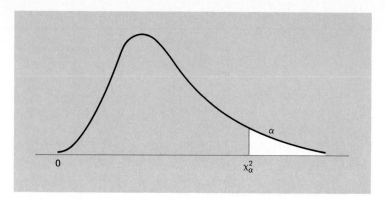

The entries in Table IV are values for which the area to their right under the chi-square distribution with given degrees of freedom (the white area in the figure shown above) is equal to α.

Table IV   Values of χ^2 [†]

d.f.	$\chi^2_{.05}$	$\chi^2_{.01}$	d.f.
1	3.841	6.635	1
2	5.991	9.210	2
3	7.815	11.345	3
4	9.488	13.277	4
5	11.070	15.086	5
6	12.592	16.812	6
7	14.067	18.475	7
8	15.507	20.090	8
9	16.919	21.666	9
10	18.307	23.209	10
11	19.675	24.725	11
12	21.026	26.217	12
13	22.362	27.688	13
14	23.685	29.141	14
15	24.996	30.578	15
16	26.296	32.000	16
17	27.587	33.409	17
18	28.869	34.805	18
19	30.144	36.191	19
20	31.410	37.566	20
21	32.671	38.932	21
22	33.924	40.289	22
23	35.172	41.638	23
24	36.415	42.980	24
25	37.652	44.314	25
26	38.885	45.642	26
27	40.113	46.963	27
28	41.337	48.278	28
29	42.557	49.588	29
30	43.773	50.892	30

[†] Based on Table 8 of *Biometrika Tables for Statisticians, Volume I* (Cambridge: Cambridge University Press, 1954) by permission of the *Biometrika* trustees.

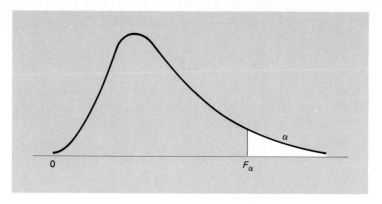

The entries in Table V are values for which the area to their right under the F distribution with given degrees of freedom (the white area in the figure shown above) is equal to α.

Table V   Values of $F_{0.05}$[†]

Degrees of freedom for numerator

$\nu_2 \backslash \nu_1$	1	2	3	4	5	6	7	8	9	10	12	15	20	24	30	40	60	120	∞
1	161	200	216	225	230	234	237	239	241	242	244	246	248	249	250	251	252	253	254
2	18.5	19.0	19.2	19.2	19.3	19.3	19.4	19.4	19.4	19.4	19.4	19.4	19.4	19.5	19.5	19.5	19.5	19.5	19.5
3	10.1	9.55	9.28	9.12	9.01	8.94	8.89	8.85	8.81	8.79	8.74	8.70	8.66	8.64	8.62	8.59	8.57	8.55	8.53
4	7.71	6.94	6.59	6.39	6.26	6.16	6.09	6.04	6.00	5.96	5.91	5.86	5.80	5.77	5.75	5.72	5.69	5.66	5.63
5	6.61	5.79	5.41	5.19	5.05	4.95	4.88	4.82	4.77	4.74	4.68	4.62	4.56	4.53	4.50	4.46	4.43	4.40	4.37
6	5.99	5.14	4.76	4.53	4.39	4.28	4.21	4.15	4.10	4.06	4.00	3.94	3.87	3.84	3.81	3.77	3.74	3.70	3.67
7	5.59	4.74	4.35	4.12	3.97	3.87	3.79	3.73	3.68	3.64	3.57	3.51	3.44	3.41	3.38	3.34	3.30	3.27	3.23
8	5.32	4.46	4.07	3.84	3.69	3.58	3.50	3.44	3.39	3.35	3.28	3.22	3.15	3.12	3.08	3.04	3.01	2.97	2.93
9	5.12	4.26	3.86	3.63	3.48	3.37	3.29	3.23	3.18	3.14	3.07	3.01	2.94	2.90	2.86	2.83	2.79	2.75	2.71
10	4.96	4.10	3.71	3.48	3.33	3.22	3.14	3.07	3.02	2.98	2.91	2.85	2.77	2.74	2.70	2.66	2.62	2.58	2.54
11	4.84	3.98	3.59	3.36	3.20	3.09	3.01	2.95	2.90	2.85	2.79	2.72	2.65	2.61	2.57	2.53	2.49	2.45	2.40
12	4.75	3.89	3.49	3.26	3.11	3.00	2.91	2.85	2.80	2.75	2.69	2.62	2.54	2.51	2.47	2.43	2.38	2.34	2.30
13	4.67	3.81	3.41	3.18	3.03	2.92	2.83	2.77	2.71	2.67	2.60	2.53	2.46	2.42	2.38	2.34	2.30	2.25	2.21
14	4.60	3.74	3.34	3.11	2.96	2.85	2.76	2.70	2.65	2.60	2.53	2.46	2.39	2.35	2.31	2.27	2.22	2.18	2.13
15	4.54	3.68	3.29	3.06	2.90	2.79	2.71	2.64	2.59	2.54	2.48	2.40	2.33	2.29	2.25	2.20	2.16	2.11	2.07
16	4.49	3.63	3.24	3.01	2.85	2.74	2.66	2.59	2.54	2.49	2.42	2.35	2.28	2.24	2.19	2.15	2.11	2.06	2.01
17	4.45	3.59	3.20	2.96	2.81	2.70	2.61	2.55	2.49	2.45	2.38	2.31	2.23	2.29	2.15	2.10	2.06	2.01	1.96
18	4.41	3.55	3.16	2.93	2.77	2.66	2.58	2.51	2.46	2.41	2.34	2.27	2.19	2.15	2.11	2.06	2.02	1.97	1.92
19	4.38	3.52	3.13	2.90	2.74	2.63	2.54	2.48	2.42	2.38	2.31	2.23	2.16	2.11	2.07	2.03	1.98	1.93	1.88
20	4.35	3.49	3.10	2.87	2.71	2.60	2.51	2.45	2.39	2.35	2.28	2.20	2.12	2.08	2.04	1.99	1.95	1.90	1.84
21	4.32	3.47	3.07	2.84	2.68	2.57	2.49	2.42	2.37	2.32	2.25	2.18	2.10	2.05	2.01	1.96	1.92	1.87	1.81
22	4.30	3.44	3.05	2.82	2.66	2.55	2.46	2.40	2.34	2.30	2.23	2.15	2.07	2.03	1.98	1.94	1.89	1.84	1.78
23	4.28	3.42	3.03	2.80	2.64	2.53	2.44	2.37	2.32	2.27	2.20	2.13	2.05	2.01	1.96	1.91	1.86	1.81	1.76
24	4.26	3.40	3.01	2.78	2.62	2.51	2.42	2.36	2.30	2.25	2.18	2.11	2.03	1.98	1.94	1.89	1.84	1.79	1.73
25	4.24	3.39	2.99	2.76	2.60	2.49	2.40	2.34	2.28	2.24	2.16	2.09	2.01	1.96	1.92	1.87	1.82	1.77	1.71
30	4.17	3.32	2.92	2.69	2.53	2.42	2.33	2.27	2.21	2.16	2.09	2.01	1.93	1.89	1.84	1.79	1.74	1.68	1.62
40	4.08	3.23	2.84	2.61	2.45	2.34	2.25	2.18	2.12	2.08	2.00	1.92	1.84	1.79	1.74	1.69	1.64	1.58	1.51
60	4.00	3.15	2.76	2.53	2.37	2.25	2.17	2.10	2.04	1.99	1.92	1.84	1.75	1.70	1.65	1.59	1.53	1.47	1.39
120	3.92	3.07	2.68	2.45	2.29	2.18	2.09	2.02	1.96	1.91	1.83	1.75	1.66	1.61	1.55	1.50	1.43	1.35	1.25
∞	3.84	3.00	2.60	2.37	2.21	2.10	2.01	1.94	1.88	1.83	1.75	1.67	1.57	1.52	1.46	1.39	1.32	1.22	1.00

Degrees of freedom of denominator

[†] Reproduced from M. Merrington and C. M. Thompson, "Tables of percentage points of the inverted beta (F) distribution," *Biometrika*, vol. 33 (1943), by permission of the *Biometrika* trustees.

Table V Values of $F_{0.01}$ [†]

Degrees of freedom for numerator

	1	2	3	4	5	6	7	8	9	10	12	15	20	24	30	40	60	120	∞
1	4,052	5,000	5,403	5,625	5,764	5,859	5,928	5,982	6,023	6,056	6,106	6,157	6,209	6,235	6,261	6,287	6,313	6,339	6,366
2	98.5	99.0	99.2	99.2	99.3	99.3	99.4	99.4	99.4	99.4	99.4	99.4	99.4	99.5	99.5	99.5	99.5	99.5	99.5
3	34.1	30.8	29.5	28.7	28.2	27.9	27.7	27.5	27.3	27.2	27.1	26.9	26.7	26.6	26.5	26.4	26.3	26.2	26.1
4	21.2	18.0	16.7	16.0	15.5	15.2	15.0	14.8	14.7	14.5	14.4	14.2	14.0	13.9	13.8	13.7	13.7	13.6	13.5
5	16.3	13.3	12.1	11.4	11.0	10.7	10.5	10.3	10.2	10.1	9.89	9.72	9.55	9.47	9.38	9.29	9.20	9.11	9.02
6	13.7	10.9	9.78	9.15	8.75	8.47	8.26	8.10	7.98	7.87	7.72	7.56	7.40	7.31	7.23	7.14	7.05	6.97	6.88
7	12.2	9.55	8.45	7.85	7.46	7.19	6.99	6.84	6.72	6.62	6.47	6.31	6.16	6.07	5.99	5.91	5.82	5.74	5.65
8	11.3	8.65	7.59	7.01	6.63	6.37	6.18	6.03	5.91	5.81	5.67	5.52	5.36	5.28	5.20	5.12	5.03	4.95	4.86
9	10.6	8.02	6.99	6.42	6.06	5.80	5.61	5.47	5.35	5.26	5.11	4.96	4.81	4.73	4.65	4.57	4.48	4.40	4.31
10	10.0	7.56	6.55	5.99	5.64	5.39	5.20	5.06	4.94	4.85	4.71	4.56	4.41	4.33	4.25	4.17	4.08	4.00	3.91
11	9.65	7.21	6.22	5.67	5.32	5.07	4.89	4.74	4.63	4.54	4.40	4.25	4.10	4.02	3.94	3.86	3.78	3.69	3.60
12	9.33	6.93	5.95	5.41	5.06	4.82	4.64	4.50	4.39	4.30	4.16	4.01	3.86	3.78	3.70	3.62	3.54	3.45	3.36
13	9.07	6.70	5.74	5.21	4.86	4.62	4.44	4.30	4.19	4.10	3.96	3.82	3.66	3.59	3.51	3.43	3.34	3.25	3.17
14	8.86	6.51	5.56	5.04	4.70	4.46	4.28	4.14	4.03	3.94	3.80	3.66	3.51	3.43	3.35	3.27	3.18	3.09	3.00
15	8.68	6.36	5.42	4.89	4.56	4.32	4.14	4.00	3.89	3.80	3.67	3.52	3.37	3.29	3.21	3.13	3.05	2.96	2.87
16	8.53	6.23	5.29	4.77	4.44	4.20	4.03	3.89	3.78	3.69	3.55	3.41	3.26	3.18	3.10	3.02	2.93	2.84	2.75
17	8.40	6.11	5.19	4.67	4.34	4.10	3.93	3.79	3.68	3.59	3.46	3.31	3.16	3.08	3.00	2.92	2.83	2.75	2.65
18	8.29	6.01	5.09	4.58	4.25	4.01	3.84	3.71	3.60	3.51	3.37	3.23	3.08	3.00	2.92	2.84	2.75	2.66	2.57
19	8.19	5.93	5.01	4.50	4.17	3.94	3.77	3.63	3.52	3.43	3.30	3.15	3.00	2.92	2.84	2.76	2.67	2.58	2.49
20	8.10	5.85	4.94	4.43	4.10	3.87	3.70	3.56	3.46	3.37	3.23	3.09	2.94	2.86	2.78	2.69	2.61	2.52	2.42
21	8.02	5.78	4.87	4.37	4.04	3.81	3.64	3.51	3.40	3.31	3.17	3.03	2.88	2.80	2.72	2.64	2.55	2.46	2.36
22	7.95	5.72	4.82	4.31	3.99	3.76	3.59	3.45	3.35	3.26	3.12	2.98	2.83	2.75	2.67	2.58	2.50	2.40	2.31
23	7.88	5.66	4.76	4.26	3.94	3.71	3.54	3.41	3.30	3.21	3.07	2.93	2.78	2.70	2.62	2.54	2.45	2.35	2.26
24	7.82	5.61	4.72	4.22	3.90	3.67	3.50	3.36	3.26	3.17	3.03	2.89	2.74	2.66	2.58	2.49	2.40	2.31	2.21
25	7.77	5.57	4.68	4.18	3.86	3.63	3.46	3.32	3.22	3.13	2.99	2.85	2.70	2.62	2.53	2.45	2.36	2.27	2.17
30	7.56	5.39	4.51	4.02	3.70	3.47	3.30	3.17	3.07	2.98	2.84	2.70	2.55	2.47	2.39	2.30	2.21	2.11	2.01
40	7.31	5.18	4.31	3.83	3.51	3.29	3.12	2.99	2.89	2.80	2.66	2.52	2.37	2.29	2.20	2.11	2.02	1.92	1.80
60	7.08	4.98	4.13	3.65	3.34	3.12	2.95	2.82	2.72	2.63	2.50	2.35	2.20	2.12	2.03	1.94	1.84	1.73	1.60
120	6.85	4.79	3.95	3.48	3.17	2.96	2.79	2.66	2.56	2.47	2.34	2.19	2.03	1.95	1.86	1.76	1.66	1.53	1.38
∞	6.63	4.61	3.78	3.32	3.02	2.80	2.64	2.51	2.41	2.32	2.18	2.04	1.88	1.79	1.70	1.59	1.47	1.32	1.00

Degrees of freedom for denominator

[†] Reproduced from M. Merrington and C. M. Thompson, "Tables of percentage points of the inverted beta (F) distribution," *Biometrika*, vol. 33 (1943), by permission of the *Biometrika* trustees.

n	$r_{.025}$	$r_{.005}$	n	$r_{.025}$	$r_{.005}$
3	0.997		18	0.468	0.590
4	0.950	0.999	19	0.456	0.575
5	0.878	0.959	20	0.444	0.561
6	0.811	0.917	21	0.433	0.549
7	0.754	0.875	22	0.423	0.537
8	0.707	0.834	27	0.381	0.487
9	0.666	0.798	32	0.349	0.449
10	0.632	0.765	37	0.325	0.418
11	0.602	0.735	42	0.304	0.393
12	0.576	0.708	47	0.288	0.372
13	0.553	0.684	52	0.273	0.354
14	0.532	0.661	62	0.250	0.325
15	0.514	0.641	72	0.232	0.302
16	0.497	0.623	82	0.217	0.283
17	0.482	0.606	92	0.205	0.267

Table VII   Critical values of U[†]

Values of $U'_{0.05}$

n_1 \ n_2	2	3	4	5	6	7	8	9	10	11	12	13	14	15
2							0	0	0	0	1	1	1	1
3				0	1	1	2	2	3	3	4	4	5	5
4			0	1	2	3	4	4	5	6	7	8	9	10
5		0	1	2	3	5	6	7	8	9	11	12	13	14
6		1	2	3	5	6	8	10	11	13	14	16	17	19
7		1	3	5	6	8	10	12	14	16	18	20	22	24
8	0	2	4	6	8	10	13	15	17	19	22	24	26	29
9	0	2	4	7	10	12	15	17	20	23	26	28	31	34
10	0	3	5	8	11	14	17	20	23	26	29	30	36	39
11	0	3	6	9	13	16	19	23	26	30	33	37	40	44
12	1	4	7	11	14	18	22	26	29	33	37	41	45	49
13	1	4	8	12	16	20	24	28	30	37	41	45	50	54
14	1	5	9	13	17	22	26	31	36	40	45	50	55	59
15	1	5	10	14	19	24	29	34	39	44	49	54	59	64

Values of $U'_{0.01}$

n_1 \ n_2	3	4	5	6	7	8	9	10	11	12	13	14	15
3							0	0	0	1	1	1	2
4				0	0	1	1	2	2	3	3	4	5
5			0	1	1	2	3	4	5	6	7	7	8
6		0	1	2	3	4	5	6	7	9	10	11	12
7		0	1	3	4	6	7	9	10	12	13	15	16
8		1	2	4	6	7	9	11	13	15	17	18	20
9	0	1	3	5	7	9	11	13	16	18	20	22	24
10	0	2	4	6	9	11	13	16	18	21	24	26	29
11	0	2	5	7	10	13	16	18	21	24	27	30	33
12	1	3	6	9	12	15	18	21	24	27	31	34	37
13	1	3	7	10	13	17	20	24	27	31	34	38	42
14	1	4	7	11	15	18	22	26	30	34	38	42	46
15	2	5	8	12	16	20	24	29	33	37	42	46	51

[†] This table is based on Table 11.4 of Donald B. Owen, *Handbook of Statistical Tables*, © 1962. Addison-Wesley, Reading, Massachusetts. Reprinted with permission.

Table **VIII**   *Critical values of u[†]*

Values of $u_{0.025}$

n_1 \ n_2	4	5	6	7	8	9	10	11	12	13	14	15
4		9	9									
5	9	10	10	11	11							
6	9	10	11	12	12	13	13	13	13			
7		11	12	13	13	14	14	14	14	15	15	15
8		11	12	13	14	14	15	15	16	16	16	16
9			13	14	14	15	16	16	16	17	17	18
10			13	14	15	16	16	17	17	18	18	18
11			13	14	15	16	17	17	18	19	19	19
12			13	14	16	16	17	18	19	19	20	20
13				15	16	17	18	19	19	20	20	21
14				15	16	17	18	19	20	20	21	22
15				15	16	18	18	19	20	21	22	22

Values of $u'_{0.025}$

n_1 \ n_2	2	3	4	5	6	7	8	9	10	11	12	13	14	15
2											2	2	2	2
3					2	2	2	2	2	2	2	2	2	3
4				2	2	2	3	3	3	3	3	3	3	3
5			2	2	3	3	3	3	3	4	4	4	4	4
6		2	2	3	3	3	3	4	4	4	4	5	5	5
7		2	2	3	3	3	4	4	5	5	5	5	5	6
8		2	3	3	3	4	4	5	5	5	6	6	6	6
9		2	3	3	4	4	5	5	5	6	6	6	7	7
10		2	3	3	4	5	5	5	6	6	7	7	7	7
11		2	3	4	4	5	5	6	6	7	7	7	8	8
12	2	2	3	4	4	5	6	6	7	7	7	8	8	8
13	2	2	3	4	5	5	6	6	7	7	8	8	9	9
14	2	2	3	4	5	5	6	7	7	8	8	9	9	9
15	2	3	3	4	5	6	6	7	7	8	8	9	9	10

[†] This table is adapted, by permission, from F. S. Swed and C. Eisenhart, "Tables for testing randomness of grouping in a sequence of alternatives," *Annals of Mathematical of Statistics*, Vol. 14.

Table VIII   Critical values of u (continued)

Values of $u_{0.005}$

n_1 \ n_2	5	6	7	8	9	10	11	12	13	14	15
5		11									
6	11	12	13	13							
7		13	13	14	15	15	15				
8		13	14	15	15	16	16	17	17	17	
9			15	15	16	17	17	18	18	18	19
10			15	16	17	17	18	19	19	19	20
11			15	16	17	18	19	19	20	20	21
12				17	18	19	19	20	21	21	22
13				17	18	19	20	21	21	22	22
14				17	18	19	20	21	22	23	23
15					19	20	21	22	22	23	24

Values of $u'_{0.005}$

n_1 \ n_2	3	4	5	6	7	8	9	10	11	12	13	14	15
3										2	2	2	2
4						2	2	2	2	2	2	2	3
5				2	2	2	2	3	3	3	3	3	3
6			2	2	2	3	3	3	3	3	3	4	4
7			2	2	3	3	3	3	4	4	4	4	4
8		2	2	3	3	3	3	4	4	4	5	5	5
9		2	2	3	3	3	4	4	5	5	5	5	6
10		2	3	3	3	4	4	5	5	5	5	6	6
11		2	3	3	4	4	5	5	5	6	6	6	7
12	2	2	3	3	4	4	5	5	6	6	6	7	7
13	2	2	3	3	4	5	5	5	6	6	7	7	7
14	2	2	3	4	4	5	5	6	6	7	7	7	8
15	2	3	3	4	4	5	6	6	7	7	7	8	8

Table IX   Factorials

n	n!
0	1
1	1
2	2
3	6
4	24
5	120
6	720
7	5,040
8	40,320
9	362,880
10	3,628,800
11	39,916,800
12	479,001,600
13	6,227,020,800
14	87,178,291,200
15	1,307,674,368,000

Table X   Binomial coeffcients

n	$\binom{n}{0}$	$\binom{n}{1}$	$\binom{n}{2}$	$\binom{n}{3}$	$\binom{n}{4}$	$\binom{n}{5}$	$\binom{n}{6}$	$\binom{n}{7}$	$\binom{n}{8}$	$\binom{n}{9}$	$\binom{n}{10}$
0	1										
1	1	1									
2	1	2	1								
3	1	3	3	1							
4	1	4	6	4	1						
5	1	5	10	10	5	1					
6	1	6	15	20	15	6	1				
7	1	7	21	35	35	21	7	1			
8	1	8	28	56	70	56	28	8	1		
9	1	9	36	84	126	126	84	36	9	1	
10	1	10	45	120	210	252	210	120	45	10	1
11	1	11	55	165	330	462	462	330	165	55	11
12	1	12	66	220	495	792	924	792	495	220	66
13	1	13	78	286	715	1287	1716	1716	1287	715	286
14	1	14	91	364	1001	2002	3003	3432	3003	2002	1001
15	1	15	105	455	1365	3003	5005	6435	6435	5005	3003
16	1	16	120	560	1820	4368	8008	11440	12870	11440	8008
17	1	17	136	680	2380	6188	12376	19448	24310	24310	19448
18	1	18	153	816	3060	8568	18564	31824	43758	48620	43758
19	1	19	171	969	3876	11628	27132	50388	75582	92378	92378
20	1	20	190	1140	4845	15504	38760	77520	125970	167960	184756

If necessary, use the identity $\binom{n}{r} = \binom{n}{n-r}$

Table XI   Values of e^{-x}

x	e^{-x}	x	e^{-x}	x	e^{-x}	x	e^{-x}
0.0	1.000	2.5	0.082	5.0	0.0067	7.5	0.00055
0.1	0.905	2.6	0.074	5.1	0.0061	7.6	0.00050
0.2	0.819	2.7	0.067	5.2	0.0055	7.7	0.00045
0.3	0.741	2.8	0.061	5.3	0.0050	7.8	0.00041
0.4	0.670	2.9	0.055	5.4	0.0045	7.9	0.00037
0.5	0.607	3.0	0.050	5.5	0.0041	8.0	0.00034
0.6	0.549	3.1	0.045	5.6	0.0037	8.1	0.00030
0.7	0.497	3.2	0.041	5.7	0.0033	8.2	0.00028
0.8	0.449	3.3	0.037	5.8	0.0030	8.3	0.00025
0.9	0.407	3.4	0.033	5.9	0.0027	8.4	0.00023
1.0	0.368	3.5	0.030	6.0	0.0025	8.5	0.00020
1.1	0.333	3.6	0.027	6.1	0.0022	8.6	0.00018
1.2	0.301	3.7	0.025	6.2	0.0020	8.7	0.00017
1.3	0.273	3.8	0.022	6.3	0.0018	8.8	0.00015
1.4	0.247	3.9	0.020	6.4	0.0017	8.9	0.00014
1.5	0.223	4.0	0.018	6.5	0.0015	9.0	0.00012
1.6	0.202	4.1	0.017	6.6	0.0014	9.1	0.00011
1.7	0.183	4.2	0.015	6.7	0.0012	9.2	0.00010
1.8	0.165	4.3	0.014	6.8	0.0011	9.3	0.00009
1.9	0.150	4.4	0.012	6.9	0.0010	9.4	0.00008
2.0	0.135	4.5	0.011	7.0	0.0009	9.5	0.00008
2.1	0.122	4.6	0.010	7.1	0.0008	9.6	0.00007
2.2	0.111	4.7	0.009	7.2	0.0007	9.7	0.00006
2.3	0.100	4.8	0.008	7.3	0.0007	9.8	0.00006
2.4	0.091	4.9	0.007	7.4	0.0006	9.9	0.00005

Table XII contains the square roots of the numbers from 1.00 to 9.99, and also the square roots of these numbers multiplied by 10, spaced at intervals of 0.01. The square roots are all rounded to four decimals. To find the square root of any positive number rounded to three significant digits, we use the following rule in deciding whether to take the entry of the $\sqrt{n}$ or $\sqrt{10n}$ column:

> **Move the decimal point an even number of places to the right or to the left until a number greater than or equal to 1 but less than 100 is reached. If the resulting number is less than 10 go to the $\sqrt{n}$ column; if it is 10 or more go to the $\sqrt{10n}$ column.**

Thus, to find the square roots of 12,800, 379, and 0.0812, we go to the $\sqrt{n}$ column since the decimal point has to be moved, respectively, four places to the left, two places to the left, and two places to the right, to give 1.28, 3.79, and 8.12. Similarly, to find the square roots of 5,240, 0.281, and 0.0000259, we go to the $\sqrt{10n}$ column since the decimal point has to be moved, respectively, two places to the left, two places to the right, and six places to the right, to give 52.4, 28.1, and 25.9.

After we locate a square root in the appropriate column of Table XII, we must be sure to get the decimal point in the right place. Here it will help to use the following rule:

> **Having previously moved the decimal point an even number of places to the left or right to get a number greater than or equal to 1 but less than 100, move the decimal point of the entry of the appropriate column in Table XII half as many places in the opposite direction.**

For example, to determine the square root of 12,800, we first note that the decimal point has to be moved *four places to the left* to give 1.28. We then take the entry of the $\sqrt{n}$ column corresponding to 1.28, move its decimal point *two places to the right*, and get $\sqrt{12,800} = 113.14$. Similarly, to find the square root of 0.0000259, we note that the decimal point has to be moved *six places to the right* to give 25.9. We thus take the entry of the $\sqrt{10n}$ column corresponding to 2.59, move the decimal point *three places to the left*, and get $\sqrt{0.0000259} = 0.0050892$. In actual practice, if a number whose square root we want to find is rounded, the square root will have to be rounded to as many significant digits as the original number.

Table XII   Square roots

n	$\sqrt{n}$	$\sqrt{10n}$	n	$\sqrt{n}$	$\sqrt{10n}$	n	$\sqrt{n}$	$\sqrt{10n}$
1.00	1.0000	3.1623	1.50	1.2247	3.8730	2.00	1.4142	4.4721
1.01	1.0050	3.1780	1.51	1.2288	3.8859	2.01	1.4177	4.4833
1.02	1.0100	3.1937	1.52	1.2329	3.8987	2.02	1.4213	4.4944
1.03	1.0149	3.2094	1.53	1.2369	3.9115	2.03	1.4248	4.5056
1.04	1.0198	3.2249	1.54	1.2410	3.9243	2.04	1.4283	4.5166
1.05	1.0247	3.2404	1.55	1.2450	3.9370	2.05	1.4318	4.5277
1.06	1.0296	3.2558	1.56	1.2490	3.9497	2.06	1.4353	4.5387
1.07	1.0344	3.2711	1.57	1.2530	3.9623	2.07	1.4387	4.5497
1.08	1.0392	3.2863	1.58	1.2570	3.9749	2.08	1.4422	4.5607
1.09	1.0440	3.3015	1.59	1.2610	3.9875	2.09	1.4457	4.5717
1.10	1.0488	3.3166	1.60	1.2649	4.0000	2.10	1.4491	4.5826
1.11	1.0536	3.3317	1.61	1.2689	4.0125	2.11	1.4526	4.5935
1.12	1.0583	3.3466	1.62	1.2728	4.0249	2.12	1.4560	4.6043
1.13	1.0630	3.3615	1.63	1.2767	4.0373	2.13	1.4595	4.6152
1.14	1.0677	3.3764	1.64	1.2806	4.0497	2.14	1.4629	4.6260
1.15	1.0724	3.3912	1.65	1.2845	4.0620	2.15	1.4663	4.6368
1.16	1.0770	3.4059	1.66	1.2884	4.0743	2.16	1.4697	4.6476
1.17	1.0817	3.4205	1.67	1.2923	4.0866	2.17	1.4731	4.6583
1.18	1.0863	3.4351	1.68	1.2961	4.0988	2.18	1.4765	4.6690
1.19	1.0909	3.4496	1.69	1.3000	4.1110	2.19	1.4799	4.6797
1.20	1.0954	3.4641	1.70	1.3038	4.1231	2.20	1.4832	4.6904
1.21	1.1000	3.4785	1.71	1.3077	4.1352	2.21	1.4866	4.7011
1.22	1.1045	3.4928	1.72	1.3115	4.1473	2.22	1.4900	4.7117
1.23	1.1091	3.5071	1.73	1.3153	4.1593	2.23	1.4933	4.7223
1.24	1.1136	3.5214	1.74	1.3191	4.1713	2.24	1.4967	4.7329
1.25	1.1180	3.5355	1.75	1.3229	4.1833	2.25	1.5000	4.7434
1.26	1.1225	3.5496	1.76	1.3266	4.1952	2.26	1.5033	4.7539
1.27	1.1269	3.5637	1.77	1.3304	4.2071	2.27	1.5067	4.7645
1.28	1.1314	3.5777	1.78	1.3342	4.2190	2.28	1.5100	4.7749
1.29	1.1358	3.5917	1.79	1.3379	4.2308	2.29	1.5133	4.7854
1.30	1.1402	3.6056	1.80	1.3416	4.2426	2.30	1.5166	4.7958
1.31	1.1446	3.6194	1.81	1.3454	4.2544	2.31	1.5199	4.8062
1.32	1.1489	3.6332	1.82	1.3491	4.2661	2.32	1.5232	4.8166
1.33	1.1533	3.6469	1.83	1.3528	4.2778	2.33	1.5264	4.8270
1.34	1.1576	3.6606	1.84	1.3565	4.2895	2.34	1.5297	4.8374
1.35	1.1619	3.6742	1.85	1.3601	4.3012	2.35	1.5330	4.8477
1.36	1.1662	3.6878	1.86	1.3638	4.3128	2.36	1.5362	4.8580
1.37	1.1705	3.7014	1.87	1.3675	4.3243	2.37	1.5395	4.8683
1.38	1.1747	3.7148	1.88	1.3711	4.3359	2.38	1.5427	4.8785
1.39	1.1790	3.7283	1.89	1.3748	4.3474	2.39	1.5460	4.8888
1.40	1.1832	3.7417	1.90	1.3784	4.3589	2.40	1.5492	4.8990
1.41	1.1874	3.7550	1.91	1.3820	4.3704	2.41	1.5524	4.9092
1.42	1.1916	3.7683	1.92	1.3856	4.3818	2.42	1.5556	4.9193
1.43	1.1958	3.7815	1.93	1.3892	4.3932	2.43	1.5588	4.9295
1.44	1.2000	3.7947	1.94	1.3928	4.4045	2.44	1.5620	4.9396
1.45	1.2042	3.8079	1.95	1.3964	4.4159	2.45	1.5652	4.9497
1.46	1.2083	3.8210	1.96	1.4000	4.4272	2.46	1.5684	4.9598
1.47	1.2124	3.8341	1.97	1.4036	4.4385	2.47	1.5716	4.9699
1.48	1.2166	3.8471	1.98	1.4071	4.4497	2.48	1.5748	4.9800
1.49	1.2207	3.8601	1.99	1.4107	4.4609	2.49	1.5780	4.9900

Table XII   *Square roots (continued)*

n	$\sqrt{n}$	$\sqrt{10n}$	n	$\sqrt{n}$	$\sqrt{10n}$	n	$\sqrt{n}$	$\sqrt{10n}$
2.50	1.5811	5.0000	3.00	1.7321	5.4772	3.50	1.8708	5.9161
2.51	1.5843	5.0100	3.01	1.7349	5.4863	3.51	1.8735	5.9245
2.52	1.5875	5.0200	3.02	1.7378	5.4955	3.52	1.8762	5.9330
2.53	1.5906	5.0299	3.03	1.7407	5.5045	3.53	1.8788	5.9414
2.54	1.5937	5.0398	3.04	1.7436	5.5136	3.54	1.8815	5.9498
2.55	1.5969	5.0498	3.05	1.7464	5.5227	3.55	1.8841	5.9582
2.56	1.6000	5.0596	3.06	1.7493	5.5317	3.56	1.8868	5.9666
2.57	1.6031	5.0695	3.07	1.7521	5.5408	3.57	1.8894	5.9749
2.58	1.6062	5.0794	3.08	1.7550	5.5498	3.58	1.8921	5.9833
2.59	1.6093	5.0892	3.09	1.7578	5.5588	3.59	1.8947	5.9917
2.60	1.6125	5.0990	3.10	1.7607	5.5678	3.60	1.8974	6.0000
2.61	1.6155	5.1088	3.11	1.7635	5.5767	3.61	1.9000	6.0083
2.62	1.6186	5.1186	3.12	1.7664	5.5857	3.62	1.9026	6.0166
2.63	1.6217	5.1284	3.13	1.7692	5.5946	3.63	1.9053	6.0249
2.64	1.6248	5.1381	3.14	1.7720	5.6036	3.64	1.9079	6.0332
2.65	1.6279	5.1478	3.15	1.7748	5.6125	3.65	1.9105	6.0415
2.66	1.6310	5.1575	3.16	1.7776	5.6214	3.66	1.9131	6.0498
2.67	1.6340	5.1672	3.17	1.7804	5.6303	3.67	1.9157	6.0581
2.68	1.6371	5.1769	3.18	1.7833	5.6391	3.68	1.9183	6.0663
2.69	1.6401	5.1865	3.19	1.7861	5.6480	3.69	1.9209	6.0745
2.70	1.6432	5.1962	3.20	1.7889	5.6569	3.70	1.9235	6.0828
2.71	1.6462	5.2058	3.21	1.7916	5.6657	3.71	1.9261	6.0910
2.72	1.6492	5.2154	3.22	1.7944	5.6745	3.72	1.9287	6.0992
2.73	1.6523	5.2249	3.23	1.7972	5.6833	3.73	1.9313	6.1074
2.74	1.6553	5.2345	3.24	1.8000	5.6921	3.74	1.9339	6.1156
2.75	1.6583	5.2440	3.25	1.8028	5.7009	3.75	1.9365	6.1237
2.76	1.6613	5.2536	3.26	1.8055	5.7096	3.76	1.9391	6.1319
2.77	1.6643	5.2631	3.27	1.8083	5.7184	3.77	1.9416	6.1400
2.78	1.6673	5.2726	3.28	1.8111	5.7271	3.78	1.9442	6.1482
2.79	1.6703	5.2820	3.29	1.8138	5.7359	3.79	1.9468	6.1563
2.80	1.6733	5.2915	3.30	1.8166	5.7446	3.80	1.9494	6.1644
2.81	1.6763	5.3009	3.31	1.8193	5.7533	3.81	1.9519	6.1725
2.82	1.6793	5.3104	3.32	1.8221	5.7619	3.82	1.9545	6.1806
2.83	1.6823	5.3198	3.33	1.8248	5.7706	3.83	1.9570	6.1887
2.84	1.6852	5.3292	3.34	1.8276	5.7793	3.84	1.9596	6.1968
2.85	1.6882	5.3385	3.35	1.8303	5.7879	3.85	1.9621	6.2048
2.86	1.6912	5.3479	3.36	1.8330	5.7966	3.86	1.9647	6.2129
2.87	1.6941	5.3572	3.37	1.8358	5.8052	3.87	1.9672	6.2209
2.88	1.6971	5.3666	3.38	1.8385	5.8138	3.88	1.9698	6.2290
2.89	1.7000	5.3759	3.39	1.8412	5.8224	3.89	1.9723	6.2370
2.90	1.7029	5.3852	3.40	1.8439	5.8310	3.90	1.9748	6.2450
2.91	1.7059	5.3944	3.41	1.8466	5.8395	3.91	1.9774	6.2530
2.92	1.7088	5.4037	3.42	1.8493	5.8481	3.92	1.9799	6.2610
2.93	1.7117	5.4129	3.43	1.8520	5.8566	3.93	1.9824	6.2690
2.94	1.7146	5.4222	3.44	1.8547	5.8652	3.94	1.9849	6.2769
2.95	1.7176	5.4314	3.45	1.8574	5.8737	3.95	1.9875	6.2849
2.96	1.7205	5.4406	3.46	1.8601	5.8822	3.96	1.9900	6.2929
2.97	1.7234	5.4498	3.47	1.8628	5.8907	3.97	1.9925	6.3008
2.98	1.7263	5.4589	3.48	1.8655	5.8992	3.98	1.9950	6.3087
2.99	1.7292	5.4681	3.49	1.8682	5.9076	3.99	1.9975	6.3166

Table XII   Square roots (continued)

n	$\sqrt{n}$	$\sqrt{10n}$	n	$\sqrt{n}$	$\sqrt{10n}$	n	$\sqrt{n}$	$\sqrt{10n}$
4.00	2.0000	6.3246	4.50	2.1213	6.7082	5.00	2.2361	7.0711
4.01	2.0025	6.3325	4.51	2.1237	6.7157	5.01	2.2383	7.0781
4.02	2.0050	6.3403	4.52	2.1260	6.7231	5.02	2.2405	7.0852
4.03	2.0075	6.3482	4.53	2.1284	6.7305	5.03	2.2428	7.0922
4.04	2.0100	6.3561	4.54	2.1307	6.7380	5.04	2.2450	7.0993
4.05	2.0125	6.3640	4.55	2.1331	6.7454	5.05	2.2472	7.1063
4.06	2.0149	6.3718	4.56	2.1354	6.7528	5.06	2.2494	7.1134
4.07	2.0174	6.3797	4.57	2.1378	6.7602	5.07	2.2517	7.1204
4.08	2.0199	6.3875	4.58	2.1401	6.7676	5.08	2.2539	7.1274
4.09	2.0224	6.3953	4.59	2.1424	6.7750	5.09	2.2561	7.1344
4.10	2.0248	6.4031	4.60	2.1448	6.7823	5.10	2.2583	7.1414
4.11	2.0273	6.4109	4.61	2.1471	6.7897	5.11	2.2605	7.1484
4.12	2.0298	6.4187	4.62	2.1494	6.7971	5.12	2.2627	7.1554
4.13	2.0322	6.4265	4.63	2.1517	6.8044	5.13	2.2650	7.1624
4.14	2.0347	6.4343	4.64	2.1541	6.8118	5.14	2.2672	7.1694
4.15	2.0372	6.4420	4.65	2.1564	6.8191	5.15	2.2694	7.1764
4.16	2.0396	6.4498	4.66	2.1587	6.8264	5.16	2.2716	7.1833
4.17	2.0421	6.4576	4.67	2.1610	6.8337	5.17	2.2738	7.1903
4.18	2.0445	6.4653	4.68	2.1633	6.8411	5.18	2.2760	7.1972
4.19	2.0469	6.4730	4.69	2.1656	6.8484	5.19	2.2782	7.2042
4.20	2.0494	6.4807	4.70	2.1679	6.8557	5.20	2.2804	7.2111
4.21	2.0518	6.4885	4.71	2.1703	6.8629	5.21	2.2825	7.2180
4.22	2.0543	6.4962	4.72	2.1726	6.8702	5.22	2.2847	7.2250
4.23	2.0567	6.5038	4.73	2.1749	6.8775	5.23	2.2869	7.2319
4.24	2.0591	6.5115	4.74	2.1772	6.8848	5.24	2.2891	7.2388
4.25	2.0616	6.5192	4.75	2.1794	6.8920	5.25	2.2913	7.2457
4.26	2.0640	6.5269	4.76	2.1817	6.8993	5.26	2.2935	7.2526
4.27	2.0664	6.5345	4.77	2.1840	6.9065	5.27	2.2956	7.2595
4.28	2.0688	6.5422	4.78	2.1863	6.9138	5.28	2.2978	7.2664
4.29	2.0712	6.5498	4.79	2.1886	6.9210	5.29	2.3000	7.2732
4.30	2.0736	6.5574	4.80	2.1909	6.9282	5.30	2.3022	7.2801
4.31	2.0761	6.5651	4.81	2.1932	6.9354	5.31	2.3043	7.2870
4.32	2.0785	6.5727	4.82	2.1954	6.9426	5.32	2.3065	7.2938
4.33	2.0809	6.5803	4.83	2.1977	6.9498	5.33	2.3087	7.3007
4.34	2.0833	6.5879	4.84	2.2000	6.9570	5.34	2.3108	7.3075
4.35	2.0857	6.5955	4.85	2.2023	6.9642	5.35	2.3130	7.3144
4.36	2.0881	6.6030	4.86	2.2045	6.9714	5.36	2.3152	7.3212
4.37	2.0905	6.6106	4.87	2.2068	6.9785	5.37	2.3173	7.3280
4.38	2.0928	6.6182	4.88	2.2091	6.9857	5.38	2.3195	7.3348
4.39	2.0952	6.6257	4.89	2.2113	6.9929	5.39	2.3216	7.3417
4.40	2.0976	6.6332	4.90	2.2136	7.0000	5.40	2.3238	7.3485
4.41	2.1000	6.6408	4.91	2.2159	7.0071	5.41	2.3259	7.3553
4.42	2.1024	6.6483	4.92	2.2181	7.0143	5.42	2.3281	7.3621
4.43	2.1048	6.6558	4.93	2.2204	7.0214	5.43	2.3302	7.3689
4.44	2.1071	6.6633	4.94	2.2226	7.0285	5.44	2.3324	7.3756
4.45	2.1095	6.6708	4.95	2.2249	7.0356	5.45	2.3345	7.3824
4.46	2.1119	6.6783	4.96	2.2271	7.0427	5.46	2.3367	7.3892
4.47	2.1142	6.6858	4.97	2.2293	7.0498	5.47	2.3388	7.3959
4.48	2.1166	6.6933	4.98	2.2316	7.0569	5.48	2.3409	7.4027
4.49	2.1190	6.7007	4.99	2.2338	7.0640	5.49	2.3431	7.4095

Table XII   Square roots (continued)

n	$\sqrt{n}$	$\sqrt{10n}$	n	$\sqrt{n}$	$\sqrt{10n}$	n	$\sqrt{n}$	$\sqrt{10n}$
5.50	2.3452	7.4162	6.00	2.4495	7.7460	6.50	2.5495	8.0623
5.51	2.3473	7.4229	6.01	2.4515	7.7524	6.51	2.5515	8.0685
5.52	2.3495	7.4297	6.02	2.4536	7.7589	6.52	2.5534	8.0747
5.53	2.3516	7.4364	6.03	2.4556	7.7653	6.53	2.5554	8.0808
5.54	2.3537	7.4431	6.04	2.4576	7.7717	6.54	2.5573	8.0870
5.55	2.3558	7.4498	6.05	2.4597	7.7782	6.55	2.5593	8.0932
5.56	2.3580	7.4565	6.06	2.4617	7.7846	6.56	2.5612	8.0994
5.57	2.3601	7.4632	6.07	2.4637	7.7910	6.57	2.5632	8.1056
5.58	2.3622	7.4699	6.08	2.4658	7.7974	6.58	2.5652	8.1117
5.59	2.3643	7.4766	6.09	2.4678	7.8038	6.59	2.5671	8.1179
5.60	2.3664	7.4833	6.10	2.4698	7.8102	6.60	2.5690	8.1240
5.61	2.3685	7.4900	6.11	2.4718	7.8166	6.61	2.5710	8.1302
5.62	2.3707	7.4967	6.12	2.4739	7.8230	6.62	2.5729	8.1363
5.63	2.3728	7.5033	6.13	2.4759	7.8294	6.63	2.5749	8.1425
5.64	2.3749	7.5100	6.14	2.4779	7.8358	6.64	2.5768	8.1486
5.65	2.3770	7.5166	6.15	2.4799	7.8422	6.65	2.5788	8.1548
5.66	2.3791	7.5233	6.16	2.4819	7.8486	6.66	2.5807	8.1609
5.67	2.3812	7.5299	6.17	2.4839	7.8549	6.67	2.5826	8.1670
5.68	2.3833	7.5366	6.18	2.4860	7.8613	6.68	2.5846	8.1731
5.69	2.3854	7.5432	6.19	2.4880	7.8677	6.69	2.5865	8.1792
5.70	2.3875	7.5498	6.20	2.4900	7.8740	6.70	2.5884	8.1854
5.71	2.3896	7.5565	6.21	2.4920	7.8804	6.71	2.5904	8.1915
5.72	2.3917	7.5631	6.22	2.4940	7.8867	6.72	2.5923	8.1976
5.73	2.3937	7.5697	6.23	2.4960	7.8930	6.73	2.5942	8.2037
5.74	2.3958	7.5763	6.24	2.4980	7.8994	6.74	2.5962	8.2098
5.75	2.3979	7.5829	6.25	2.5000	7.9057	6.75	2.5981	8.2158
5.76	2.4000	7.5895	6.26	2.5020	7.9120	6.76	2.6000	8.2219
5.77	2.4021	7.5961	6.27	2.5040	7.9183	6.77	2.6019	8.2280
5.78	2.4042	7.6026	6.28	2.5060	7.9246	6.78	2.6038	8.2341
5.79	2.4062	7.6092	6.29	2.5080	7.9310	6.79	2.6058	8.2401
5.80	2.4083	7.6158	6.30	2.5100	7.9373	6.80	2.6077	8.2462
5.81	2.4104	7.6223	6.31	2.5120	7.9436	6.81	2.6096	8.2523
5.82	2.4125	7.6289	6.32	2.5140	7.9498	6.82	2.6115	8.2583
5.83	2.4145	7.6354	6.33	2.5159	7.9561	6.83	2.6134	8.2644
5.84	2.4166	7.6420	6.34	2.5179	7.9624	6.84	2.6153	8.2704
5.85	2.4187	7.6485	6.35	2.5199	7.9687	6.85	2.6173	8.2765
5.86	2.4207	7.6551	6.36	2.5219	7.9750	6.86	2.6192	8.2825
5.87	2.4228	7.6616	6.37	2.5239	7.9812	6.87	2.6211	8.2885
5.88	2.4249	7.6681	6.38	2.5259	7.9875	6.88	2.6230	8.2946
5.89	2.4269	7.6746	6.39	2.5278	7.9937	6.89	2.6249	8.3006
5.90	2.4290	7.6811	6.40	2.5298	8.0000	6.90	2.6268	8.3066
5.91	2.4310	7.6877	6.41	2.5318	8.0062	6.91	2.6287	8.3126
5.92	2.4331	7.6942	6.42	2.5338	8.0125	6.92	2.6306	8.3187
5.93	2.4352	7.7006	6.43	2.5357	8.0187	6.93	2.6325	8.3247
5.94	2.4372	7.7071	6.44	2.5377	8.0250	6.94	2.6344	8.3307
5.95	2.4393	7.7136	6.45	2.5397	8.0312	6.95	2.6363	8.3367
5.96	2.4413	7.7201	6.46	2.5417	8.0374	6.96	2.6382	8.3427
5.97	2.4434	7.7266	6.47	2.5436	8.0436	6.97	2.6401	8.3487
5.98	2.4454	7.7330	6.48	2.5456	8.0498	6.98	2.6420	8.3546
5.99	2.4474	7.7395	6.49	2.5475	8.0561	6.99	2.6439	8.3606

Table XII   Square roots (continued)

n	$\sqrt{n}$	$\sqrt{10n}$	n	$\sqrt{n}$	$\sqrt{10n}$	n	$\sqrt{n}$	$\sqrt{10n}$
7.00	2.6458	8.3666	7.50	2.7386	8.6603	8.00	2.8284	8.9443
7.01	2.6476	8.3726	7.51	2.7404	8.6660	8.01	2.8302	8.9499
7.02	2.6495	8.3785	7.52	2.7423	8.6718	8.02	2.8320	8.9554
7.03	2.6514	8.3845	7.53	2.7441	8.6776	8.03	2.8337	8.9610
7.04	2.6533	8.3905	7.54	2.7459	8.6833	8.04	2.8355	8.9666
7.05	2.6552	8.3964	7.55	2.7477	8.6891	8.05	2.8373	8.9722
7.06	2.6571	8.4024	7.56	2.7495	8.6948	8.06	2.8390	8.9778
7.07	2.6589	8.4083	7.57	2.7514	8.7006	8.07	2.8408	8.9833
7.08	2.6608	8.4143	7.58	2.7532	8.7063	8.08	2.8425	8.9889
7.09	2.6627	8.4202	7.59	2.7550	8.7121	8.09	2.8443	8.9944
7.10	2.6646	8.4261	7.60	2.7568	8.7178	8.10	2.8460	9.0000
7.11	2.6665	8.4321	7.61	2.7586	8.7235	8.11	2.8478	9.0056
7.12	2.6683	8.4380	7.62	2.7604	8.7293	8.12	2.8496	9.0111
7.13	2.6702	8.4439	7.63	2.7622	8.7350	8.13	2.8513	9.0167
7.14	2.6721	8.4499	7.64	2.7641	8.7407	8.14	2.8531	9.0222
7.15	2.6739	8.4558	7.65	2.7659	8.7464	8.15	2.8548	9.0277
7.16	2.6758	8.4617	7.66	2.7677	8.7521	8.16	2.8566	9.0333
7.17	2.6777	8.4676	7.67	2.7695	8.7579	8.17	2.8583	9.0388
7.18	2.6796	8.4735	7.68	2.7713	8.7636	8.18	2.8601	9.0443
7.19	2.6814	8.4794	7.69	2.7731	8.7693	8.19	2.8618	9.0499
7.20	2.6833	8.4853	7.70	2.7749	8.7750	8.20	2.8636	9.0554
7.21	2.6851	8.4912	7.71	2.7767	8.7807	8.21	2.8653	9.0609
7.22	2.6870	8.4971	7.72	2.7785	8.7864	8.22	2.8671	9.0664
7.23	2.6889	8.5029	7.73	2.7803	8.7920	8.23	2.8688	9.0719
7.24	2.6907	8.5088	7.74	2.7821	8.7977	8.24	2.8705	9.0774
7.25	2.6926	8.5147	7.75	2.7839	8.8034	8.25	2.8723	9.0830
7.26	2.6944	8.5206	7.76	2.7857	8.8091	8.26	2.8740	9.0885
7.27	2.6963	8.5264	7.77	2.7875	8.8148	8.27	2.8758	9.0940
7.28	2.6981	8.5323	7.78	2.7893	8.8204	8.28	2.8775	9.0995
7.29	2.7000	8.5381	7.79	2.7911	8.8261	8.29	2.8792	9.1049
7.30	2.7019	8.5440	7.80	2.7928	8.8318	8.30	2.8810	9.1104
7.31	2.7037	8.5499	7.81	2.7946	8.8374	8.31	2.8827	9.1159
7.32	2.7055	8.5557	7.82	2.7964	8.8431	8.32	2.8844	9.1214
7.33	2.7074	8.5615	7.83	2.7982	8.8487	8.33	2.8862	9.1269
7.34	2.7092	8.5674	7.84	2.8000	8.8544	8.34	2.8879	9.1324
7.35	2.7111	8.5732	7.85	2.8018	8.8600	8.35	2.8896	9.1378
7.36	2.7129	8.5790	7.86	2.8036	8.8657	8.36	2.8914	9.1433
7.37	2.7148	8.5849	7.87	2.8054	8.8713	8.37	2.8931	9.1488
7.38	2.7166	8.5907	7.88	2.8071	8.8769	8.38	2.8948	9.1542
7.39	2.7185	8.5965	7.89	2.8089	8.8826	8.39	2.8965	9.1597
7.40	2.7203	8.6023	7.90	2.8107	8.8882	8.40	2.8983	9.1652
7.41	2.7221	8.6081	7.91	2.8125	8.8938	8.41	2.9000	9.1706
7.42	2.7240	8.6139	7.92	2.8142	8.8994	8.42	2.9017	9.1761
7.43	2.7258	8.6197	7.93	2.8160	8.9051	8.43	2.9034	9.1815
7.44	2.7276	8.6255	7.94	2.8178	8.9107	8.44	2.9052	9.1869
7.45	2.7295	8.6313	7.95	2.8196	8.9163	8.45	2.9069	9.1924
7.46	2.7313	8.6371	7.96	2.8213	8.9219	8.46	2.9086	9.1978
7.47	2.7331	8.6429	7.97	2.8231	8.9275	8.47	2.9103	9.2033
7.48	2.7350	8.6487	7.98	2.8249	8.9331	8.48	2.9120	9.2087
7.49	2.7368	8.6545	7.99	2.8267	8.9387	8.49	2.9138	9.2141

Table XII   Square roots (continued)

n	$\sqrt{n}$	$\sqrt{10n}$	n	$\sqrt{n}$	$\sqrt{10n}$	n	$\sqrt{n}$	$\sqrt{10n}$
8.50	2.9155	9.2195	9.00	3.0000	9.4868	9.50	3.0822	9.7468
8.51	2.9172	9.2250	9.01	3.0017	9.4921	9.51	3.0838	9.7519
8.52	2.9189	9.2304	9.02	3.0033	9.4974	9.52	3.0854	9.7570
8.53	2.9206	9.2358	9.03	3.0050	9.5026	9.53	3.0871	9.7622
8.54	2.9223	9.2412	9.04	3.0067	9.5079	9.54	3.0887	9.7673
8.55	2.9240	9.2466	9.05	3.0083	9.5131	9.55	3.0903	9.7724
8.56	2.9257	9.2520	9.06	3.0100	9.5184	9.56	3.0919	9.7775
8.57	2.9275	9.2574	9.07	3.0116	9.5237	9.57	3.0935	9.7826
8.58	2.9292	9.2628	9.08	3.0133	9.5289	9.58	3.0952	9.7877
8.59	2.9309	9.2682	9.09	3.0150	9.5341	9.59	3.0968	9.7929
8.60	2.9326	9.2736	9.10	3.0166	9.5394	9.60	3.0984	9.7980
8.61	2.9343	9.2790	9.11	3.0183	9.5446	9.61	3.1000	9.8031
8.62	2.9360	9.2844	9.12	3.0199	9.5499	9.62	3.1016	9.8082
8.63	2.9377	9.2898	9.13	3.0216	9.5551	9.63	3.1032	9.8133
8.64	2.9394	9.2952	9.14	3.0232	9.5603	9.64	3.1048	9.8184
8.65	2.9411	9.3005	9.15	3.0249	9.5656	9.65	3.1064	9.8234
8.66	2.9428	9.3059	9.16	3.0265	9.5708	9.66	3.1081	9.8285
8.67	2.9445	9.3113	9.17	3.0282	9.5760	9.67	3.1097	9.8336
8.68	2.9462	9.3167	9.18	3.0299	9.5812	9.68	3.1113	9.8387
8.69	2.9479	9.3220	9.19	3.0315	9.5864	9.69	3.1129	9.8438
8.70	2.9496	9.3274	9.20	3.0332	9.5917	9.70	3.1145	9.8489
8.71	2.9513	9.3327	9.21	3.0348	9.5969	9.71	3.1161	9.8539
8.72	2.9530	9.3381	9.22	3.0364	9.6021	9.72	3.1177	9.8590
8.73	2.9547	9.3434	9.23	3.0381	9.6073	9.73	3.1193	9.8641
8.74	2.9563	9.3488	9.24	3.0397	9.6125	9.74	3.1209	9.8691
8.75	2.9580	9.3541	9.25	3.0414	9.6177	9.75	3.1225	9.8742
8.76	2.9597	9.3595	9.26	3.0430	9.6229	9.76	3.1241	9.8793
8.77	2.9614	9.3648	9.27	3.0447	9.6281	9.77	3.1257	9.8843
8.78	2.9631	9.3702	9.28	3.0463	9.6333	9.78	3.1273	9.8894
8.79	2.9648	9.3755	9.29	3.0480	9.6385	9.79	3.1289	9.8944
8.80	2.9665	9.3808	9.30	3.0496	9.6437	9.80	3.1305	9.8995
8.81	2.9682	9.3862	9.31	3.0512	9.6488	9.81	3.1321	9.9045
8.82	2.9698	9.3915	9.32	3.0529	9.6540	9.82	3.1337	9.9096
8.83	2.9715	9.3968	9.33	3.0545	9.6592	9.83	3.1353	9.9146
8.84	2.9732	9.4021	9.34	3.0561	9.6644	9.84	3.1369	9.9197
8.85	2.9749	9.4074	9.35	3.0578	9.6695	9.85	3.1385	9.9247
8.86	2.9766	9.4128	9.36	3.0594	9.6747	9.86	3.1401	9.9298
8.87	2.9783	9.4181	9.37	3.0610	9.6799	9.87	3.1417	9.9348
8.88	2.9799	9.4234	9.38	3.0627	9.6850	9.88	3.1432	9.9398
8.89	2.9816	9.4287	9.39	3.0643	9.6902	9.89	3.1448	9.9448
8.90	2.9833	9.4340	9.40	3.0659	9.6954	9.90	3.1464	9.9499
8.91	2.9850	9.4393	9.41	3.0676	9.7005	9.91	3.1480	9.9549
8.92	2.9866	9.4446	9.42	3.0692	9.7057	9.92	3.1496	9.9599
8.93	2.9883	9.4499	9.43	3.0708	9.7108	9.93	3.1512	9.9649
8.94	2.9900	9.4552	9.44	3.0725	9.7160	9.94	3.1528	9.9700
8.95	2.9917	9.4604	9.45	3.0741	9.7211	9.95	3.1544	9.9750
8.96	2.9933	9.4657	9.46	3.0757	9.7263	9.96	3.1559	9.9800
8.97	2.9950	9.4710	9.47	3.0773	9.7314	9.97	3.1575	9.9850
8.98	2.9967	9.4763	9.48	3.0790	9.7365	9.98	3.1591	9.9900
8.99	2.9983	9.4816	9.49	3.0806	9.7417	9.99	3.1607	9.9950

Table XIII   Random numbers[†]

04433	80674	24520	18222	10610	05794	37515
60298	47829	72648	37414	75755	04717	29899
67884	59651	67533	68123	17730	95862	08034
89512	32155	51906	61662	64130	16688	37275
32653	01895	12506	88535	36553	23757	34209
95913	15405	13772	76638	48423	25018	99041
55864	21694	13122	44115	01601	50541	00147
35334	49810	91601	40617	72876	33967	73830
57729	32196	76487	11622	96297	24160	09903
86648	13697	63677	70119	94739	25875	38829
30574	47609	07967	32422	76791	39725	53711
81307	43694	83580	79974	45929	85113	72268
02410	54905	79007	54939	21410	86980	91772
18969	75274	52233	62319	08598	09066	95288
87863	82384	66860	62297	80198	19347	73234
68397	71708	15438	62311	72844	60203	46412
28529	54447	58729	10854	99058	18260	38765
44285	06372	15867	70418	57012	72122	36634
86299	83430	33571	23309	57040	29285	67870
84842	68668	90894	61658	15001	94055	36308
56970	83609	52098	04184	54967	72938	56834
83125	71257	60490	44369	66130	72936	69848
55503	52423	02464	26141	68779	66388	75242
47019	76273	33203	29608	54553	25971	69573
84828	32592	79526	29554	84580	37859	28504
68921	08141	79227	05748	51276	57143	31926
36458	96045	30424	98420	72925	40729	22337
95752	59445	36847	87729	81679	59126	59437
26768	47323	58454	56958	20575	76746	49878
42613	37056	43636	58085	06766	60227	96414
95457	30566	65482	25596	02678	54592	63607
95276	17894	63564	95958	39750	64379	46059
66954	52324	64776	92345	95110	59448	77249
17457	18481	14113	62462	02798	54977	48349
03704	36872	83214	59337	01695	60666	97410
21538	86497	33210	60337	27976	70661	08250
57178	67619	98310	70348	11317	71623	55510
31048	97558	94953	55866	96283	46620	52087
69799	55380	16498	80733	96422	58078	99643
90595	61867	59231	17772	67831	33317	00520
33570	04981	98939	78784	09977	29398	93896
15340	93460	57477	13898	48431	72936	78160
64079	42483	36512	56186	99098	48850	72527
63491	05546	67118	62063	74958	20946	28147
92003	63868	41034	28260	79708	00770	88643
52360	46658	66511	04172	73085	11795	52594
74622	12142	68355	65635	21828	39539	18988
04157	50079	61343	64315	70836	82857	35335
86003	60070	66241	32836	27573	11479	94114
41268	80187	20351	09636	84668	42486	71303

[†] Based on parts of *Tables of 105,000 Random Decimal Digits*. Interstate Commerce Commission, Bureau of Transport Economics and Statistics, Washington D.C.

Table XIII   Random numbers (continued)

48611	62866	33963	14045	79451	04934	45576
78812	03509	78673	73181	29973	18664	04555
19472	63971	37271	31445	49019	49405	46925
51266	11569	08697	91120	64156	40365	74297
55806	96275	26130	47949	14877	69594	83041
77527	81360	18180	97421	55541	90275	18213
77680	58788	33016	61173	93049	04694	43534
15404	96554	88265	34537	38526	67924	40474
14045	22917	60718	66487	46346	30949	03173
68376	43918	77653	04127	69930	43283	35766
93385	13421	67957	20384	58731	53396	59723
09858	52104	32014	53115	03727	98624	84616
93307	34116	49516	42148	57740	31198	70336
04794	01534	92058	03157	91758	80611	45357
86265	49096	97021	92582	61422	75890	86442
65943	79232	45702	67055	39024	57383	44424
90038	94209	04055	27393	61517	23002	96560
97283	95943	78363	36498	40662	94188	18202
21913	72958	75637	99936	58715	07943	23748
41161	37341	81838	19389	80336	46346	91895
23777	98392	31417	98547	92058	02277	50315
59973	08144	61070	73094	27059	69181	55623
82690	74099	77885	23813	10054	11900	44653
83854	24715	48866	65745	31131	47636	45137
61980	34997	41825	11623	07320	15003	56774
99915	45821	97702	87125	44488	77613	56823
48293	86847	43186	42951	37804	85129	28993
33225	31280	41232	34750	91097	60752	69783
06846	32828	24425	30249	78801	26977	92074
32671	45587	79620	84831	38156	74211	82752
82096	21913	75544	55228	89796	05694	91552
51666	10433	10945	55306	78562	89630	41230
54044	67942	24145	42294	27427	84875	37022
66738	60184	75679	38120	17640	36242	99357
55064	17427	89180	74018	44865	53197	74810
69599	60264	84549	78007	88450	06488	72274
64756	87759	92354	78694	63638	80939	98644
80817	74533	68407	55862	32476	19326	95558
39847	96884	84657	33697	39578	90197	80532
90401	41700	95510	61166	33757	23279	85523
78227	90110	81378	96659	37008	04050	04228
87240	52716	87697	79433	16336	52862	69149
08486	10951	26832	39763	02485	71688	90936
39338	32169	03713	93510	61244	73774	01245
21188	01850	69689	49426	49128	14660	14143
13287	82531	04388	64693	11934	35051	68576
53609	04001	19648	14053	49623	10840	31915
87900	36194	31567	53506	34304	39910	79630
81641	00496	36058	75899	46620	70024	88753
19512	50277	71508	20116	79520	06269	74173

Table XIII   *Random numbers (continued)*

24418	23508	91507	76455	54941	72711	39406
57404	73678	08272	62941	02349	71389	45605
77644	98489	86268	73652	98210	44546	27174
68366	65614	01443	07607	11826	91326	29664
64472	72294	95432	53555	96810	17100	35066
88205	37913	98633	81009	81060	33449	68055
98455	78685	71250	10329	56135	80647	51404
48977	36794	56054	59243	57361	65304	93258
93077	72941	92779	23581	24548	56415	61927
84533	26564	91583	83411	66504	02036	02922
11338	12903	14514	27585	45068	05520	56321
23853	68500	92274	87026	99717	01542	72990
94096	74920	25822	98026	05394	61840	83089
83160	82362	09350	98536	38155	42661	02363
97425	47335	69709	01386	74319	04318	99387
83951	11954	24317	20345	18134	90062	10761
93085	35203	05740	03206	92012	42710	34650
33762	83193	58045	89880	78101	44392	53767
49665	85397	85137	30496	23469	42846	94810
37541	82627	80051	72521	35342	56119	97190
22145	85304	35348	82854	55846	18076	12415
27153	08662	61078	52433	22184	33998	87436
00301	49425	66682	25442	83668	66236	79655
43815	43272	73778	63469	50083	70696	13558
14689	86482	74157	46012	97765	27552	49617
16680	55936	82453	19532	49988	13176	94219
86938	60429	01137	86168	78257	86249	46134
33944	29219	73161	46061	30946	22210	79302
16045	67736	18608	18198	19468	76358	69203
37044	52523	25627	63107	30806	80857	84383
61471	45322	35340	35132	42163	69332	98851
47422	21296	16785	66393	39249	51463	95963
24133	39719	14484	58613	88717	29289	77360
67253	67064	10748	16006	16767	57345	42285
62382	76941	01635	35829	77516	98468	51686
98011	16503	09201	03523	87192	66483	55649
37366	24386	20654	85117	74078	64120	04643
73587	83993	54176	05221	94119	20108	78101
33583	68291	50547	96085	62180	27453	18567
02878	33223	39199	49536	56199	05993	71201
91498	41673	17195	33175	04994	09879	70337
91127	19815	30219	55591	21725	43827	78862
12997	55013	18662	81724	24305	37661	18956
96098	13651	15393	69995	14762	69734	89150
97627	17837	10472	18983	28387	99781	52977
40064	47981	31484	76603	54088	91095	00010
16239	68743	71374	55863	22672	91609	51514
58354	24913	20435	30965	17453	65623	93058
52567	65085	60220	84641	18273	49604	47418
06236	29052	91392	07551	83532	68130	56970

Table XIII   *Random numbers (continued)*

94620	27963	96478	21559	19246	88097	44926
60947	60775	73181	43264	56895	04232	59604
27499	53523	63110	57106	20865	91683	80688
01603	23156	89223	43429	95353	44662	59433
00815	01552	06392	31437	70385	45863	75971
83844	90942	74857	52419	68723	47830	63010
06626	10042	93629	37609	57215	08409	81906
56760	63348	24949	11859	29793	37457	59377
64416	29934	00755	09418	14230	62887	92683
63569	17906	38076	32135	19096	96970	75917
22693	35089	72994	04252	23791	60249	83010
43413	59744	01275	71326	91382	45114	20245
09224	78530	50566	49965	04851	18280	14039
67625	34683	03142	74733	63558	09665	22610
86874	12549	98699	54952	91579	26023	81076
54548	49505	62515	63903	13193	33905	66936
73236	66167	49728	03581	40699	10396	81827
15220	66319	13543	14071	59148	95154	72852
16151	08029	36954	03891	38313	34016	18671
43635	84249	88984	80993	55431	90793	62603
30193	42776	85611	57635	51362	79907	77364
37430	45246	11400	20986	43996	73122	88474
88312	93047	12088	86937	70794	01041	74867
98995	58159	04700	90443	13168	31553	67891
51734	20849	70198	67906	00880	82899	66065
88698	41755	56216	66852	17748	04963	54859
51865	09836	73966	65711	41699	11732	17173
40300	08852	27528	84648	79589	95295	72895
02760	28625	70476	76410	32988	10194	94917
78450	26245	91763	73117	33047	03577	62599
50252	56911	62693	73817	98693	18728	94741
07929	66728	47761	81472	44806	15592	71357
09030	39605	87507	85446	51257	89555	75520
56670	88445	85799	76200	21795	38894	58070
48140	13583	94911	13318	64741	64336	95103
36764	86132	12463	28385	94242	32063	45233
14351	71381	28133	68269	65145	28152	39087
81276	00835	63835	87174	42446	08882	27067
55524	86088	00069	59254	24654	77371	26409
78852	65889	32719	13758	23937	90740	16866
11861	69032	51915	23510	32050	52052	24004
67699	01009	07050	73324	06732	27510	33761
50064	39500	17450	18030	63124	48061	59412
93126	17700	94400	76075	08317	27324	72723
01657	92602	41043	05686	15650	29970	95877
13800	76690	75133	60456	28491	03845	11507
98135	42870	48578	29036	69876	86563	61729
08313	99293	00990	13595	77457	79969	11339
90974	83965	62732	85161	54330	22406	86253
33273	61993	88407	69399	17301	70975	99129

Answers to Odd-Numbered Exercises

Chapter 1

1.3 (a) Generalization; (b) description; (c) description; (d) generalization.

1.5 (a) Since $\dfrac{9 + 8 + 13}{3} = \dfrac{30}{3} = 10$, the conclusion is obtained by purely descriptive methods.

(b) The conclusion does not necessarily follow from the data; it requires a generalization.

1.7 (a) Convenience foods are more likely to be used by persons who work, and hence are not home during weekday mornings.

(b) Persons coming out of the building are more likely to favor the political party having its headquarters there.

(c) Expenses on a luxury cruise are not typical of what the average person spends on his or her vacation.

1.9 The conclusion is not valid, of course. If an elevator goes up and down more or less continuously in a very tall building, it is more likely to be above the third floor than below it.

Chapter 2

2.3 One possibility is $200.00–219.99, 220.00–239.99, 240.00–259.99, 260.00–279.99, 280.00–299.99, and 300.00–319.99.

2.5 (a) No; (b) yes; (c) no; (d) yes.

2.9 (a) 0, 20, 40, 60, 80, 100, 120, and 140;
(b) 19, 39, 59, 79, 99, 119, 139, and 159;

(c) 9.5, 29.5, 49.5, 69.5, 89.5, 109.5, 129.5, and 149.5;

(d) 20.

2.13 (a) 31.5, 40.5, 49.5, 58.5, 67.5, 76.5, 85.5, and 94.5;

(b) 32–40, 41–49, 50–58, 59–67, 68–76, 77–85, and 86–94.

2.15 There is no provision for 11 and 31. Also, there is an ambiguity because 23 can be put into the fourth class or the fifth.

2.17 (a) The percentages are 4, 6, 18, 26, 26, 14, and 6%.

(b) The cumulative "or less" percentages are 0, 4, 10, 28, 54, 80, 94, and 100%.

2.19 (a) The cumulative frequencies less than 20, less than 25, ..., and less than 55 are 0, 129, 350, 660, 823, 928, 990, and 1,000.

(b) The cumulative frequencies more than 19, more than 24, ..., and more than 54 are 1,000, 871, 650, 340, 177, 72, 10, and 0.

2.21 0, 1, 2, 3, 4, 5, or 6 students were absent 6, 14, 9, 6, 3, 1, and 1 times.

2.23 The frequencies corresponding to excellent, very good, good, fair, poor, and very poor are 3, 9, 20, 6, 1, and 1.

2.25 The cumulative "less than" frequencies are 0, 18, 80, 143, 186, and 200.

2.27 The cumulative "less than" percentages, rounded to one decimal, are 0.0, 2.9, 19.3, 54.3, 81.4, 93.6, 97.9, and 100.0.

2.29 The percentages corresponding to the six categories are 13.0, 16.7, 37.3, 8.0, 11.3, and 13.7. Multiplying these percentages by 3.6, we get central angles of 46.8, 60.1, 134.3, 28.8, 40.7, and 49.3 degrees.

2.31 The class frequencies are 3, 5, 7, 11, 15, 7, and 2.

2.35

16	9							
17	0	5						
18	6	1	3	7	7			
19	4	4	0	8	6	9	2	
20	5	4	7	3				
21	6	2	8					
22	3	6						

2.37 (a) 10, 12, 17, 15, 11, 11, and 18;

(b) 125, 123, 123, 120, and 122;

(c) 345, 318, 366, and 301;

(d) 1.50, 1.57, 1.52, 1.52, and 1.59.

2.39

5·	7	9						
6*	1	2	0	4	1	0		
6·	8	5	9	8	6	6	5	7	7
7*	3	4	0	1	2	3	2	
7·	6	8	6					
8*	1	3						
8·	5							

2.41

	x			
	26–30	*31–35*	*36–40*	*41–45*
21–25				1
26–30	3	2		
y *31–35*	1	5	1	1
36–40		2	4	3
41–45			1	

Chapter 3

3.3 (a) The data would constitute a population if the executives have to prepare a report on their expenses during the first six months of 1985.

(b) The data would constitute a sample if they are used to prepare budgets for future six-month periods.

3.5 3.0 pounds.

3.7 335.1 calories.

3.9 The mean is 2, but among the 15 criminals only three had two foreign-born grandparents. As it is used here, the term "average criminal" is too vague.

3.11 The total weight of the 36 vehicles is 166,680 pounds, which is less than the maximum load of 180,000 pounds. There is no real danger that the bridge might collapse.

3.13 $\bar{x} = 7.3$ hours. It is not a very useful average because it is inflated by the one large value.

3.15 $\bar{x} = 85$ degrees. It is absurd to speak of the mean as a "comfortable" 85 degrees, as it is extremely hot there in the summer and fairly cool in the winter.

3.19 $\bar{x}_w = 43.3\%$.

3.21 0.295.

3.23 $6,108.

3.27 (a) The median is the value of the 20th item.

(b) The median is the mean of the values of the 75th and 76th items.

3.29 The median is 5.

3.31 The median is 55.

3.33 The median is 2.

3.35 The median is 68.5.

3.41 The midranges are 29.8, 30.0, and 30.3, so that the manufacturers of car C can use the midrange to substantiate the claim that their car performed best in the test.

3.43 The Q_1, median, and Q_3 positions are 5, 9.5, and 14.

3.47 (a) $Q_1 = 129$ and $Q_3 = 150.5$; (b) $Q_1 = 129$ and $Q_3 = 150.5$.

3.51 The mode is 0.

3.53 The mode is 142 minutes.

3.55 Blue is the modal choice.

3.59 For stock A the range is $\frac{7}{8}$ and for stock B the range is $\frac{5}{8}$. Thus, stock B is less variable.

3.61 The range is 66.

3.63 (a) $s = 3.70$; (b) $s = 3.70$.

3.65 $s = 2.26$.

3.67 $s = 4.42$.

3.69 $\sigma = 35.4$.

3.73 $\bar{x} = 9.029$ and $s = 0.503$.

3.77 (a) At least 96%; (b) at least 98.44%.

3.79 (a) At least 88.89%; (b) at least 91.84%.

3.81 At least 88.89%. For a bell-shaped distribution it is about 99.7%.

3.83 It would be wiser to dispose of stock D, which, at that time, is selling at a higher level within its range.

3.85 (a) The student is in a relatively better position with respect to the first university.

(b) The student is in a relatively better position with respect to the second university.

3.87 The first student is relatively more consistent.

3.91 $\bar{x} = \$2,170.50$ and $s = \$246.30$.

3.95 The median is 31.92.

3.99 $Q_1 = 27.24$ and $Q_3 = 37.26$.

3.103 $P_{20} = \$1,980.15$ and $P_{80} = \$2,353.89$.

3.107 $SK = -0.28$.

3.109 The distribution is U-shaped because the number of H's is apt to stay ahead of the number of T's once it gets ahead, and vice versa.

3.111 (a) $x_1 + x_2 + x_3 + x_4 + x_5 + x_6$;

(b) $y_1 + y_2 + y_3 + y_4 + y_5$;

(c) $x_1 y_1 + x_2 y_2 + x_3 y_3$;

(d) $x_1 f_1 + x_2 f_2 + x_3 f_3 + x_4 f_4 + x_5 f_5 + x_6 f_6 + x_7 f_7 + x_8 f_8$;

(e) $x_3^2 + x_4^2 + x_5^2 + x_6^2 + x_7^2$;

(f) $(x_1 + y_1) + (x_2 + y_2) + (x_3 + y_3) + (x_4 + y_4)$.

3.115 (a) 27; (b) 35; (c) 137; (d) 587.

3.117 (a) 8, −1, and 0; (b) 4, 5, 2, and −4; (c) 7.

3.119 No.

R.1 (a) $\bar{x} = 16.6$; (b) 16.05.

R.3 (a) Description; (b) description; (c) generalization; (d) generalization.

R.5 (a) The data would constitute a population if the dean wants to determine the average number of failing grades given by faculty members in the academic year 1984–1985.

(b) The data would constitute a sample if the dean wants to predict how many failing grades the faculty members will give in future years.

R7 $\bar{x} = \$712.50$.

R.11

12	4					
13	5	0				
14	2	6	9	1		
15	1	3	5	8	9	6
16	5	2	2			
17	3	7				
18	2					
19						
20	4					

R.13 (a) $\bar{x} = 0.066$ and $s = 0.004$; (b) 0.0655 and 0.012.

R.15 (a) 25; (b) 151.

R.17 $SK = 1.98$.

R.19 (a) 18 and 17, so that bus A averaged more passengers per run;

(b) 4.64 and 2.65;

(c) 25.8% and 15.6%, so that the number of passengers is relatively more variable for bus A.

R.23 (a) 50, 100, 150, 200, 250, 300, and 350;

(b) 99, 149, 199, 249, 299, 349, and 399;

(c) 49.5, 99.5, 149.5, 199.5, 249.5, 299.5, 349.5, and 399.5;

(d) 74.5, 124.5, 174.5, 224.5, 274.5, 324.5, and 374.5;

(e) 50.

R.25 $s = 0.65$ second.

R.27 (a) 52, 14, and 5.89; (b) 2, 14, and 5.89; (c) 26, 7, and 2.94;

(d) If we add a constant to each value, the same constant is added to the mean, but the range and the standard deviation remain unchanged. If we multiply each value by a constant, the mean, the range, and the standard deviation are all multiplied by the same constant.

R.29 (a) 16; (b) 31.5.

R.31 (a) 2.46; (b) 1.17 and 3.86.

R.33 (a) $P_{35} = 1.74$ and $P_{65} = 3.18$.

R.35 The range is 12.

R.37 (a) The class frequencies are 7, 10, 16, 19, 5, 2, and 1.

(b) The cumulative frequencies less than 0, less than 5, less than 10, ... , and less than 35 are 0, 7, 17, 33, 52, 57, 59, and 60.

R.39 $\tilde{x} = 3.5$.

Chapter 4

4.3 In two cases he will be exactly $1 ahead.

4.5 In five of the nine outcomes, the two union officials will not be of the same sex.

4.9 In 24 different ways.

4.11 In ten different ways.

4.13 In 1,080 different ways.

4.15 In 32,768 different ways.

4.17 In 288 different ways.

4.21 (a) True; (b) false; (c) false.

4.25 (a) 210; (b) 343.

4.27 In 1,320 different ways.

4.29 In 30,240 different ways.

4.31 (a) 720; (b) 6; (c) 48; (d) 72.

4.35 In 330 different ways.

4.37 In 792 different ways.

4.39 (a) 120; (b) 90.

4.41 In 2,450 different ways.

4.43 In 2,362,080 different ways.

4.45 1, 6, 15, 20, 15, 6, and 1;   1, 7, 21, 35, 35, 21, 7, and 1;   1, 8, 28, 56, 70, 56, 28, 8, and 1.

4.49 1/4, 1/2, and 1/4.

4.51 (a) 9/25; (b) 9/25; (c) 13/25.

4.53 (a) 1/17; (b) 1/221; (c) 8/663.

4.55 (a) 35/46; (b) 21/92; (c) 1/92.

4.57 2/3.

4.59 0.78.

4.61 0.26.

4.69 2 cents.

4.71 $1.00.

4.73 1.92 times.

4.77 His assessment of the probability is 0.16.

4.81 Continuing the operation will maximize the company's expected profit.

Chapter 5

5.3 (a) $S = \{2, 3, 4, 5, 6, 7, 8, 9, 10, 11, 12\}$;

(b) $A' = \{5, 6, 11, 12\}$ is the event that we roll a 5, 6, 11, or 12; $A \cup B = \{2, 3, 4, 5, 6, 7, 8, 9, 10\}$ is the event that we roll a 10 or less; $A \cap B = \{4, 7, 8\}$ is the event that we roll a 4, 7, or 8.

5.5 (a) $M \cup N = \{Q_c, K_c, Q_d, K_d, Q_h, K_h, 10_s, J_s, Q_s, K_s\}$ is the event that we draw a queen, a king, or the 10 or jack of spades.

(b) $M \cap N = \{Q_s, K_s\}$ is the event that we draw the queen or king of spades.

(c) $M' = \{A_c, 2_c, \ldots, J_c, A_d, 2_d, \ldots, J_d, A_h, 2_h, \ldots, J_h, A_s, 2_s, \ldots, J_s\}$ is the event that we do not draw a queen or a king.

(d) $N' = \{A_c, 2_c, \ldots, K_c, A_d, 2_d, \ldots, K_d, A_h, 2_h, \ldots, K_h, A_s, 2_s, \ldots, 9_s\}$ is the event that we do not draw the 10, jack, queen, or king of spades.

(e) $M' \cup N' = \{A_c, 2_c, \ldots, K_c, A_d, 2_d, \ldots, K_d, A_h, 2_h, \ldots, K_h, A_s, 2_s, \ldots, J_s\}$ is the event that we do not draw the queen or king of spades.

(f) $M' \cap N' = \{A_c, 2_c, \ldots, J_c, A_d, 2_d, \ldots, J_d, A_h, 2_h, \ldots, J_h, A_s, 2_s, \ldots, 9_s\}$ is the event that we do not draw a queen, a king, or the 10 or jack of spades.

5.7 A and B are not mutually exclusive; A and C are mutually exclusive; B and C are not mutually exclusive.

5.9 T and U are not mutually exclusive; T and V are not mutually exclusive; T and W are mutually exclusive; U and V are mutually exclusive; U and W are mutually exclusive, V and W are not mutually exclusive.

5.11 (a) The events are not mutually exclusive, since a driver can speed through a red light.

(b) The events are mutually exclusive, since, by law, the President of the United States cannot be foreign-born.

(c) The events are mutually exclusive.

(d) The events are not mutually exclusive, since a baseball player can get a walk in one at bat and a home run in another at bat in the same game.

(e) The events are not mutually exclusive.

5.13 (a) Regions 1 and 2 together represent the event that a person vacationing in Southern California visits Disneyland.

(b) Regions 2 and 3 together represent the event that a person vacationing in Southern California visits Disneyland or Universal Studios, but not both.

(c) Regions 2 and 4 together represent the event that a person vacationing in Southern California does not visit Universal Studios.

5.15 (a) Regions 1 and 3 together represent the event that the burglar is found not guilty.

(b) Regions 1 and 4 together represent the event that the burglar is caught and found guilty or not caught and not found guilty.

(c) Regions 3 and 4 together represent the event that the burglar is not caught.

5.17 (a) Someone will bring beer, someone will bring pretzels, and someone will bring cheese.

(b) Someone will bring pretzels, someone will bring cheese, but no one will bring beer.

(c) Someone will bring cheese, but no one will bring beer and no one will bring pretzels.

(d) No one will bring beer, no one will bring pretzels, and no one will bring cheese.

(e) Someone will bring beer and someone will bring cheese.

(f) Someone will bring pretzels, but no one will bring beer.

(g) Someone will bring cheese.

(h) No one will bring cheese.

5.21 (a) $P(J')$; (b) $P(J \cap M)$; (c) $P(J' \cap M')$; (d) $P(J \cup M)$.

5.23 (a) A probability cannot exceed 1.

(b) The sum of the two probabilities should be 1.

(c) The third probability should equal the sum of the other two.

(d) The sum of the two probabilities should not exceed 1.

5.25 (a) 0.59; (b) 0.64; (c) 0.77; (d) 0.

5.27 (a) 0.82; (b) 0.73; (c) 0.27.

5.29 1. Since $0 \le s \le n$, division by n yields $0 \le \dfrac{s}{n} \le 1$.

2. If an event is certain to occur, $s = n$, and its probability is $\dfrac{s}{n} = \dfrac{n}{n} = 1$.

3. The respective probabilities are $\dfrac{s_1}{n}$, $\dfrac{s_2}{n}$, and $\dfrac{s_1 + s_2}{n}$, and the sum of the first two is equal to the third.

4. $\dfrac{s}{n} + \dfrac{n-s}{n} = \dfrac{n}{n} = 1$.

5.33 The odds cannot all be right, since they correspond to probabilities of 2/3, 1/5, and 1/10, whose sum does not equal 1.

5.35 The probabilities are consistent.

5.39 0.83.

5.41 (a) 0.60; (b) 0.25; (c) 0.34.

5.45 (a) 0.34; (b) 0.44; (c) 0.22.

5.49 0.09.

5.51 1/16.

5.55 (a) $P(Q \mid W)$; (b) $P(W' \mid Q)$; (c) $P(Q' \mid W')$.

5.57 (a) $P(W \mid E)$; (b) $P(E \mid H')$; (c) $P(W \mid H \cap E)$.

5.61 (a) 3/4; (b) 5/16; (c) 11/20; (d) 11/80; (e) 4/5; (f) 11/20.

5.63 (a) 5/12; (b) 3/23; (c) 3/20; (d) 4/11; (e) 19/42; (f) 8/23.

5.65 0.80.

5.67 0.85.

5.71 (a) 1/36; (b) 5/36.

5.73 (a) 64/225; (b) 8/29.

5.75 (a) 1/256; (b) 625/1,296.

5.77 1/55.

5.79 (a) 0.196; (b) 0.084.

5.83 0.59.

5.85 0.92.

5.87 0.14.

Review Exercises: Chapters 4 and 5

R.41 24.

R.45 1.63 dresses.

R.47 (a) $\{A, D\}$; (b) $\{C, E\}$; (c) $\{B\}$.

R.49 It is not worthwhile to pay 50 cents.

R.51 In 1,120 different ways.

R.53 (a) 3; (b) 2.

R.55 (a) 48; (b) 12; (c) 24.

R.57 (a) 0.57; (b) 0.64; (c) 0.36.

R.59 In 336 different ways.

R.61 0.64.

R.65 The odds are 16 to 9 that next year's inflation rate will not exceed this year's.

R.67 Region 1 represents the event that the school's football team is rated among the top twenty by both AP and UPI. Region 2 represents the event that the school's football team is rated among the top twenty by AP but not by UPI. Region 3 represents the event that the school's football team is rated among the top twenty by UPI but not by AP. Region 4 represents the event that the school's football team is rated among the top twenty by neither AP nor UPI.

R.69 0.68.

R.71 (a) 28; (b) 6; (c) 16.

R.73 (a) Region 5; (b) regions 1 and 2 together; (c) regions 3, 5, and 6 together; (d) regions 1, 3, 4, and 6 together.

R.75 (a) $\{1, 6\}$ is the event that the program will be rated terrible or excellent.
(b) $\{2, 3, 4, 5, 6\}$ is the event that the program will not be rated terrible.
(c) $\{4, 5\}$ is the event that the program will be rated good or very good.
(d) $\{2, 3\}$ is the event that the program will be rated poor or fair.

R.77 (a) 0.43; (b) 0.67; (c) 0.11; (d) 0.59.

R.79 0.25.

6.3 (a) Yes;   (b) no, since $f(4)$ is greater than 1;   (c) no, since $f(4)$ is negative.

6.9 0.057.

6.11 0.181.

6.13 (a) 0.164;   (b) 0.164.

6.15 (a) 0.943;   (b) 0.004;   (c) 0.497.

6.17 (a) 0.196;   (b) 0.059;   (c) 0.153;   (d) 0.304.

6.19 (a) 0.099;   (b) 0.545;   (c) 0.179.

6.21 (a) 0.9596;   (b) 0.3414.

6.25 0.117.

6.27 (a) 0.682;   (b) 0.576.

6.31 (a) No;   (b) yes;   (c) no.

6.33 0.206.

6.35 (a) No;   (b) yes;   (c) yes.

6.37 0.264.

6.39 0.879.

6.41 0.201.

6.43 (a) 0.449;   (b) 0.359;   (c) 0.144;   (d) 0.048.

6.45 0.0011.

6.47 (a) 0.5803;   (b) 0.2375;   (c) 0.7905.

6.51 0.117.

6.55 $\mu = 2$ and $\sigma = 1.37$.

6.57 $\sigma = 1.708$.

6.61 (a) $\sigma^2 = 1.25$;   (b) $\sigma^2 = 1.25$.

6.63 (a) $\sigma = 1.266$;   (b) $\sigma = 1.265$.

6.65 (a) $\mu = 450$ and $\sigma = 15$;   (b) $\mu = 67.5$ and $\sigma = 7.5$;
(c) $\mu = 226.8$ and $\sigma = 12.6$.

6.67 $\mu = 0.937$ and $\mu = 0.9375$.

6.71 (a) The probability is at least 63/64 that between 86 and 206 marriage licenses
will be issued.
(b) The probability is at least 0.99.

7.3 (a) 0.2823;   (b) 0.9938;   (c) 0.0606;   (d) 0.1161.

7.5 (a) 0.1974;   (b) 0.8320;   (c) 0.0099;   (d) 0.0873.

7.7 (a) $z = 0.53$ or $z = -0.53$;   (b) $z = -1.18$;   (c) $z = 1.83$;
(d) $z = 0.82$ or $z = -0.82$.

7.9 (a) 0.9332;   (b) 0.7734;   (c) 0.0987;   (d) 0.9198.

7.11 (a) 0.6826;   (b) 0.9544;   (c) 0.9974;   (d) 0.99994.

7.15 $\mu = 68.1$.

7.17 (a) 0.1587; (b) 0.0668.

7.19 (a) 0.3707; (b) 0.7314.

7.21 4.96 inches.

7.25 0.0661.

7.27 (a) 0.0080; (b) 0.5944.

7.29 (a) Not satisfied; (b) satisfied; (c) not satisfied.

7.31 0.1662; the error of the approximation is 0.0001.

7.33 0.8365.

7.35 0.1112.

7.37 (a) 0.2358; (b) 0.4908; (c) 0.9556.

7.39 (a) 0.0056; (b) 0.0034.

Chapter 8

8.3 (a) 28; (b) 66; (c) 190.

8.5 $\dfrac{1}{3,060}$.

8.7 a and b, a and c, a and d, a and e, a and f, b and c, b and d, b and e, b and f, c and d, c and e, c and f, d and e, d and f, and e and f.

8.9 TWA, American, Western, and Continental; TWA, American, Western, and PSA; TWA, American, Western, and Delta; TWA, American, Continental, and PSA; TWA, American, Continental, and Delta; TWA, American, PSA, and Delta; TWA, Western, Continental, and PSA; TWA, Western, Continental, and Delta; TWA, Western, PSA, and Delta; TWA, Continental, PSA, and Delta; American, Western, Continental, and PSA; American, Western, Continental, and Delta; American, Western, PSA, and Delta; American, Continental, PSA, and Delta; and Western, Continental, PSA, and Delta. (a) 1/15; (b) 2/3.

8.11 495, 661, 663, 080, 427, 452, 581, and 208.

8.13 042, 446, 458, 478, 084, 374, 451, 182, 096, 260, 339, and 103.

8.17 (a) $\mu = 5$ and $\sigma = \sqrt{5}$.
(b) 2 and 2, 2 and 4, 2 and 6, 2 and 8, 4 and 2, 4 and 4, 4 and 6, 4 and 8, 6 and 2, 6 and 4, 6 and 6, 6 and 8, 8 and 2, 8 and 4, 8 and 6, and 8 and 8; the means are 2, 3, 4, 5, 3, 4, 5, 6, 4, 5, 6, 7, 5, 6, 7, and 8.

(c)

$\bar{x}$	Probability
2	1/16
3	2/16
4	3/16
5	4/16
6	3/16
7	2/16
8	1/16

(d) $\mu_{\bar{x}} = 5$ and $\sigma_{\bar{x}} = \sqrt{\dfrac{5}{2}}$.

8.19 (a) It is divided by 1.5;   (b) it is multiplied by 2.5.

8.27 (a) 8/9;   (b) 0.9974.

8.29 (a) 0.9342;   (b) 0.9978.

8.31 (a) 0.9772;   (b) 0.1587.

8.35 $n = 625$.

Review Exercises: Chapters 6, 7, and 8

R.83 0.1008.

R.85 (a) 0.1262;   (b) 3156.

R.87 (a) 0.125;   (b) 0.24.

R.89 0.150; the value in the table is 0.153.

R.91 (a) 0.1379;   (b) 0.5255.

R.93 (a) 0.091;   (b) 0.262;   (c) 0.060.

R.95 (a) 1/3, 8/15, and 2/15;   (b) $\mu = 0.80$ and $\sigma = 0.65$;   (c) $\mu = 0.80$.

R.97 (a) The probability is at least 5/9;   (b) 0.8664.

R.99 0.7123.

R.101 (a) 0.396;   (b) 0.183.

R.103 (a) No, since the sum of the probabilities is less than 1.
(b) No, since $f(4)$ is negative.
(c) Yes, since the values are all on the interval from 0 to 1 and their sum is 1.

R.105 0.2514.

R.107 (a) Yes;   (b) no;   (c) no;   (d) yes.

R.109 (a) $z = 2.36$ or $z = -2.36$;   (b) $z = -0.52$;   (c) $z = 2.37$;
(d) $z = 0.22$ or $z = -0.22$.

R.111 (a) Yes;   (b) no;   (c) yes.

R.113 (a) 0.3644;   (b) 0.3631.

R.115 (a) Yes;   (b) yes.

R.117 0.192.

Chapter 9

9.3 $E = 17.75$.

9.5 $E = 1.26$ pounds.

9.7 $E = \$1.34$.

9.9 81.3%.

9.11 $n = 250$.

9.13 $n = 50$.

Answers to Odd-Numbered Exercises

9.15 $21.95 < \mu < 23.75$.

9.17 $\$91.81 < \mu < \$95.33$.

9.19 $\$24.76 < \mu < \$26.00$.

9.21 $62.35 < \mu < 65.45$.

9.23 $E = 0.003$ cm.

9.25 $5.7 < \mu < 12.3$.

9.27 $13.1 < \mu < 16.9$.

9.29 $14.19 < \mu < 14.41$.

9.33 $2.04 < \sigma < 3.24$.

9.35 $1.58 < \sigma < 2.21$.

9.37 We would commit a Type I error if we erroneously reject the hypothesis that the average noise level of the vacuum cleaner meets specifications. We would commit a Type II error if we erroneously accept the hypothesis that the average noise level of the vacuum cleaner meets specifications.

9.39 (a) She is testing the hypothesis that the method of computer-assisted instruction is not effective.

(b) She is testing the hypothesis that the method of computer-assisted instruction is effective.

9.45 (a) The manufacturer should use the alternative hypothesis $\mu < 20$ and make the modification only if the null hypothesis can be rejected.

(b) The manufacturer should use the alternative hypothesis $\mu > 20$ and make the modification unless the null hypothesis can be rejected.

9.47 $z = -2.47$; we can conclude that eighth graders from the given school district can be expected to average less than the norm of 81.7.

9.49 $z = 1.50$; the null hypothesis cannot be rejected.

9.53 $t = -1.98$; the data support the employer's claim.

9.55 $t = -5.07$; we conclude that the truck is not operating at an average of 11.5 miles per gallon with the gasoline.

9.57 $t = 3.08$; we conclude that on the average it takes more than 9.5 days to fill such orders.

9.59 The probability of getting a value of t less than or equal to that calculated for the given data is 0.0358.

9.63 $z = 2.38$; the difference between the two sample means is significant.

9.67 $t = -1.88$; the null hypothesis cannot be rejected.

9.69 $t = 1.11$; the difference between the means of the two samples is not significant.

9.71 $t = 2.20$; the difference between the means of the weights obtained with the two scales is not significant.

9.77 (a) 34, 5.2, and $F = 6.54$;

(b) The differences among the sample means are significant.

9.79

Source of variation	Degrees of freedom	Sum of squares	Mean square	F
Treatments	2	10	5	0.86
Error	12	70	5.83	
Total	14	80		

The differences among the three sample means can be attributed to chance.

9.81

Source of variation	Degrees of freedom	Sum of squares	Mean square	F
Treatments	4	111.04	27.76	0.84
Error	20	660.00	33.00	
Total	24	771.04		

The differences among the five sample means can be attributed to chance.

9.85

Source of variation	Degrees of freedom	Sum of squares	Mean square	F
Treatments	2	78.55	39.28	2.97
Error	12	158.78	13.23	
Total	14	237.33		

The differences among the three sample means are not significant.

Chapter 10

10.3 $E = 0.137$.

10.5 $0.245 < p < 0.295$.

10.7 $0.625 < p < 0.775$.

10.9 $0.341 < p < 0.559$.

10.13 $n = 4{,}145$.

10.15 $n = 267$.

10.17 $n = 2,653$.

10.21 The probability of 9 or fewer successes is 0.061; the claim cannot be rejected.

10.23 The probability of 6 or fewer successes is 0.094 and that of 6 or more successes is 0.966; the null hypothesis cannot be rejected.

10.27 $z = 1.49$; the null hypothesis cannot be rejected.

10.29 $z = 1.85$; the null hypothesis cannot be rejected.

10.33 $z = -1.21$; the null hypothesis cannot be rejected.

10.35 $z = -2.78$; the difference between the two sample proportions is significant.

10.37 $\chi^2 = 6.55$; the differences among the sample proportions are significant.

10.39 $\chi^2 = 8.96$; the differences among the sample proportions are significant.

10.43 $\chi^2 = 7.71$; the difference between the two sample proportions is significant. Also, $(-2.78)^2 = 7.73$, and the difference is due to rounding.

10.45 $\chi^2 = 20.13$; we conclude that the therapy is effective.

10.47 $\chi^2 = 25.5$; the null hypothesis must be rejected.

10.49 $\chi^2 = 52.8$; we conclude that fidelity is not independent of selectivity.

10.57 $\chi^2 = 4.30$; the null hypothesis that the coins are balanced and randomly tossed cannot be rejected.

10.59 $\chi^2 = 4.56$; the hypothesis that the number of calls received by the switchboard in a five-minute interval is a random variable having the Poisson distribution with $\lambda = 1.5$ cannot be rejected.

Review Exercises: Chapters 9 and 10

R.119 $t = 1.84$; the difference between the two sample means is not significant.

R.121 $n = 1,068$.

R.123 $\chi^2 = 10.125$; the null hypothesis that the coins are balanced and randomly tossed cannot be rejected.

R.125 $n = 718$.

R.127 $\chi^2 = 42.87$; the null hypothesis must be rejected.

R.129 $t = 5.66$; the difference between $\bar{x} = 14.4$ and $\mu = 14.0$ is significant.

R.131 $18.5 < \sigma < 29.8$.

R.133 $\chi^2 = 4.32$; the null hypothesis cannot be rejected.

R.135 $\chi^2 = 9.23$; the differences among the proportions of "yes" answers are significant.

R.137 $5.56 < \mu < 7.84$.

R.139 $F = 1.52$; the differences among the four sample means can be attributed to chance.

R.141 $t = -2.87$; the difference between the mean repair costs is not significant.

R.143 We would be committing a Type I error if we erroneously reject the hypothesis that unit A is more efficient than unit B. We would be committing a Type II error if we erroneously accept the hypothesis that unit A is more efficient than unit B.

R.145 $z = 2.67$; reject the editor's claim.

R.147 $t = -1.98$; method B is more effective.

R.149 $32.7 < \mu < 36.9$.

R.151 $15.95 < \mu < 24.39$.

R.153 $16.88 < \mu < 19.12$.

R.155 2.34, 3.66, and $F = 0.64$.

Chapter 11

11.3 (a) $\hat{y} = 325.44 - 14.14x$; (b) $\hat{y} = 42.64$ or approximately 43,000 units.

11.5 (a) $\hat{y} = -87,177 + 1,338.8x$; (b) $\hat{y} = 20,596.4$ or approximately $20,596,000.

11.7 $\hat{y} = 0.66 + 0.60x$, where the years are numbered $x = 1, x = 2, x = 3, x = 4$, and $x = 5$; $\hat{y} = 4.26$ or $4,260,000.

11.9 $\hat{y} = 70.95 + 1.126x$.

11.13 $t = -2.14$; the null hypothesis cannot be rejected.

11.15 $-20.70 < \beta < -7.58$.

11.17 $93,214 to $97,998.

11.19 $t = 0.26$; the null hypothesis cannot be rejected.

11.21 $0.175 < \beta < 2.077$.

11.25 $r = 0.35$.

11.27 $r = 0.885$.

11.29 (a) $r = 0.916$;
(b) Regardless of what number we subtract, the result should be $r = 0.916$.

11.33 45%.

11.35 Neither answer should come as a surprise, since we can always draw a straight line passing through two given points, and, hence, get a perfect fit.
(a) $r = 1$ since the larger value of x corresponds to the larger value of y.
(b) $r = -1$ since the larger value of x corresponds to the smaller value of y.

11.37 (a) Significant; (b) significant; (c) not significant; (d) not significant.

11.39 The total and residual sums of squares are 2,660.64 and 1,858.33; $r = 0.549$.

11.41 $r = 0.916$; 83.9%.

Chapter 12

12.3 $x = 5$; the null hypothesis cannot be rejected.

12.5 $x = 9$; the null hypothesis cannot be rejected.

12.9 $z = 1.73$; the null hypothesis cannot be rejected.

12.11 $z = 2.75$; the null hypothesis must be rejected.

12.13 $z = -2.04$; the null hypothesis must be rejected.

548

12.15 $U = 5.5$; the null hypothesis must be rejected.

12.17 $U = 6$; the null hypothesis cannot be rejected.

12.21 $z = -3.20$; the null hypothesis must be rejected.

12.23 $z = 0.10$; the null hypothesis cannot be rejected.

12.25 $H = 4.51$; the null hypothesis cannot be rejected.

12.27 $H = 0.245$; the null hypothesis cannot be rejected.

12.29 $u = 4$; the null hypothesis of randomness cannot be rejected.

12.31 $u = 17$; the null hypothesis of randomness must be rejected.

12.33 $z = -0.88$; the null hypothesis of randomness cannot be rejected.

12.35 $z = 0.38$; the signal may be regarded as random noise.

12.39 $z = -2.88$; the null hypothesis of randomness must be rejected. There seems to be a trend, with the grades decreasing as the students take longer to finish the examination.

12.43 $r_S = 0.65$.

12.45 $r_S = 0.99$.

12.47 $r_S = 0.893$ compared to $r = 0.885$.

12.49 $r_S = 0.75$.

Review Exercises: Chapters 11 and 12

R.157 $\hat{y} = -1.07 + 1.22x$; 3.2.

R.159 (a) $r = 0.33$ is significant;   (b) $r = 0.33$ is not significant.

R.161 $U = 9$; the null hypothesis cannot be rejected.

R.163 $x = 2$; the null hypothesis must be rejected.

R.165 $r = -0.65$ is significant.

R.167 (a) Negative correlation;   (b) positive correlation;   (c) zero correlation;   (d) negative correlation;   (e) zero correlation;   (f) positive correlation.

R.169 $x = 9$; the null hypothesis must be rejected.

R.171 $\hat{y} = 14.253 + 0.637x$; 39.733 minutes.

R.173 $H = 5.03$; the null hypothesis cannot be rejected.

R.175 $z = 2.29$; the null hypothesis must be rejected.

R.177 $u = 11$; the null hypothesis of randomness cannot be rejected.

R.179 $U = 35$; the null hypothesis must be rejected.

R.181 $t = -0.804$; the null hypothesis cannot be rejected.

R.183 $z = 2.38$; it is not significant.

Index

A

Addition rules, 159

a (alpha), probability of Type I error, 314

a (alpha), regression coefficient, 429

Alternative hypothesis, 310
  one-sided and two-sided, 317

Analysis of variance, 343
  one-way, 344
    unequal sample sizes, 349
  table, 346

Arithmetic mean (*see* Mean)

Average (*see* Mean)

B

Bar chart, 25

Bayesian analysis, 132

Bayesian inference, 180

Bayes' theorem, 179

Bell-shaped distribution, 80

B (beta), probability of Type II error, 314

B (beta), regression coefficient, 429
  confidence interval for, 433
  test for, 431

Betting odds, 154

Binomial coefficients, 116
  table, 520

Binomial distribution, 200
  and hypergeometric distribution, 211
  mean, 222
  and normal distribution, 253
  and Poisson distribution, 312
  standard deviation, 224
  table, 502

Binomial population, 204

Boundary, class, 17

Box-and-whisker plot, 55

C

Categorical distribution, 13

Cell, 379

Central limit theorem, 279

Chebyshev's theorem, 67, 227, 278

Chi-square distribution, 379
  degrees of freedom, 379, 386
  table, 511

χ^2(chi-square) statistic, 378, 385, 393

Class:
  boundary, 17
  frequency, 16
  interval, 17
  limit, 16
  mark, 17
  open, 15
  real limits, 17

Classical probability concept, 119

Coding, 74

Coefficient, regression (*see* Regression coefficients)

Coefficient of correlation, 440
  computing formula, 441
  interpretation, 442
  population, 444
  rank, 478
  significance test for, 444

Coefficient of skewness, Pearsonian, 82

Coefficient of variation, 69

Coefficients, binomial, 116
  table, 520

Combinations, 115

Complement, 143

Composite hypothesis, 316

Computer simulation, 272

Conditional probability, 168

Confidence, degree of, 302

Confidence and probability, 298

Confidence interval, 302
  mean, 302, 305
  mean of y for given x, 434
  proportion, 363
  regression coefficient B, 433
  standard deviation, 308

D

Confidence limits (*see* Confidence interval)

Consistency criterion, 156

Contingency table, 383
  chi-square statistics, 385
  expected cell frequencies, 383, 384

Continuity correction, 250

Continous distribution (*see* Probability density)

Correction factor, finite population, 275

Correlation, 413
  coefficient, 440
  negative, 441
  positive, 441
  rank, 478

Correlation coefficient (*see* Coefficient of correlation)

Count data, 361

Cumulative distribution, 18

Curve fitting, 414

Data:
  count, 361
  paired, 33
  points, 419
  raw, 12

Decision theory, 8

Degrees of freedom, 302
  chi-square distribution, 379, 386, 393
  F distribution, 342
  t distribution, 303, 326, 431

Dependent events, 170

Dependent samples, 330

Descriptive statistics, 4

Deviation from mean, 61

Difference between means:
  paired data, 334
  standard error, 330
  test for, 331, 332

Difference between proportions:

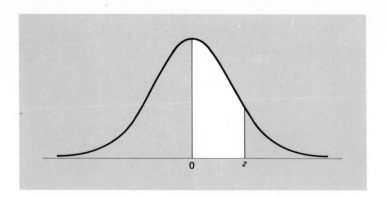

Normal-curve areas

z	.00	.01	.02	.03	.04	.05	.06	.07	.08	.09
0.0	.0000	.0040	.0080	.0120	.0160	.0199	.0239	.0279	.0319	.0359
0.1	.0398	.0438	.0478	.0517	.0557	.0596	.0636	.0675	.0714	.0753
0.2	.0793	.0832	.0871	.0910	.0948	.0987	.1026	.1064	.1103	.1141
0.3	.1179	.1217	.1255	.1293	.1331	.1368	.1406	.1443	.1480	.1517
0.4	.1554	.1591	.1628	.1664	.1700	.1736	.1772	.1808	.1844	.1879
0.5	.1915	.1950	.1985	.2019	.2054	.2088	.2123	.2157	.2190	.2224
0.6	.2257	.2291	.2324	.2357	.2389	.2422	.2454	.2486	.2517	.2549
0.7	.2580	.2611	.2642	.2673	.2704	.2734	.2764	.2794	.2823	.2852
0.8	.2881	.2910	.2939	.2967	.2995	.3023	.3051	.3078	.3106	.3133
0.9	.3159	.3186	.3212	.3238	.3264	.3289	.3315	.3340	.3365	.3389
1.0	.3413	.3438	.3461	.3485	.3508	.3531	.3554	.3577	.3599	.3621
1.1	.3643	.3665	.3686	.3708	.3729	.3749	.3770	.3790	.3810	.3830
1.2	.3849	.3869	.3888	.3907	.3925	.3944	.3962	.3980	.3997	.4015
1.3	.4032	.4049	.4066	.4082	.4099	.4115	.4131	.4147	.4162	.4177
1.4	.4192	.4207	.4222	.4236	.4251	.4265	.4279	.4292	.4306	.4319
1.5	.4332	.4345	.4357	.4370	.4382	.4394	.4406	.4418	.4429	.4441
1.6	.4452	.4463	.4474	.4484	.4495	.4505	.4515	.4525	.4535	.4545
1.7	.4554	.4564	.4573	.4582	.4591	.4599	.4608	.4616	.4625	.4633
1.8	.4641	.4649	.4656	.4664	.4671	.4678	.4686	.4693	.4699	.4706
1.9	.4713	.4719	.4726	.4732	.4738	.4744	.4750	.4756	.4761	.4767
2.0	.4772	.4778	.4783	.4788	.4793	.4798	.4803	.4808	.4812	.4817
2.1	.4821	.4826	.4830	.4834	.4838	.4842	.4846	.4850	.4854	.4857
2.2	.4861	.4864	.4868	.4871	.4875	.4878	.4881	.4884	.4887	.4890
2.3	.4893	.4896	.4898	.4901	.4904	.4906	.4909	.4911	.4913	.4916
2.4	.4918	.4920	.4922	.4925	.4927	.4929	.4931	.4932	.4934	.4936
2.5	.4938	.4940	.4941	.4943	.4945	.4946	.4948	.4949	.4951	.4952
2.6	.4953	.4955	.4956	.4957	.4959	.4960	.4961	.4962	.4963	.4964
2.7	.4965	.4966	.4967	.4968	.4969	.4970	.4971	.4972	.4973	.4974
2.8	.4974	.4975	.4976	.4977	.4977	.4978	.4979	.4979	.4980	.4981
2.9	.4981	.4982	.4982	.4983	.4984	.4984	.4985	.4985	.4986	.4986
3.0	.4987	.4987	.4987	.4988	.4988	.4989	.4989	.4989	.4990	.4990

Also, for $z = 4.0$, 5.0, and 6.0, the areas are 0.49997, 0.49999997, and 0.499999999.